A Friendly Introduction to Number Theory

Third Edition

Joseph H. Silverman

Brown University

Upper Saddle River, New Jersey 07458

Library of Congress Cataloging-in-Publication Data

Silverman, Joseph H.
A friendly introduction to number theory / Joseph H. Silverman.–3rd ed.
p. cm.
Includes bibliographical references
ISBN 0-13-186137-9
1. Number theory–Textbooks. I. Title

QA241.S497 2006
512.7–dc22 2005042950

Executive Acquisitions Editor: *George Lobell*
Editor-in-Chief: *Sally Yagan*
Production Editor: *Bob Walters, Prepress Management, Inc.*
Senior Managing Editor: *Linda Mihatov Behrens*
Executive Managing Editor: *Kathleen Schiaparelli*
Assistant Manufacturing Manager/Buyer: *Michael Bell*
Marketing Manager: *Halee Dinsey*
Marketing Assistant: *Joon Moon*
Editorial Assistant: *Jennifer Urban*
Art Director: *Jayne Conte*
Director of Creative Services: *Paul Belfanti*
Cover Designer: *Bruce Kenselaar*
Cover Photo: *Steve Gorton* ©*Darling Kindersley*

Pearson Prentice Hall
Pearson Education, Inc.
Upper Saddle River, New Jersey 07458

Printed in the United States of America
10 9 8 7 6 5 4 3 2 1

ISBN: 0-13-186137-9

Pearson Education LTD., *London*
Pearson Education Australia PTY, Limited, *Sydney*
Pearson Education, Singapore, Pte. Ltd.
Pearson Education North Asia Ltd., *Hong Kong*
Pearson Education Canada, Ltd., *Toronto*
Pearson Educaciûn de Mexico S.A. de C.V.
Pearson Education — Japan, *Tokyo*
Pearson Education Malaysia, Pte. Ltd.

Contents

Preface

The 1990s saw a wave of calculus reform whose aim was to teach students to think for themselves and to solve substantial problems, rather than merely memorizing formulas and performing rote algebraic manipulations. This book has a similar, albeit somewhat more ambitious, goal; to lead you to think mathematically and to experience the thrill of independent intellectual discovery. Our chosen subject, Number Theory, is particularly well suited for this purpose. The natural numbers 1, 2, 3, ... satisfy a multitude of beautiful patterns and relationships, many of which can be discerned at a glance; others are so subtle that one marvels they were noticed at all. Experimentation requires nothing more than paper and pencil, but many false alleys beckon to those who make conjectures on too scanty evidence. It is only by rigorous demonstration that one is finally convinced that the numerical evidence reflects a universal truth. This book will lead you through the groves wherein lurk some of the brightest flowers of Number Theory, as it simultaneously encourages you to investigate, analyze, conjecture, and ultimately prove your own beautiful number theoretic results.

This book was originally written to serve as a text for Math 42, a course created by Jeff Hoffstein at Brown University in the early 1990s. Math 42 was designed to attract nonscience majors, those with little interest in pursuing the standard calculus sequence, and to convince them to study some college mathematics. The intent was to create a course similar to one on, say, "The Music of Mozart" or "Elizabethan Drama," wherein an audience is introduced to the overall themes and methodology of an entire discipline through the detailed study of a particular facet of the subject. Math 42 has been extremely successful, attracting both its intended audience and also scientifically oriented undergraduates interested in a change of pace from their large-lecture, cookbook-style courses.

The prerequisites for reading this book are few. Some facility with high school algebra is required, and those who know how to program a computer will have fun generating reams of data and implementing assorted algorithms, but in truth the reader needs nothing more than a simple calculator. Concepts from calculus are mentioned in passing, but are not used in an essential way. However, and the reader is hereby forewarned, it is not possible to truly appreciate Number Theory without an eager and questioning mind and a spirit that is not afraid to experiment, to make mistakes and profit from them, to accept frustration and persevere to the ultimate triumph. Readers who are able to cultivate these qualities will find themselves richly rewarded, both in their study of Number Theory and their appreciation of all that life has to offer.

Acknowledgments for the First Edition

There are many people I would like to thank for their assistance—Jeff Hoffstein, Karen Bender, and Rachel Pries for their pioneering work in Math 42; Bill Amend for kindly permitting me to use some of his wonderful FoxTrot cartoons; the creators of PARI for providing the ultimate in number theory computational power; Nick Fiori, Daniel Goldston, Rob Gross, Matt Holford, Alan Landman, Paul Lockhart, Matt Marcy, Patricia Pacelli, Rachel Pries (again), Michael Schlessinger, Thomas Shemanske, Jeffrey Stopple, Chris Towse, Roger Ware, Larry Washington, Yangbo Ye, and Karl Zimmerman for looking at the initial draft and offering invaluable suggestions; Michael Artin, Richard Guy, Marc Hindry, Mike Rosen, Karl Rubin, Ed Scheinerman, John Selfridge, and Sam Wagstaff for much helpful advice; and George Lobell and Gale Epps at Prentice Hall for their excellent advice and guidance during the publication process.

Finally, and most important, I want to thank my wife Susan and children Debby, Daniel, and Jonathan for their patience and understanding while this book was being written.

Acknowledgments for the Second Edition

I would like to thank all those who took the time to send me corrections and suggestions that were invaluable in preparing this second edition, including Arthur Baragar, Aaron Bertram, Nigel Boston, David Boyd, Seth Braver, Michael Catalano-Johnson, L. Chang, Robin Chapman, Miguel Cordero, John Cremona, Jim Delany, Lisa Fastenberg, Nicholas Fiori, Fumiyasu Funami, Jim Funderburk, Andrew Granville, Rob Gross, Shamita Dutta Gupta, Tom Hagedorn, Ron Jacobowitz, Jerry S. Kelly, Hershy Kisilevsky, Hendrik Lenstra, Gordon S. Lessells, Ken Levasseur, Stephen Lichtenbaum, Nidia Lopez Jerry Metzger, Jukka Pihko, Carl Pomerance, Rachel Pries, Ken Ribet, John Robeson, David Rohrlich, Daniel Silverman, Alfred Tang, and Wenchao Zhou.

Acknowledgments for the Third Edition

I would like to thank Jiro Suzuki for his beautiful translation of my book into Japanese. I would also like to thank all those who took the time to send me corrections and suggestions that were invaluable in preparing this third edition, including Bill Adams, Autumn Alden, Robert Altshuler, Avner Ash, Joe Auslander, Dave Benoit, Jürgen Bierbrauer, Andrew Clifford, Keith Conrad, Sarah DeGooyer, Amartya Kumar Dutta, Laurie Fanning, Benji Fisher, Joe Fisher, Jon Graff, Eric Gutman, Edward Hinson, Bruce Hugo, Ole Jensen, Peter Kahn, Avinash Kalra, Jerry Kelly, Yukio Kikuchi, Amartya Kumar, Andrew Lenard, Sufatrio Liu, Troy Madsen, Russ Mann, Gordon Mason, Farley Mawyer, Mike McConnell, Jerry Metzger, Steve Paik, Nicole Perez, Dinakar Ramakrishnan, Cecil Rousseau, Marc Roth, Ehud Schreiber, Tamina Stephenson, Jiro Suzuki, James Tanton, James Tong, Chris Towse, Roger Turton, Fernando Villegas, and Chung Yi.

Email and Electronic Resources

All the people listed above have helped me to correct numerous mistakes and to greatly refine the exposition, but no book is ever free from error or incapable of being improved. I would be delighted to receive comments, good or bad, and corrections from my readers. You can send mail to me at

`jhs@math.brown.edu`

Additional material, including an errata sheet, links to interesting number theoretic sites, and downloadable versions of various computer exercises, are available on the *Friendly Introduction to Number Theory* Home Page:

`www.math.brown.edu/~jhs/frint.html`

Joseph H. Silverman

Introduction

Euclid alone
Has looked on Beauty bare. Fortunate they
Who, though once only and then but far away,
Have heard her massive sandal set on stone.
Edna St. Vincent Millay (1923)

The origins of the natural numbers 1, 2, 3, 4, 5, 6, ... are lost in the mists of time. We have no knowledge of who first realized that there is a certain concept of "threeness" that applies equally well to three rocks, three stars, and three people. From the very beginnings of recorded history, numbers have inspired an endless fascination—mystical, aesthetic, and practical as well. It is not just the numbers themselves, of course, that command attention. Far more intriguing are the relationships that numbers exhibit, one with another. It is within these profound and often subtle relationships that one finds the Beauty[1] so strikingly described in Edna St. Vincent Millay's poem. Here is another description by a celebrated twentieth-century philosopher.

> Mathematics, rightly viewed, possesses not only truth, but supreme beauty—a beauty cold and austere, like that of sculpture, without appeal to any part of our weaker nature, without the gorgeous trappings of paintings or music, yet sublimely pure, and capable of a stern perfection such as only the greatest art can show. (Bertrand Russell, 1902)

The Theory of Numbers is that area of mathematics whose aim is to uncover the many deep and subtle relationships between different sorts of numbers. To take a simple example, many people through the ages have been intrigued by the square numbers 1, 4, 9, 16, 25, ... If we perform the experiment of adding together pairs

[1]Euclid, indeed, has looked on Beauty bare, and not merely the beauty of geometry that most people associate with his name. Number theory is prominently featured in Books VII, VIII, and IX of Euclid's famous *Elements*.

of square numbers, we will find that occasionally we get another square. The most famous example of this phenomenon is

$$3^2 + 4^2 = 5^2,$$

but there are many others, such as

$$5^2 + 12^2 = 13^2, \quad 20^2 + 21^2 = 29^2, \quad 28^2 + 45^2 = 53^2.$$

Triples like $(3, 4, 5)$, $(5, 12, 13)$, $(20, 21, 29)$, and $(28, 45, 53)$ have been given the name Pythagorean triples.[2] Based on this experiment, anyone with a lively curiosity is bound to pose various questions, such as "Are there infinitely many Pythagorean triples?" and "If so, can we find a formula that describes all of them?" These are the sorts of questions dealt with by number theory.

As another example, consider the problem of finding the remainder when the huge number

$$32478543^{743921429837645}$$

is divided by 54817263. Here's one way to solve this problem. Take the number 32478543, multiply it by itself 743921429837645 times, use long division to divide by 54817263, and take the remainder. In principle, this method will work, but in practice it would take far longer than a lifetime, even on the world's fastest computers. Number theory provides a means for solving this problem, too. "Wait a minute," I hear you say, "Pythagorean triples have a certain elegance that is pleasing to the eye, but where is the beauty in long division and remainders?" The answer is not in the remainders themselves, but in the use to which such remainders can be put. In a striking turn of events, mathematicians have shown how the solution of this elementary remainder problem (and its inverse) leads to the creation of simple codes that are so secure that even the National Security Agency[3] is unable to break them. So much for G.H. Hardy's singularly unprophetic remark that "no one has yet discovered any warlike purpose to be served by the theory of numbers or relativity, and it seems very unlikely that anyone will do so for many years."[4]

The land of Number Theory is populated by a variety of exotic flora and fauna. There are square numbers and prime numbers and odd numbers and perfect numbers (but no square-prime numbers and, as far as anyone knows, no odd-perfect

[2]In fairness, it should be mentioned that the Babylonians compiled large tables of "Pythagorean" triples many centuries before Pythagoras was born.

[3]The National Security Agency (NSA) is the arm of the United States government charged with data collection, code making, and code breaking. The NSA, with a budget larger than that of the CIA, is the single largest employer of mathematicians in the world.

[4]*A Mathematician's Apology*, §28, G.H. Hardy, Camb. Univ. Press, 1940.

numbers). There are Fermat equations and Pell equations, Pythagorean triples and elliptic curves, Fibonacci's rabbits, unbreakable codes, and much, much more. You will meet all these creatures, and many others, as we journey through the Theory of Numbers.

Guide for the Instructor

This book is designed to be used as a text for a one-semester or full-year course in undergraduate number theory or for an independent study or reading course. It contains approximately two semesters' worth of material, so the instructor of a one-semester course will have some flexibility in the choice of topics. The first 11 chapters are basic, and probably most instructors will want to continue through the RSA cryptosystem in Chapter 18, since in my experience this is one of the students' favorite topics.

There are now many ways to proceed. Here are a few possibilities that seem to fit comfortably into one semester, but feel free to slice-and-dice the later chapters to fit your own tastes.

Chapters 20–32. Primitive roots, quadratic reciprocity, sums of squares, Pell's equation, and Diophantine approximation. (Add Chapters 39 and 40 on continued fractions if time permits.)

Chapters 28–32 & 43–48. Fermat's equation for exponent 4, Pell's equation, Diophantine approximation, elliptic curves, and Fermat's Last Theorem.

Chapters 29–37 & 39–40. Pell's equation, Diophantine approximation, Gaussian integers transcendental numbers, binomial coefficients, linear recurrences, and continued fractions.

Chapters 19–25 & 36–38. Primality testing, primitive roots, quadratic reciprocity, binomial coefficients, linear recurrences, big-Oh notation. (This syllabus is designed in particular for students planning further work in computer science or cryptography.)

In any case, a good final project is to have the students read a few of the omitted chapters and do the exercises.

Most of the nonnumerical nonprogramming exercises in this book are designed to foster discussion and experimentation. They do not necessarily have "correct" or "complete" answers. Many students will find this extremely disconcerting at first, so it must be stressed repeatedly. You can make your students feel more at ease by prefacing such questions with the phrase "Tell me as much as you can about ...." Tell your students that accumulating data and solving special cases are not merely acceptable, but encouraged. On the other hand, tell them that there is no such thing as a complete solution, since the solution of a good problem always raises additional questions. So if they can fully answer the specific question given in the text, their next task is to look for generalizations and for limitations on the validity of their solution.

Aside from a few clearly marked exercises, calculus is required only in two late chapters (Big-Oh notation in Chapter 38 and Generating Functions in Chapter 41). If the class has not taken calculus, these chapters may be omitted with no harm to the flow of the material.

Number theory is not easy, so there's no point in trying to convince the students that it is. Instead, this book will show your students that they are capable of mastering a difficult subject and experiencing the intense satisfaction of intellectual discovery. Your reward as the instructor is to bask in the glow of their endeavors.

Computers, Number Theory, and This Book

At this point I would like to say a few words about the use of computers in conjunction with this book. I neither expect nor desire that the reader make use of a high-level computer package such as Maple, Mathematica, PARI, or Derive, and most exercises (except as otherwised indicated) can be done with a simple pocket calculator. To take a concrete example, studying greatest common divisors (Chapter 5) by typing `GCD[M,N]` into a computer is akin to studying electronics by turning on a television set. Admittedly, computers allow one to do examples with large numbers, and you will find such computer-generated examples scattered through the text, but our ultimate goal is always to understand concepts and relationships. So if I were forced to make a firm ruling, yea or nay, regarding computers, I would undoubtedly forbid their use.

However, just as with any good rule, certain exceptions will be admitted. First, one of the best ways to understand a subject is to explain it to someone else; so if you know a little bit of how to write computer programs, you will find it extremely enlightening to explain to a computer how to perform the algorithms described in this book. In other words, don't rely on a canned computer package; do the programming yourself. Good candidates for such treatment are the Euclidean algorithm (Chapters 5–6) the RSA cryptosystem (Chapters 16–18), quadratic reciprocity (Chapter 25), writing numbers as sums of two squares (Chapters 26–27), primality testing (Chapter 19), and generating rational points on elliptic curves (Chapter 43).

The second exception to the "no computer rule" is generation of data. Discovery in number theory is usually based on experimentation, which may involve examining reams of data to try to distinguish underlying patterns. Computers are well suited to generating such data and also sometimes to assist in searching for patterns, and I have no objection to their being used for these purposes.

I have included a number of computer exercises and computer projects to encourage you to use computers properly as tools to help understand and investigate the theory of numbers. Some of these exercises can be implemented on a small computer (or even a programmable calculator), while others require more sophisticated machines and/or programming languages. Exercises and projects requiring a computer are marked by the symbol 💻.

For many of the projects I have not given a precise formulation, since part of the project is to decide exactly what the user should input and exactly what form the output should take. Note that a good computer program must include all the following features:

- Clearly written documentation explaining what the program does, how to use it, what quantities it takes as input, and what quantities it returns as output.
- Extensive internal comments explaining how the program works.
- Complete error handling with informative error messages. For example, if $a = b = 0$, then the $\gcd(a, b)$ routine should return the error message "`gcd(0,0) is undefined`" instead of going into an infinite loop or returning a "division by zero" error.

As you write your own programs, try to make them user friendly and as versatile as possible, since ultimately you will want to link the pieces together to form your own package of number theoretic routines.

The moral is that computers are useful as a tool for experimentation and that you can learn a lot by teaching a computer how to perform number theoretic calculations, but when you are first learning a subject, prepackaged computer programs merely provide a crutch that prevent you from learning to walk on your own.

Chapter 1

What Is Number Theory?

Number theory is the study of the set of positive whole numbers

$$1, 2, 3, 4, 5, 6, 7, \ldots,$$

which are often called the set of *natural numbers*. We will especially want to study the *relationships* between different sorts of numbers. Since ancient times, people have separated the natural numbers into a variety of different types. Here are some familiar and not-so-familiar examples:

odd	$1, 3, 5, 7, 9, 11, \ldots$
even	$2, 4, 6, 8, 10, \ldots$
square	$1, 4, 9, 16, 25, 36, \ldots$
cube	$1, 8, 27, 64, 125, \ldots$
prime	$2, 3, 5, 7, 11, 13, 17, 19, 23, 29, 31, \ldots$
composite	$4, 6, 8, 9, 10, 12, 14, 15, 16, \ldots$
1 (modulo 4)	$1, 5, 9, 13, 17, 21, 25, \ldots$
3 (modulo 4)	$3, 7, 11, 15, 19, 23, 27, \ldots$
triangular	$1, 3, 6, 10, 15, 21, \ldots$
perfect	$6, 28, 496, \ldots$
Fibonacci	$1, 1, 2, 3, 5, 8, 13, 21, \ldots$

Many of these types of numbers are undoubtedly already known to you. Others, such as the "modulo 4" numbers, may not be familiar. A number is said to be congruent to 1 (modulo 4) if it leaves a remainder of 1 when divided by 4, and similarly for the 3 (modulo 4) numbers. A number is called triangular if that number of pebbles can be arranged in a triangle, with one pebble at the top, two pebbles in the next row, and so on. The Fibonacci numbers are created by starting with 1 and 1. Then, to get the next number in the list, just add the previous two. Finally, a number is perfect if the sum of all its divisors, other than itself, adds back up to the original number. Thus, the numbers dividing 6 are 1, 2, and 3, and $1 + 2 + 3 = 6$. Similarly, the divisors of 28 are 1, 2, 4, 7, and 14, and

$$1 + 2 + 4 + 7 + 14 = 28.$$

We will encounter all these types of numbers, and many others, in our excursion through the Theory of Numbers.

Some Typical Number Theoretic Questions

The main goal of number theory is to discover interesting and unexpected relationships between different sorts of numbers and to prove that these relationships are true. In this section we will describe a few typical number theoretic problems, some of which we will eventually solve, some of which have known solutions too difficult for us to include, and some of which remain unsolved to this day.

Sums of Squares I. Can the sum of two squares be a square? The answer is clearly "YES"; for example $3^2 + 4^2 = 5^2$ and $5^2 + 12^2 = 13^2$. These are examples of *Pythagorean triples*. We will describe all Pythagorean triples in Chapter 2.

Sums of Higher Powers. Can the sum of two cubes be a cube? Can the sum of two fourth powers be a fourth power? In general, can the sum of two n^{th} powers be an nth power? The answer is "NO." This famous problem, called *Fermat's Last Theorem*, was first posed by Pierre de Fermat in the seventeenth century, but was not completely solved until 1994 by Andrew Wiles. Wiles's proof uses sophisticated mathematical techniques that we will not be able to describe in detail, but in Chapter 28 we will prove that no fourth power is a sum of two fourth powers, and in Chapter 48 we will sketch some of the ideas that go into Wiles's proof.

Infinitude of Primes. A *prime number* is a number p whose only factors are 1 and p.

- Are there infinitely many prime numbers?
- Are there infinitely many primes that are 1 modulo 4 numbers?
- Are there infinitely many primes that are 3 modulo 4 numbers?

The answer to all these questions is "YES." We will prove these facts in Chapters 12 and 24 and also discuss a much more general result proved by Lejeune Dirichlet in 1837.

Sums of Squares II. Which numbers are sums of two squares? It often turns out that questions of this sort are easier to answer first for primes, so we ask which (odd) prime numbers are a sum of two squares. For example,

$$\begin{array}{llll} 3 = \text{NO}, & 5 = 1^2 + 2^2, & 7 = \text{NO}, & 11 = \text{NO}, \\ 13 = 2^2 + 3^2, & 17 = 1^2 + 4^2, & 19 = \text{NO}, & 23 = \text{NO}, \\ 29 = 2^2 + 5^2, & 31 = \text{NO}, & 37 = 1^2 + 6^2, & \ldots \end{array}$$

Do you see a pattern? Possibly not, since this is only a short list, but a longer list leads to the conjecture that p is a sum of two squares if it is congruent to 1 (modulo 4). In other words, p is a sum of two squares if it leaves a remainder of 1 when divided by 4, and it is not a sum of two squares if it leaves a remainder of 3. We will prove that this is true in Chapter 26.

Number Shapes. The square numbers are the numbers 1, 4, 9, 16, ... that can be arranged in the shape of a square. The triangular numbers are the numbers 1, 3, 6, 10, ... that can be arranged in the shape of a triangle. The first few triangular and square numbers are illustrated in Figure 1.1.

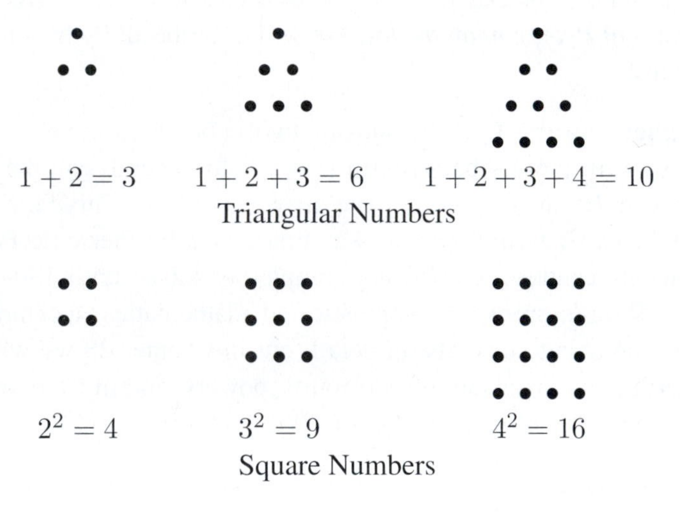

Figure 1.1: Numbers that form interesting shapes

A natural question to ask is whether there are any triangular numbers that are also square numbers (other than 1). The answer is "YES," the smallest example being

$$36 = 6^2 = 1 + 2 + 3 + 4 + 5 + 6 + 7 + 8.$$

So we might ask whether there are more examples and, if so, are there infinitely many? To search for examples, the following formula is helpful:

$$1 + 2 + 3 + \cdots + (n-1) + n = \frac{n(n+1)}{2}.$$

> There is an amusing anecdote associated with this formula. One day when the young Karl Friedrich Gauss (1777–1855) was in grade school, his teacher became so incensed with the class that he set them the task of adding up all the numbers from 1 to 100. As Gauss's classmates dutifully began to add, Gauss walked up to the teacher and presented the answer, 5050. The story goes that the teacher was neither impressed nor amused, but there's no record of what the next make-work assignment was!

There is an easy geometric way to verify Gauss's formula, which may be the way he discovered it himself. The idea is to take two triangles consisting of $1 + 2 + \cdots + n$ pebbles and fit them together with one additional diagonal of $n + 1$ pebbles. Figure 1.2 illustrates this idea for $n = 6$.

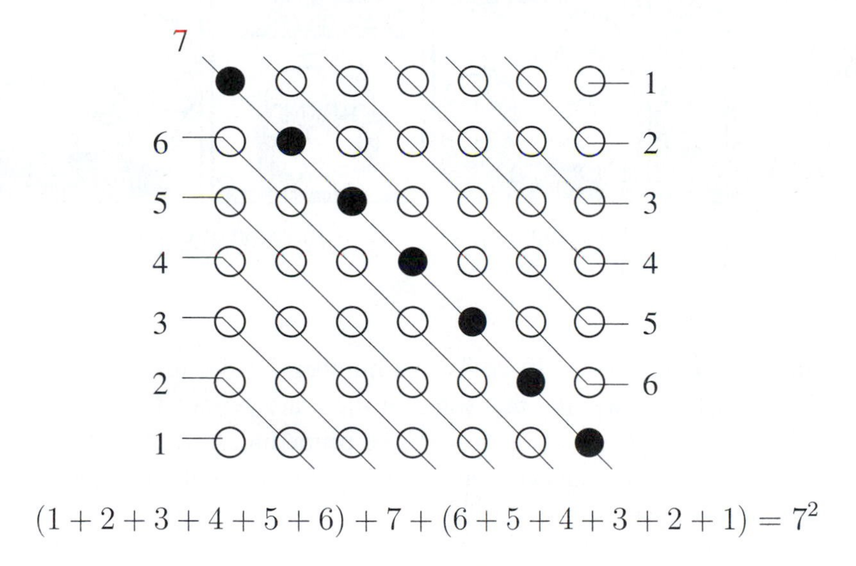

Figure 1.2: The sum of the first n integers

In the figure, we have marked the extra $n + 1 = 7$ pebbles on the diagonal with black dots. The resulting square has sides consisting of $n + 1$ pebbles, so in mathematical terms we obtain the formula

$$\begin{array}{ccccc} 2(1+2+3+\cdots+n) & + & (n+1) & = & (n+1)^2, \\ \text{two triangles} & + & \text{diagonal} & = & \text{square.} \end{array}$$

Now we can subtract $n + 1$ from each side and divide by 2 to get Gauss's formula.

Twin Primes. In the list of primes it is sometimes true that consecutive odd numbers are both prime. We have boxed these *twin primes* in the following list of primes less than 100:

$$\boxed{3}, \boxed{5}, \boxed{7}, \quad \boxed{11}, \boxed{13}, \quad \boxed{17}, \boxed{19}, \quad 23, \quad \boxed{29}, \boxed{31}, \quad 37$$
$$\boxed{41}, \boxed{43}, \quad 47, 53, \quad \boxed{59}, \boxed{61}, \quad 67, \quad \boxed{71}, \boxed{73}, \quad 79, 83, 89, 97.$$

Are there infinitely many twin primes? That is, are there infinitely many prime numbers p such that $p + 2$ is also a prime? At present, no one knows the answer to this question.

FOXTROT ©Bill Amend. Reprinted with permission of UNIVERSAL SYNDICATE.

Primes of the Form $N^2 + 1$. If we list the numbers of the form $N^2 + 1$ taking $N = 1, 2, 3, \ldots$, we find that some of them are prime. Of course, if N is odd, then $N^2 + 1$ is even, so it won't be prime unless $N = 1$. So it's really only interesting to take even values of N. We've highlighted the primes in the following list:

$$2^2 + 1 = \mathbf{5} \qquad 4^2 + 1 = \mathbf{17} \qquad 6^2 + 1 = \mathbf{37} \qquad 8^2 + 1 = 65 = 5 \cdot 13$$
$$10^2 + 1 = \mathbf{101} \qquad 12^2 + 1 = 145 = 5 \cdot 29 \qquad 14^2 + 1 = \mathbf{197}$$
$$16^2 + 1 = \mathbf{257} \qquad 18^2 + 1 = 325 = 5^2 \cdot 13 \qquad 20^2 + 1 = \mathbf{401}.$$

It looks like there are quite a few prime values, but if you take larger values of N you will find that they become much rarer. So we ask whether there are infinitely many primes of the form $N^2 + 1$. Again, no one presently knows the answer to this question.

We have now seen some of the types of questions that are studied in the Theory of Numbers. How does one attempt to answer these questions? The answer is that Number Theory is partly experimental and partly theoretical. The experimental part normally comes first; it leads to questions and suggests ways to answer them. The theoretical part follows; in this part one tries to devise an argument that gives a conclusive answer to the questions. In summary, here are the steps to follow:

1. Accumulate data, usually numerical, but sometimes more abstract in nature.
2. Examine the data and try to find patterns and relationships.
3. Formulate conjectures (that is, guesses) that explain the patterns and relationships. These are frequently given by formulas.
4. Test your conjectures by collecting additional data and checking whether the new information fits your conjectures.
5. Devise an argument (that is, a proof) that your conjectures are correct.

All five steps are important in number theory and in mathematics. More generally, the scientific method always involves at least the first four steps. Be wary of any purported "scientist" who claims to have "proved" something using only the first three. Given any collection of data, it's generally not too difficult to devise numerous explanations. The true test of a scientific theory is its ability to predict the outcome of experiments that have not yet taken place. In other words, a scientific theory only becomes plausible when it has been tested against new data. This is true of all real science. In mathematics one requires the further step of a proof, that is, a logical sequence of assertions, starting from known facts and ending at the desired statement.

Exercises

1.1. The first two numbers that are both squares and triangles are 1 and 36. Find the next one and, if possible, the one after that. Can you figure out an efficient way to find triangular–square numbers? Do you think that there are infinitely many?

1.2. Try adding up the first few odd numbers and see if the numbers you get satisfy some sort of pattern. Once you find the pattern, express it as a formula. Give a geometric verification that your formula is correct.

1.3. The consecutive odd numbers 3, 5, and 7 are all primes. Are there infinitely many such "prime triplets"? That is, are there infinitely many prime numbers p so that $p + 2$ and $p + 4$ are also primes?

1.4. It is generally believed that infinitely many primes have the form $N^2 + 1$, although no one knows for sure.

(a) Do you think that there are infinitely many primes of the form $N^2 - 1$?
(b) Do you think that there are infinitely many primes of the form $N^2 - 2$?
(c) How about of the form $N^2 - 3$? How about $N^2 - 4$?
(d) Which values of a do you think give infinitely many primes of the form $N^2 - a$?

1.5. The following two lines indicate another way to derive the formula for the sum of the first n integers by rearranging the terms in the sum. Fill in the details.

$$\begin{aligned} 1 + 2 + 3 + \cdots + n &= (1 + n) + (2 + (n - 1)) + (3 + (n - 2)) + \cdots \\ &= (1 + n) + (1 + n) + (1 + n) + \cdots . \end{aligned}$$

In particular, how many copies of $n + 1$ are in there in the second line? (You may need to consider the cases of odd n and even n separately. If that's not clear, try first writing it out explicitly for $n = 6$ and $n = 7$.)

Chapter 2

Pythagorean Triples

The Pythagorean Theorem, that "beloved" formula of all high school geometry students, says that the sum of the squares of the sides of a right triangle equals the square of the hypotenuse. In symbols,

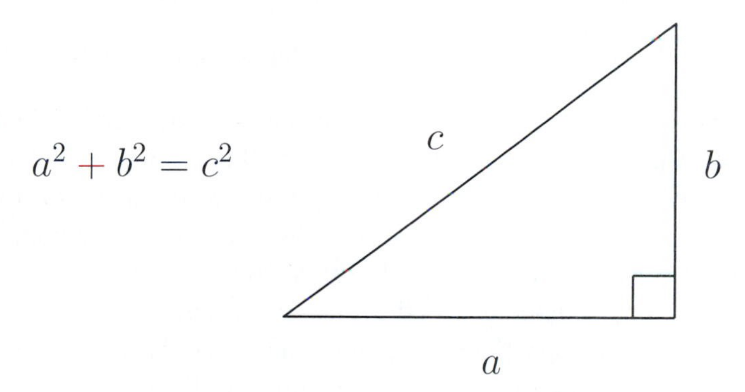

Figure 2.1: A Pythagorean Triangle

Since we're interested in number theory, that is, the theory of the natural numbers, we will ask whether there are any Pythagorean triangles all of whose sides are natural numbers. There are many such triangles. The most famous has sides 3, 4, and 5. Here are the first few examples:

$$3^2 + 4^2 = 5^2, \quad 5^2 + 12^2 = 13^2, \quad 8^2 + 15^2 = 17^2, \quad 28^2 + 45^2 = 53^2.$$

The study of these *Pythagorean triples* began long before the time of Pythagoras. There are Babylonian tablets that contain lists of such triples, including quite large ones, indicating that the Babylonians probably had a systematic method for producing them. Pythagorean triples were also used in ancient Egypt. For example, a rough-and-ready way to produce a right angle is to take a piece of string, mark it into 12 equal segments, tie it into a loop, and hold it taut in the form of a 3-4-5 triangle, as illustrated in Figure 2.2. This provides an inexpensive right angle tool for use on small construction projects (such as marking property boundaries or building pyramids). Even more amazing is the fact that the Babylonians created tables of quite large Pythagorean triples, which they may have used as primitive trigonometric tables.

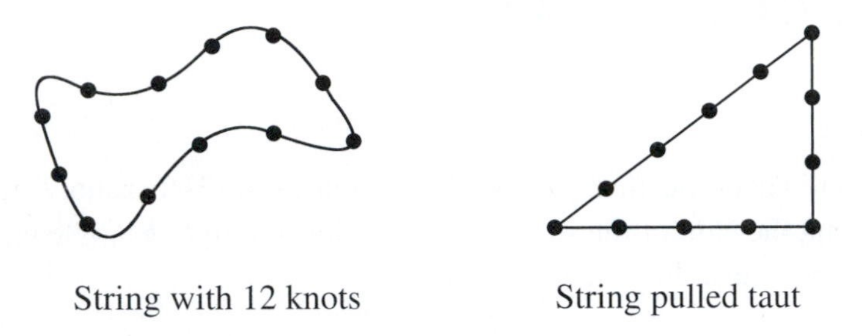

String with 12 knots String pulled taut

Figure 2.2: Using a knotted string to create a right triangle

The Babylonians and Egyptians had practical reasons for studying Pythagorean triples. Do such practical reasons still exist? For this particular problem, the answer is "probably not." However, there is at least one good reason to study Pythagorean triples, and it's the same reason why it is worthwhile studying the art of Rembrandt and the music of Beethoven. There is a beauty to the ways in which numbers interact with one another, just as there is a beauty in the composition of a painting or a symphony. To appreciate this beauty, one has to be willing to expend a certain amount of mental energy. But the end result is well worth the effort. Our goal in this book is to understand and appreciate some truly beautiful mathematics, to learn how this mathematics was discovered and proved, and maybe even to make some original contributions of our own.

Enough blathering, you are undoubtedly thinking. Let's get to the real stuff. Our first naive question is whether there are infinitely many *Pythagorean triples*, that is triples of natural numbers (a, b, c) satisfying the equation $a^2 + b^2 = c^2$. The answer is "YES" for a very silly reason. If we take a Pythagorean triple (a, b, c) and multiply it by some other number d, then we obtain a new Pythagorean triple (da, db, dc). This is true because

$$(da)^2 + (db)^2 = d^2(a^2 + b^2) = d^2c^2 = (dc)^2.$$

Clearly these new Pythagorean triples are not very interesting. So we will concentrate our attention on triples with no common factors. We will even give them a name:

> A *primitive Pythagorean triple* (or PPT for short) is a triple of numbers (a, b, c) so that a, b, and c have no common factors[1] and satisfy
>
> $$a^2 + b^2 = c^2.$$

Recall our checklist from Chapter 1. The first step is to accumulate some data. I used a computer to substitute in values for a and b and checked if $a^2 + b^2$ is a square. Here are some primitive Pythagorean triples that I found:

$$\begin{array}{llll} (3,4,5), & (5,12,13), & (8,15,17), & (7,24,25), \\ (20,21,29), & (9,40,41), & (12,35,37), & (11,60,61), \\ (28,45,53), & (33,56,65), & (16,63,65). & \end{array}$$

A few conclusions can easily be drawn even from such a short list. For example, it certainly looks like one of a and b is odd and the other even. It also seems that c is always odd.

It's not hard to prove that these conjectures are correct. First, if a and b are both even, then c would also be even. This means that a, b, and c would have a common factor of 2, so the triple would not be primitive. Next, suppose that a and b are both odd, which means that c would have to be even. This means that there are numbers x, y, and z so that

$$a = 2x + 1, \qquad b = 2y + 1, \qquad \text{and} \qquad c = 2z.$$

We can substitute these into the equation $a^2 + b^2 = c^2$ to get

$$(2x + 1)^2 + (2y + 1)^2 = (2z)^2,$$
$$4x^2 + 4x + 4y^2 + 4y + 2 = 4z^2.$$

Now divide by 2,

$$2x^2 + 2x + 2y^2 + 2y + 1 = 2z^2.$$

This last equation says that an odd number is equal to an even number, which is impossible, so a and b cannot both be odd. Since we've just checked that they cannot both be even and cannot both be odd, it must be true that one is even and

[1] A *common factor* of a, b, and c is a number d so that each of a, b and c is a multiple of d . For example, 3 is a common factor of 30, 42, and 105, since $30 = 3 \cdot 10$, $42 = 3 \cdot 14$, and $105 = 3 \cdot 35$, and indeed it is their largest common factor. On the other hand, the numbers 10, 12, and 15 have no common factor (other than 1). Since our goal in this chapter is to explore some interesting and beautiful number theory without getting bogged down in formalities, we will use common factors and divisibility informally and trust our intuition. In Chapter 5 we will return to these questions and develop the theory of divisibility more carefully.

the other is odd. It's then obvious from the equation $a^2 + b^2 = c^2$ that c is also odd.

We can always switch a and b, so our problem now is to find all solutions in natural numbers to the equation

$$a^2 + b^2 = c^2 \qquad \text{with} \qquad \begin{cases} a \text{ odd}, \\ b \text{ even}, \\ a, b, c \text{ having no common factors.} \end{cases}$$

The tools we will use are *factorization* and *divisibility*.

Our first observation is that if (a, b, c) is a primitive Pythagorean triple, then we can factor

$$a^2 = c^2 - b^2 = (c - b)(c + b).$$

Here are a few examples from the list given earlier, where note that we always take a to be odd and b to be even:

$$\begin{aligned} 3^2 &= 5^2 - 4^2 = (5 - 4)(5 + 4) = 1 \cdot 9, \\ 15^2 &= 17^2 - 8^2 = (17 - 8)(17 + 8) = 9 \cdot 25, \\ 35^2 &= 37^2 - 12^2 = (37 - 12)(37 + 12) = 25 \cdot 49, \\ 33^2 &= 65^2 - 56^2 = (65 - 56)(65 + 56) = 9 \cdot 121. \end{aligned}$$

It looks like $c - b$ and $c + b$ are themselves always squares. We check this observation with a couple more examples:

$$\begin{aligned} 21^2 &= 29^2 - 20^2 = (29 - 20)(29 + 20) = 9 \cdot 49, \\ 63^2 &= 65^2 - 16^2 = (65 - 16)(65 + 16) = 49 \cdot 81. \end{aligned}$$

How can we prove that $c - b$ and $c + b$ are squares? Another observation apparent from our list of examples is that $c - b$ and $c + b$ seem to have no common factors. We can prove this last assertion as follows. Suppose that d is a common factor of $c - b$ and $c + b$; that is, d divides both $c - b$ and $c + b$. Then d also divides

$$(c + b) + (c - b) = 2c \qquad \text{and} \qquad (c + b) - (c - b) = 2b.$$

Thus, d divides $2b$ and $2c$. But b and c have no common factor because we are assuming that (a, b, c) is a primitive Pythagorean triple. So d must equal 1 or 2. But d also divides $(c - b)(c + b) = a^2$, and a is odd, so d must be 1. In other words, the only number dividing both $c - b$ and $c + b$ is 1, so $c - b$ and $c + b$ have no common factor.

We now know that $c - b$ and $c + b$ have no common factor, and that their product is a square since $(c - b)(c + b) = a^2$. The only way that this can happen is if $c - b$ and $c + b$ are themselves squares.[2] So we can write

$$c + b = s^2 \qquad \text{and} \qquad c - b = t^2,$$

where $s > t \geq 1$ are odd integers with no common factors. Solving these two equations for b and c yields

$$c = \frac{s^2 + t^2}{2} \quad \text{and} \quad b = \frac{s^2 - t^2}{2},$$

and then

$$a = \sqrt{(c - b)(c + b)} = st.$$

We have finished our first proof! The following theorem records our accomplishment.

Theorem 2.1 (Pythagorean Triples Theorem). *You will get every primitive Pythagorean triple (a, b, c) with a odd and b even by using the formulas*

$$a = st, \qquad b = \frac{s^2 - t^2}{2}, \qquad c = \frac{s^2 + t^2}{2},$$

where $s > t \geq 1$ are chosen to be any odd integers with no common factors.

For example, if we take $t = 1$, then we get a triple $\left(s, \frac{s^2-1}{2}, \frac{s^2+1}{2}\right)$ whose b and c entries differ by 1. This explains many of the examples we listed above. The following table gives all possible triples with $s \leq 9$.

s	t	$a = st$	$b = \frac{s^2 - t^2}{2}$	$c = \frac{s^2 + t^2}{2}$
3	1	3	4	5
5	1	5	12	13
7	1	7	24	25
9	1	9	40	41
5	3	15	8	17
7	3	21	20	29
7	5	35	12	37
9	5	45	28	53
9	7	63	16	65

[2] This is intuitively clear if you consider the factorization of $c - b$ and $c + b$ into primes, since the primes in the factorization of $c - b$ will be distinct from the primes in the factorization of $c + b$. However, the existence and uniqueness of the factorization into primes is by no means as obvious as it appears. We will discuss this further in Chapter 7.

A Notational Interlude

Mathematicians have created certain standard notations as a shorthand for various quantities. We will keep our use of such notation to a minimum, but there are a few symbols that are so commonly used and are so useful that it is worthwhile to introduce them here. They are

$$\mathbb{N} = \text{the set of natural numbers} = 1, 2, 3, 4, \dots,$$
$$\mathbb{Z} = \text{the set of integers} = \dots - 3, -2, -1, 0, 1, 2, 3, \dots,$$
$$\mathbb{Q} = \text{the set of rational numbers (i.e., fractions).}$$

In addition, mathematicians often use $\mathbb{R}$ to denote the real numbers and $\mathbb{C}$ for the complex numbers, but we will not need these. Why were these letters chosen? The choice of $\mathbb{N}$, $\mathbb{R}$, and $\mathbb{C}$ needs no explanation. The letter $\mathbb{Z}$ for the set of integers comes from the German word "Zahlen," which means numbers. Similarly, $\mathbb{Q}$ comes from the German "Quotient" (which is the same as the English word). We will also use the standard mathematical symbol $\in$ to mean "is an element of the set." So, for example, $a \in \mathbb{N}$ means that a is a natural number, and $x \in \mathbb{Q}$ means that x is a rational number.

Exercises

2.1. (a) We showed that in any primitive Pythagorean triple (a, b, c), either a or b is even. Use the same sort of argument to show that either a or b must be a multiple of 3.
(b) By examining the above list of primitive Pythagorean triples, make a guess about when a, b, or c is a multiple of 5. Try to show that your guess is correct.

2.2. A nonzero integer d is said to *divide* an integer m if $m = dk$ for some number k. Show that if d divides both m and n, then d also divides $m - n$ and $m + n$.

2.3. For each of the following questions, begin by compiling some data; next examine the data and formulate a conjecture; and finally try to prove that your conjecture is correct. (But don't worry if you can't solve every part of this problem; some parts are quite difficult.)
(a) Which odd numbers a can appear in a primitive Pythagorean triple (a, b, c)?
(b) Which even numbers b can appear in a primitive Pythagorean triple (a, b, c)?
(c) Which numbers c can appear in a primitive Pythagorean triple (a, b, c)?

2.4. In our list of examples are the two primitive Pythagorean triples

$$33^2 + 56^2 = 65^2 \qquad \text{and} \qquad 16^2 + 63^2 = 65^2.$$

Find at least one more example of two primitive Pythagorean triples with the same value of c. Can you find three primitive Pythagorean triples with the same c? Can you find more than three?

2.5. In Chapter 1 we saw that the nth triangular number T_n is given by the formula

$$T_n = 1 + 2 + 3 + \cdots + n = \frac{n(n+1)}{2}.$$

The first few triangular numbers are 1, 3, 6, and 10. In the list of the first few Pythagorean triples (a, b, c), we find $(3, 4, 5)$, $(5, 12, 13)$, $(7, 24, 25)$, and $(9, 40, 41)$. Notice that in each case, the value of b is four times a triangular number.

(a) Find a primitive Pythagorean triples (a, b, c) with $b = T_5$. Do the same for $b = T_6$ and with $b = T_7$.

(b) Do you think that for every triangular number T_n, there is a primitive Pythagorean triple (a, b, c) with $b = 4T_n$? If you believe that this is true, then prove it. Otherwise, find some triangular number for which it is not true.

2.6. If you look at the table of primitive Pythagorean triples in this chapter, you will see many triples in which c is 2 greater than a. For example, the triples $(3, 4, 5)$, $(15, 8, 17)$, $(35, 12, 37)$, and $(63, 16, 65)$ all have this property.

(a) Find two more primitive Pythagorean triples (a, b, c) having $c = a + 2$.

(b) Find a primitive Pythagorean triple (a, b, c) having $c = a + 2$ and $c > 1000$.

(c) Try to find a formula that describes all primitive Pythagorean triples (a, b, c) having $c = a + 2$.

2.7. For each primitive Pythagorean triple (a, b, c) in the table in this chapter, compute the quantity $2c - 2a$. Do these values seem to have some special form? Try to prove that your observation is true for all primitive Pythagorean triples.

2.8. (a) Read about the Babylonian number system and write a short description, including the symbols for the numbers 1 to 10 and the multiples of 10 from 20 to 50.

(b) Read about the Babylonian tablet called Plimpton 322 and write a brief description, including its approximate date of origin and some of the large Pythagorean triples that it contains.

Chapter 3

Pythagorean Triples and the Unit Circle

In the previous chapter we described all solutions to

$$a^2 + b^2 = c^2$$

in whole numbers a, b, c. If we divide this equation by c^2, we obtain

$$\left(\frac{a}{c}\right)^2 + \left(\frac{b}{c}\right)^2 = 1.$$

So the pair of rational numbers $(a/c, b/c)$ is a solution to the equation

$$x^2 + y^2 = 1.$$

Everyone knows what the equation $x^2 + y^2 = 1$ looks like: It is a circle C of radius 1 with center at $(0,0)$. We are going to use the geometry of the circle C to find all the points on C whose xy-coordinates are rational numbers. Notice that the circle has four obvious points with rational coordinates, $(\pm 1, 0)$ and $(0, \pm 1)$. Suppose that we take any (rational) number m and look at the line L going through the point $(-1, 0)$ and having slope m. (See Figure 3.1.) The line L is given by the equation

$$L : y = m(x+1) \qquad \text{(point–slope formula).}$$

It is clear from the picture that the intersection $C \cap L$ consists of exactly two points, and one of those points is $(-1, 0)$. We want to find the other one.

To find the intersection of C and L, we need to solve the equations

$$x^2 + y^2 = 1 \qquad \text{and} \qquad y = m(x+1)$$

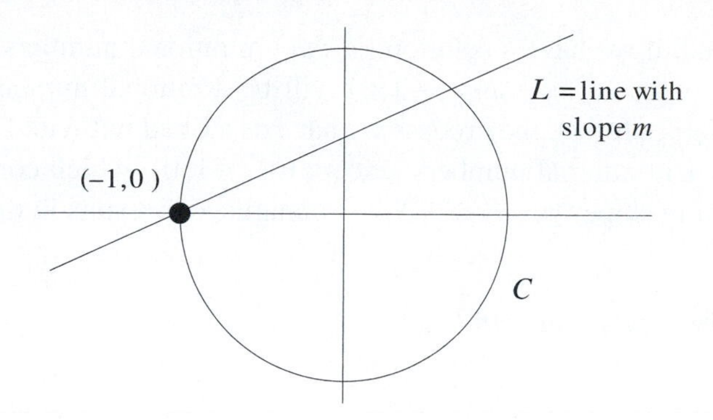

Figure 3.1: The Intersection of a Circle and a Line

for x and y. Substituting the second equation into the first and simplifying, we need to solve

$$\begin{aligned} x^2 + \left(m(x+1)\right)^2 &= 1 \\ x^2 + m^2(x^2 + 2x + 1) &= 1 \\ (m^2+1)x^2 + 2m^2x + (m^2-1) &= 0. \end{aligned}$$

This is just a quadratic equation, so we could use the quadratic formula to solve for x. But there is a much easier way to find the solution. We know that $x = -1$ must be a solution, since the point $(-1, 0)$ is on both C and L. This means that we can divide the quadratic polynomial by $x + 1$ to find the other root:

$$x+1 \overline{\smash{\big)}\,(m^2+1)x^2 + 2m^2x + (m^2-1)} \quad \text{quotient: } (m^2+1)x + (m^2-1).$$

So the other root is the solution of $(m^2 + 1)x + (m^2 - 1) = 0$, which means that

$$x = \frac{1-m^2}{1+m^2}.$$

Then we substitute this value of x into the equation $y = m(x + 1)$ of the line L to find the y-coordinate,

$$y = m(x+1) = m\left(\frac{1-m^2}{1+m^2} + 1\right) = \frac{2m}{1+m^2}.$$

Thus, for every rational number m we get a solution in rational numbers

$$\left(\frac{1-m^2}{1+m^2}, \frac{2m}{1+m^2}\right) \quad \text{to the equation} \quad x^2 + y^2 = 1.$$

On the other hand, if we have a solution (x_1, y_1) in rational numbers, then the slope of the line through (x_1, y_1) and $(-1, 0)$ will be a rational number. So by taking all possible values for m, the process we have described will yield every solution to $x^2 + y^2 = 1$ in rational numbers [except for $(-1, 0)$, which corresponds to a vertical line having slope "$m = \infty$"]. We summarize our results in the following theorem.

Theorem 3.1. *Every point on the circle*

$$x^2 + y^2 = 1$$

whose coordinates are rational numbers can be obtained from the formula

$$(x, y) = \left(\frac{1 - m^2}{1 + m^2}, \frac{2m}{1 + m^2}\right)$$

by substituting in rational numbers for m. [Except for the point $(-1, 0)$, which is the limiting value as $m \to \infty$.]

How is this formula for rational points on a circle related to our formula for Pythagorean triples? If we write the rational number m as a fraction v/u, then our formula becomes

$$(x, y) = \left(\frac{u^2 - v^2}{u^2 + v^2}, \frac{2uv}{u^2 + v^2}\right),$$

and clearing denominators gives the Pythagorean triple

$$(a, b, c) = (u^2 - v^2, 2uv, u^2 + v^2).$$

This is another way of describing all Pythagorean triples, although to describe only the primitive ones would require some restrictions on u and v. You can relate this description to the formula in Chapter 2 by setting

$$u = \frac{s + t}{2} \quad \text{and} \quad v = \frac{s - t}{2}.$$

Exercises

3.1. As we have just seen, we get every Pythagorean triple (a, b, c) with b even from the formula

$$(a, b, c) = (u^2 - v^2, 2uv, u^2 + v^2)$$

by substituting in different integers for u and v. For example, $(u, v) = (2, 1)$ gives the smallest triple $(3, 4, 5)$.

(a) If u and v have a common factor, explain why (a, b, c) will not be a primitive Pythagorean triple.
(b) Find an example of integers $u > v > 0$ that do not have a common factor, yet the Pythagorean triple $(u^2 - v^2, 2uv, u^2 + v^2)$ is not primitive.
(c) Make a table of the Pythagorean triples that arise when you substitute in all values of u and v with $1 \le v < u \le 10$.
(d) Using your table from (c), find some simple conditions on u and v that ensure that the Pythagorean triple $(u^2 - v^2, 2uv, u^2 + v^2)$ is primitive.
(e) Prove that your conditions in (d) really work.

3.2. (a) Use the lines through the point $(1, 1)$ to describe all the points on the circle

$$x^2 + y^2 = 2$$

whose coordinates are rational numbers.
(b) What goes wrong if you try to apply the same procedure to find all the points on the circle $x^2 + y^2 = 3$ with rational coordinates?

3.3. Find a formula for all the points on the hyperbola

$$x^2 - y^2 = 1$$

whose coordinates are rational numbers. [*Hint.* Take the line through the point $(-1, 0)$ having rational slope m and find a formula in terms of m for the second point where the line intersects the hyperbola.]

3.4. The curve

$$y^2 = x^3 + 8$$

contains the points $(1, -3)$ and $(-7/4, 13/8)$. The line through these two points intersects the curve in exactly one other point. Find this third point. Can you explain why the coordinates of this third point are rational numbers?

Chapter 4

Sums of Higher Powers and Fermat's Last Theorem

In the previous two chapters we discovered that the equation

$$a^2 + b^2 = c^2$$

has lots of solutions in whole numbers a, b, c. It is natural to ask whether there are solutions when the exponent 2 is replaced by a higher power. For example, do the equations

$$a^3 + b^3 = c^3 \qquad \text{and} \qquad a^4 + b^4 = c^4 \qquad \text{and} \qquad a^5 + b^5 = c^5$$

have solutions in nonzero integers a, b, c? The answer is "NO." Sometime around 1637, Pierre de Fermat showed that there is no solution for exponent 4. During the eighteenth and nineteenth centuries, Karl Friedrich Gauss and Leonhard Euler showed that there is no solution for exponent 3 and Lejeune Dirichlet and Adrien Legendre dealt with the exponent 5. The general problem of showing that the equation

$$a^n + b^n = c^n$$

has no solutions in positive integers if $n \geq 3$ is known as "Fermat's Last Theorem." It has attained almost cult status in the 350 years since Fermat scribbled the following assertion in the margin of one of his books:

> It is impossible to separate a cube into two cubes, or a fourth power into two fourth powers, or in general any power higher than the second into powers of

> like degree. I have discovered a truly remarkable proof which this margin is too small to contain.[1]

Few mathematicians today believe that Fermat had a valid proof of his "Theorem," which is called his Last Theorem because it was the last of his assertions that remained unproved. The history of Fermat's Last Theorem is fascinating, with literally hundreds of mathematicians making important contributions. Even a brief summary could easily fill a book. This is not our intent in this volume, so we will be content with a few brief remarks.

One of the first general results on Fermat's Last Theorem, as opposed to verification for specific exponents n, was given by Sophie Germain in 1823. She proved that if both p and $2p+1$ are primes then the equation $a^p + b^p = c^p$ has no solutions in integers a, b, c with p not dividing the product abc. A later result of a similar nature, due to A. Wieferich in 1909, is that the same conclusion is true if the quantity $2^p - 2$ is not divisible by p^2. Meanwhile, during the latter part of the nineteenth century a number of mathematicians, including Richard Dedekind, Leopold Kronecker, and especially Ernst Kummer, developed a new field of mathematics called algebraic number theory and used their theory to prove Fermat's Last Theorem for many exponents, although still only a finite list. Then, in 1985, L.M. Adleman, D.R. Heath-Brown, and E. Fouvry used a refinement of Germain's criterion together with difficult analytic estimates to prove that there are infinitely many primes p such that $a^p + b^p = c^p$ has no solutions with p not dividing abc.

> **Sophie Germain** (1776–1831) Sophie Germain was a French mathematician who did important work in number theory and differential equations. She is best known for her work on Fermat's Last Theorem, where she gave a simple criterion that suffices to show that the equation $a^p + b^p = c^p$ has no solutions with abc not divisible by p. She also did work on acoustics and elasticity, especially the theory of vibrating plates. As a mathematics student, she was forced to take correspondence courses from the École Polytechnique in Paris, since they did not accept women as students. For a similar reason, she began her extensive correspondence with Gauss using the pseudonym Monsieur Le Blanc; but when she eventually revealed her identity, Gauss was delighted and sufficiently impressed with her work to recommend her for an honorary degree at the University of Göttingen.

In 1986 Gerhard Frey suggested a new line of attack on Fermat's problem using a notion called modularity. Frey's idea was refined by Jean-Pierre Serre, and Ken

---

[1] Translated from the Latin: "*Cubum autem in duos cubos, aut quadrato quadratum in duos quadrato quadratos, & generaliter nullam in infinitum ultra quadratum potestatem in duos ejusdem nominis fas est dividere; cujus rei demonstrationem mirabilem sane detexi. Hanc marginis exiguitas non caperet.*"

Ribet subsequently proved that if the Modularity Conjecture is true, then Fermat's Last Theorem is true. Precisely, Ribet proved that if every semistable elliptic curve[2] is modular[3] then Fermat's Last Theorem is true. The Modularity Conjecture, which asserts that every rational elliptic curve is modular, was at that time a conjecture originally formulated by Goro Shimura and Yutaka Taniyama. Finally, in 1994, Andrew Wiles announced a proof that every semistable rational elliptic curve is modular, thereby completing the proof of Fermat's 350-year-old claim. Wiles's proof, which is a tour de force using the vast machinery of modern number theory and algebraic geometry, is far too complicated for us to describe in detail, but we will try to convey the flavor of his proof in Chapter 48.

Few mathematical or scientific discoveries arise in a vacuum. Even Sir Isaac Newton, the transcendent genius not noted for his modesty, wrote that "If I have seen further, it is by standing on the shoulders of giants." Here is a list of some of the giants, all contemporary mathematicians, whose work either directly or indirectly contributed to Wiles's brilliant proof. The diversified nationalities highlight the international character of modern mathematics. In alphabetical order: Spencer Bloch (USA), Henri Carayol (France), John Coates (Australia), Pierre Deligne (Belgium), Ehud de Shalit (Israel), Fred Diamond (USA), Gerd Faltings (Germany), Matthias Flach (Germany), Gerhard Frey (Germany), Alexander Grothendieck (Belgium), Yves Hellegouarch (France), Haruzo Hida (Japan), Kenkichi Iwasawa (Japan), Kazuya Kato (Japan), Nick Katz (USA), V.A. Kolyvagin (Russia), Ernst Kunz (Germany), Robert Langlands (Canada), Hendrik Lenstra (The Netherlands), Wen-Ch'ing Winnie Li (USA), Barry Mazur (USA), André Néron (France), Ravi Ramakrishna (USA), Michel Raynaud (France), Ken Ribet (USA), Karl Rubin (USA), Jean-Pierre Serre (France), Goro Shimura (Japan), Yutaka Taniyama (Japan), John Tate (USA), Richard Taylor (England), Jacques Tilouine (France), Jerry Tunnell (USA), André Weil (France), Andrew Wiles (England).

Exercises

4.1. Write a one- to two-page biography on one (or more) of the following mathematicians. Be sure to describe their mathematical achievements, especially in number theory, and some details of their lives. Also include a paragraph putting them into a historical context

[2]An elliptic curve is a certain sort of curve, not an ellipse, given by an equation of the form $y^2 = x^3 + ax^2 + bx + c$, where a, b, c are integers. The elliptic curve is semistable if the quantities $3b - a^2$ and $27c - 9ab + 2a^3$ have no common factors other than 2 and satisfy a few other technical conditions. We study elliptic curves in Chapters 43–48

[3]An elliptic curve is called modular if there is a map to it from another special sort of curve called a modular curve.

by describing the times (scientifically, politically, socially, etc.) during which they lived and worked: (a) Niels Abel, (b) Claude Gaspar Bachet de Meziriac, (c) Richard Dedekind, (d) Diophantus of Alexandria, (e) Lejeune Dirichlet, (f) Eratosthenes, (g) Euclid of Alexandria, (h) Leonhard Euler, (i) Pierre de Fermat, (j) Leonardo Fibonacci, (k) Karl Friedrich Gauss, (l) Sophie Germain, (m) David Hilbert, (n) Karl Jacobi, (o) Leopold Kronecker, (p) Ernst Kummer, (q) Joseph-Louis Lagrange, (r) Adrien-Marie Legendre, (s) Joseph Liouville, (t) Marin Mersenne, (u) Hermann Minkowski, (v) Sir Isaac Newton, (w) Pythagoras, (x) Srinivasa Ramanujan, (y) Bernhard Riemann, (z) P.L. Tchebychef (also spelled Chebychev).

4.2. The equation $a^2+b^2=c^2$ has lots of solutions in positive integers, while the equation $a^3+b^3=c^3$ has no solutions in positive integers. This exercise asks you to look for solutions to the equation

$$a^3+b^3=c^2 \qquad (*)$$

in integers $c \geq b \geq a \geq 1$.

(a) The equation $(*)$ has the solution $(a,b,c)=(2,2,4)$. Find three more solutions in positive integers. [*Hint*. Look for solutions of the form $(a,b,c)=(xz,yz,z^2)$. Not every choice of x,y,z will work, of course, so you'll need to figure out which ones do work.]

(b) If (A,B,C) is a solution to $(*)$ and n is any integer, show that (n^2A,n^2B,n^3C) is also a solution to $(*)$. We will say that a solution (a,b,c) to $(*)$ is *primitive* if it does not look like (n^2A,n^2B,n^3C) for any $n \geq 2$.

(c) Write down four different primitive solutions to $(*)$. [That is, redo (a) using only primitive solutions.]

(d) The solution $(2,2,4)$ has $a=b$. Find all primitive solutions that have $a=b$.

(e) Find a primitive solution to $(*)$ that has $a>10000$.

Chapter 5

Divisibility and the Greatest Common Divisor

As we have already seen in our study of Pythagorean triples, the notions of divisibility and factorizations are important tools in number theory. In this chapter we will look at these ideas more closely.

Suppose that m and n are integers with $m \neq 0$. We say that m *divides* n if n is a multiple of m, that is, if there is an integer k so that $n = mk$. If m divides n, we write $m|n$. Similarly, if m does not divide n, then we write $m \nmid n$. For example,

$$3|6 \quad \text{and} \quad 12|132, \quad \text{since} \quad 6 = 3 \cdot 2 \quad \text{and} \quad 132 = 12 \cdot 11.$$

The divisors of 6 are 1, 2, 3, and 6. On the other hand, $5 \nmid 7$, since no integer multiple of 5 is equal to 7. A number that divides n is called a *divisor of* n.

If we are given two numbers, we can look for common divisors, that is, numbers that divide both of them. For example, 4 is a common divisor of 12 and 20, since $4|12$ and $4|20$. Notice that 4 is the largest common divisor of 12 and 20. Similarly, 3 is a common divisor of 18 and 30, but it is not the largest, since 6 is also a common divisor. The largest common divisor of two numbers is an extremely important quantity that will frequently appear during our number theoretic excursions.

> The *greatest common divisor* of two numbers a and b (not both zero) is the largest number that divides both of them. It is denoted $\gcd(a, b)$. If $\gcd(a, b) = 1$, we say that a and b are *relatively prime*.

Two examples that we mentioned above are

$$\gcd(12, 20) = 4 \qquad \text{and} \qquad \gcd(18, 30) = 6.$$

Another example is

$$\gcd(225, 120) = 15.$$

We can check that this answer is correct by factoring $225 = 3^2 \cdot 5^2$ and $120 = 2^3 \cdot 3 \cdot 5$, but, in general, factoring a and b is not an efficient way to compute their greatest common divisor.[1]

The most efficient method known for finding the greatest common divisors of two numbers is called the *Euclidean algorithm.* It consists of doing a sequence of divisions with remainder until the remainder is zero. We will illustrate with two examples before describing the general method.

As our first example, we will compute $\gcd(36, 132)$. The first step is to divide 132 by 36, which gives a quotient of 3 and a remainder of 24. We write this as

$$132 = 3 \times 36 + 24.$$

The next step is to take 36 and divide it by the remainder 24 from the previous step. This gives

$$36 = 1 \times 24 + 12.$$

Next we divide 24 by 12, and we find a remainder of 0,

$$24 = 2 \times 12 + 0.$$

The Euclidean algorithm says that when you get a remainder of 0 then the remainder from the previous step is the greatest common divisor of the original two numbers. So in this case we find that $\gcd(132, 36) = 12$.

Let's do a larger example. We will compute

$$\gcd(1160718174, 316258250).$$

Our reason for doing a large example like this is to help convince you that the Euclidean algorithm gives a far more efficient way to compute gcd's than factorization. We begin by dividing 1160718174 by 316258250, which gives 3 with a remainder of 211943424. Next we take 316258250 and divide it by 211943424. This process continues until we get a remainder of 0. The calculations are given in

[1]An even less efficient way to compute the greatest common divisor of a and b is the method taught to my daughter by her fourth grade teacher, who recommended that the students make complete lists of all the divisors of a and b and then pick out the largest number that appears on both lists!

the following table:

$$\begin{aligned}
1160718174 &= 3 \times 316258250 + 211943424 \\
316258250 &= 1 \times 211943424 + 104314826 \\
211943424 &= 2 \times 104314826 + 3313772 \\
104314826 &= 31 \times 3313772 + 1587894 \\
3313772 &= 2 \times 1587894 + 137984 \\
1587894 &= 11 \times 137984 + 70070 \\
137984 &= 1 \times 70070 + 67914 \\
70070 &= 1 \times 67914 + 2156 \\
67914 &= 31 \times 2156 + \boxed{1078} \leftarrow \text{gcd} \\
2156 &= 2 \times 1078 + 0
\end{aligned}$$

Notice how at each step we divide a number A by a number B to get a quotient Q and a remainder R. In other words,

$$A = Q \times B + R.$$

Then at the next step we replace our old A and B with the numbers B and R and continue the process until we get a remainder of 0. At that point, the remainder R from the previous step is the greatest common divisor of our original two numbers. So the above calculation shows that

$$\gcd(1160718174, 316258250) = 1078.$$

We can partly check our calculation (always a good idea) by verifying that 1078 is indeed a common divisor. Thus

$$1160718174 = 1078 \times 1076733 \quad \text{and} \quad 316258250 = 1078 \times 293375.$$

There is one more practical matter to be mentioned before we undertake a theoretical analysis of the Euclidean algorithm. If we are given A and B, how can we find the quotient Q and the remainder R? Of course, you can always use long division, but that can be time consuming and subject to arithmetic errors if A and B are large. A pleasant alternative is to find a calculator or computer program that will automatically compute Q and R for you. However, even if you are only equipped with an inexpensive calculator, there is an easy three-step method to find Q and R.

Method to Compute Q and R on a Calculator So That $A = B \times Q + R$

1. Use the calculator to divide A by B. You get a number with decimals.
2. Discard all the digits to the right of the decimal point. This gives Q.
3. To find R, use the formula $R = A - B \times Q$.

For example, suppose that $A = 12345$ and $B = 417$. Then $A/B = 29.6043\ldots$, so $Q = 29$ and $R = 12345 - 417 \cdot 29 = 252$.

We're now ready to analyze the Euclidean algorithm. The general method looks like

$$\begin{aligned}
a &= q_1 \times b + r_1 \\
b &= q_2 \times r_1 + r_2 \\
r_1 &= q_3 \times r_2 + r_3 \\
r_2 &= q_4 \times r_3 + r_4 \\
&\vdots \\
r_{n-3} &= q_{n-1} \times r_{n-2} + r_{n-1} \\
r_{n-2} &= q_n \times r_{n-1} + \boxed{r_n} \leftarrow \text{gcd} \\
r_{n-1} &= q_{n+1} r_n + 0
\end{aligned}$$

If we let $r_0 = b$ and $r_{-1} = a$, then every line looks like

$$r_{i-1} = q_{i+1} \times r_i + r_{i+1}.$$

Why is the last nonzero remainder r_n a common divisor of a and b? We start from the bottom and work our way up. The last line $r_{n-1} = q_{n+1} r_n$ shows that r_n divides r_{n-1}. Then the previous line

$$r_{n-2} = q_n \times r_{n-1} + r_n$$

shows that r_n divides r_{n-2}, since it divides both r_{n-1} and r_n. Now looking at the line above that, we already know that r_n divides both r_{n-1} and r_{n-2}, so we find that r_n also divides r_{n-3}. Moving up line by line, when we reach the second line we will already know that r_n divides r_2 and r_1. Then the second line $b = q_2 \times r_1 + r_2$ tells us that r_n divides b. Finally, we move up to the top line and use the fact that r_n divides both r_1 and b to conclude that r_n also divides a. This completes our verification that the last nonzero remainder r_n is a common divisor of a and b.

But why is r_n the *greatest* common divisor of a and b? Suppose that d is any common divisor of a and b. We will work our way back down the list of equations. So from the first equation $a = q_1 \times b + r_1$ and the fact that d divides both a and b, we see that d also divides r_1. Then the second equation $b = q_2 r_1 + r_2$ shows us that d must divide r_2. Continuing down line by line, at each stage we will know that d divides the previous two remainders r_{i-1} and r_i, and then the current line $r_{i-1} = q_{i+1} \times r_i + r_{i+1}$ will tell us that d also divides the next remainder r_{i+1}. Eventually, we reach the penultimate line $r_{n-2} = q_n \times r_{n-1} + r_n$, at which point we conclude that d divides r_n. So we have shown that if d is any common divisor of a and b then d will divide r_n. Therefore, r_n must be the greatest common divisor of a and b.

This completes our verification that the Euclidean algorithm actually computes the greatest common divisor, a fact of sufficient importance to be officially recorded.

Theorem 5.1 (Euclidean Algorithm). *To compute the greatest common divisor of two numbers a and b, let $r_{-1} = a$, let $r_0 = b$, and compute successive quotients and remainders*

$$r_{i-1} = q_{i+1} \times r_i + r_{i+1}$$

for $i = 0, 1, 2, \ldots$ until some remainder r_{n+1} is 0. The last nonzero remainder r_n is then the greatest common divisor of a and b.

There remains the question of why the Euclidean algorithm always finishes. In other words, we know that the last nonzero remainder will be the desired gcd, but how do we know that we ever get a remainder that does equal 0? This is not a silly question, since it is easy to give algorithms that do not terminate; and there are even very simple algorithms for which it is not known whether or not they always terminate. Fortunately, it is easy to see that the Euclidean algorithm always terminates. The reason is simple. Each time we compute a quotient with remainder,

$$A = Q \times B + R,$$

the remainder will be between 0 and $B - 1$. This is clear, since if $R \geq B$, then we can add one more onto the quotient Q and subtract B from R. So the successive remainders in the Euclidean algorithm continually decrease:

$$b = r_0 > r_1 > r_2 > r_3 > \cdots.$$

But all the remainders are greater than or equal to 0, so we have a strictly decreasing sequence of nonnegative integers. Eventually, we must reach a remainder that equals 0; in fact, it is clear that we will reach a remainder of 0 in at most b steps. Fortunately, the Euclidean algorithm is far more efficient than this. You will show in the exercises that the number of steps in the Euclidean algorithm is at most seven times the *number of digits* in b. So, on a computer, it is quite feasible to compute $\gcd(a, b)$ when a and b have hundreds or even thousands of digits!

Exercises

5.1. Use the Euclidean algorithm to compute each of the following gcd's.
(a) $\gcd(12345, 67890)$ (b) $\gcd(54321, 9876)$

5.2. Write a program to compute the greatest common divisor $\gcd(a, b)$ of two integers a and b. Your program should work even if one of a or b is zero. Make sure that you don't go into an infinite loop if a and b are both zero!

5.3. Let $b = r_0, r_1, r_2, \ldots$ be the successive remainders in the Euclidean algorithm applied to a and b. Show that every two steps reduces the remainder by at least one half. In other words, verify that

$$r_{i+2} < \frac{1}{2} r_i \qquad \text{for every } i = 0, 1, 2, \ldots.$$

Conclude that the Euclidean algorithm terminates in at most $2\log_2(b)$ steps, where $\log_2$ is the logarithm to the base 2. In particular, show that the number of steps is at most seven times the number of digits in b. [*Hint*. What is the value of $\log_2(10)$?]

5.4. A number L is called a common multiple of m and n if both m and n divide L. The smallest such L is called the *least common multiple of* m *and* n and is denoted by $\mathrm{LCM}(m, n)$. For example, $\mathrm{LCM}(3, 7) = 21$ and $\mathrm{LCM}(12, 66) = 132$.

(a) Find the following least common multiples.
 (i) $\mathrm{LCM}(8, 12)$ (ii) $\mathrm{LCM}(20, 30)$ (iii) $\mathrm{LCM}(51, 68)$ (iv) $\mathrm{LCM}(23, 18)$.

(b) For each of the LCMs that you computed in (a), compare the value of $\mathrm{LCM}(m, n)$ to the values of m, n, and $\gcd(m, n)$. Try to find a relationship.

(c) Give an argument proving that the relationship you found is correct for all m and n.

(d) Use your result in (b) to compute $\mathrm{LCM}(301337, 307829)$.

(e) Suppose that $\gcd(m, n) = 18$ and $\mathrm{LCM}(m, n) = 720$. Find m and n. Is there more than one possibility? If so, find all of them.

5.5. The "$3n + 1$ *algorithm*" works as follows. Start with any number n. If n is even, divide it by 2. If n is odd, replace it with $3n + 1$. Repeat. So, for example, if we start with 5, we get the list of numbers

$$5, 16, 8, 4, 2, 1, 4, 2, 1, 4, 2, 1, \ldots,$$

and if we start with 7, we get

$$7, 22, 11, 34, 17, 52, 26, 13, 40, 20, 10, 5, 16, 8, 4, 2, 1, 4, 2, 1, \ldots.$$

Notice that if we ever get to 1 the list just continues to repeat with 4, 2, 1's. In general, one of the following two possibilities will occur:[2]

(i) We may end up repeating some number a that appeared earlier in our list, in which case the block of numbers between the two a's will repeat indefinitely. In this case we say that the algorithm *terminates* at the last nonrepeated value, and the number of distinct entries in the list is called the *length of the algorithm*. For example, the algorithm terminates at 1 for both 5 and 7. The length of the algorithm for 5 is 6, and the length of the algorithm for 7 is 17.

(ii) We may never repeat the same number, in which case we say that the algorithm does not terminate.

[2]There is, of course, a third possibility. We may get tired of computing and just stop working, in which case one might say that the algorithm terminates due to exhaustion of the computer!

(a) Find the length and terminating value of the $3n+1$ algorithm for each of the following starting values of n:

$$\text{(i) } n = 21 \qquad \text{(ii) } n = 13 \qquad \text{(iii) } n = 31$$

(b) Do some further experimentation and try to decide whether the $3n + 1$ algorithm always terminates and, if so, what value(s) it terminates at.

(c) Let $L(n)$ be the length of the algorithm for starting value n (assuming that it terminates, of course). For example, $L(5) = 6$ and $L(7) = 17$. Show that if $n = 8k + 4$ then $L(n) = L(n + 1)$. (*Hint.* What does the algorithm do to the starting values $8k + 4$ and $8k + 5$?)

(d) Show that if $n = 128k + 28$ then $L(n) = L(n + 1) = L(n + 2)$.

(e) Find some other conditions, similar to those in (c) and (d), for which consecutive values of n have the same length. (It might be helpful to begin by using the next exercise to accumulate some data.)

5.6. Write a program to implement the $3n + 1$ algorithm described in the previous exercise. The user will input n and your program should return the length $L(n)$ and the terminating value $T(n)$ of the $3n + 1$ algorithm. Use your program to create a table giving the length and terminating value for all starting values $1 \leq n \leq 100$.

Chapter 6

Linear Equations and the Greatest Common Divisor

Given two whole numbers a and b, we are going to look at all the possible numbers we can get by adding a multiple of a to a multiple of b. In other words, we will consider all numbers obtained from the formula

$$ax + by$$

when we substitute all possible integers for x and y. Note that we are going to allow both positive and negative values for x and y. For example, we could take $a = 42$ and $b = 30$. Some of the values of $ax + by$ for this a and b are given in the following table:

	$x = -3$	$x = -2$	$x = -1$	$x = 0$	$x = 1$	$x = 2$	$x = 3$
$y = -3$	-216	-174	-132	-90	-48	-6	36
$y = -2$	-186	-144	-102	-60	-18	24	66
$y = -1$	-156	-114	-72	-30	12	54	96
$y = \ \ 0$	-126	-84	-42	0	42	84	126
$y = \ \ 1$	-96	-54	-12	30	72	114	156
$y = \ \ 2$	-66	-24	18	60	102	144	186
$y = \ \ 3$	-36	6	48	90	132	174	216

Table of Values of $42x + 30y$

Our first observation is that every entry in the table is divisible by 6. This is not surprising, since both 42 and 30 are divisible by 6, so every number of the form $42x + 30y = 6(7x + 5y)$ is a multiple of 6. More generally, it is clear that every number of the form $ax + by$ is divisible by $\gcd(a, b)$, since both a and b are divisible by $\gcd(a, b)$.

A second observation, which is somewhat more surprising, is that the greatest common divisor of 42 and 30, that is 6, actually appears in our table. Thus from the table we see that

$$42 \cdot (-2) + 30 \cdot 3 = 6 = \gcd(42, 30).$$

Further examples suggest the following conclusion:

The smallest positive value of $ax + by$ is equal to $\gcd(a, b)$.

There are many ways to prove that this is true. We will take a constructive approach, via the Euclidean algorithm, which has the advantage of giving a procedure for finding the appropriate values of x and y. In other words, we are going to describe a method of finding integers x and y that are solutions to the equation

$$ax + by = \gcd(a, b).$$

Since, as we have already observed, every number $ax+by$ is divisible by $\gcd(a, b)$, it will follow that the smallest positive value of $ax + by$ is precisely $\gcd(a, b)$.

How might we solve the equation $ax + by = \gcd(a, b)$? If a and b are small, we might be able to guess a solution. For example, the equation

$$10x + 35y = 5$$

has the solution $x = -3$ and $y = 1$, and the equation

$$7x + 11y = 1$$

has the solution $x = -3$ and $y = 2$. We also notice that there can be more than one solution, since $x = 8$ and $y = -5$ is also a solution to $7x + 11y = 1$.

However, if a and b are large, guesswork or trial and error are not going to be helpful. We are going to start by illustrating the Euclidean algorithm method for solving $ax + by = \gcd(a, b)$ with a particular example. So we are going to try to solve

$$22x + 60y = \gcd(22, 60).$$

The first step is to perform the Euclidean algorithm to compute the gcd. We find

$$\begin{aligned} 60 &= 2 \times 22 + 16 \\ 22 &= 1 \times 16 + 6 \\ 16 &= 2 \times 6 + 4 \\ 6 &= 1 \times 4 + 2 \\ 4 &= 2 \times 2 + 0 \end{aligned}$$

This shows that $\gcd(22, 60) = 2$, a fact that is clear without recourse to the Euclidean algorithm. However, the Euclidean algorithm computation is important because we're going to use the intermediate quotients and remainders to solve the equation $22x + 60y = 2$. The first step is to rewrite the first equation as

$$16 = a - 2b, \qquad \text{where we let } a = 60 \text{ and } b = 22.$$

We next substitute this value into the 16 appearing in the second equation. This gives (remember that $b = 22$)

$$b = 1 \times 16 + 6 = 1 \times (a - 2b) + 6.$$

Rearranging this equation to isolate the remainder 6 yields

$$6 = b - (a - 2b) = -a + 3b.$$

Now substitute the values 16 and 6 into the next equation, $16 = 2 \times 6 + 4$:

$$a - 2b = 16 = 2 \times 6 + 4 = 2(-a + 3b) + 4.$$

Again we isolate the remainder 4, yielding

$$4 = (a - 2b) - 2(-a + 3b) = 3a - 8b.$$

Finally, we use the equation $6 = 1 \times 4 + 2$ to get

$$-a + 3b = 6 = 1 \times 4 + 2 = 1 \times (3a - 8b) + 2.$$

Rearranging this equation gives the desired solution

$$-4a + 11b = 2.$$

(We should check our solution: $-4 \times 60 + 11 \times 22 = -240 + 242 = 2$.)

We can summarize the above computation in the following efficient tabular form. Note that the left-hand equations are the Euclidean algorithm, and the right-hand equations compute the solution to $ax + by = \gcd(a, b)$.

$$
\begin{array}{rcl|l}
a = & 2 \times b + & 16 & 16 = a - 2b \\
b = & 1 \times 16 + & 6 & 6 = b - 1 \times 16 \\
 & & & = b - 1 \times (a - 2b) \\
 & & & = -a + 3b \\
16 = & 2 \times 6 + & 4 & 4 = 16 - 2 \times 6 \\
 & & & = (a - 2b) - 2 \times (-a + 3b) \\
 & & & = 3a - 8b \\
6 = & 1 \times 4 + & 2 & 2 = 6 - 1 \times 4 \\
 & & & = (-a + 3b) - 1 \times (3a - 8b) \\
 & & & = -4a + 11b \\
4 = & 2 \times 2 + & 0 &
\end{array}
$$

Why does this method work? As the following table makes clear, we start with the first two lines of the Euclidean algorithm, which involve the quantities a and b, and work our way down.

$$\begin{array}{r|l}
a = \ q_1 b + r_1 & r_1 = a - q_1 b \\
b = q_2 r_1 + r_2 & r_2 = b - q_2 r_1 \\
 & \quad = b - q_2(a - q_1 b) \\
 & \quad = -q_2 a + (1 + q_1 q_2) b \\
r_1 = q_3 r_2 + r_3 & r_3 = r_1 - q_3 r_2 \\
 & \quad = (a - q_1 b) - q_3\bigl(-q_2 a + (1 + q_1 q_2) b\bigr) \\
 & \quad = (1 + q_2 q_3) a - (q_1 + q_3 + q_1 q_2 q_3) b \\
\vdots & \qquad \vdots
\end{array}$$

As we move from line to line, we will continually be forming equations that look like

$$\text{latest remainder} = \text{some multiple of } a \text{ plus some multiple of } b.$$

Eventually, we get down to the last nonzero remainder, which we know is equal to $\gcd(a, b)$, and this gives the desired solution to the equation $\gcd(a, b) = ax + by$.

A larger example with $a = 12453$ and $b = 2347$ is given in tabular form on top of the next page. As before, the left-hand side is the Euclidean algorithm and the right-hand side solves $ax + by = \gcd(a, b)$. We see that $\gcd(12453, 2347) = 1$ and that the equation $12453x + 2347y = 1$ has the solution $(x, y) = (304, -1613)$.

We now know that the equation

$$ax + by = \gcd(a, b)$$

always has a solution in integers x and y. The final topic we discuss in this section is the question of how many solutions it has, and how to describe all the solutions. Let's start with the case that a and b are relatively prime, that is, $\gcd(a, b) = 1$, and suppose that (x_1, y_1) is a solution to the equation

$$ax + by = 1.$$

We can create additional solutions by subtracting a multiple of b from x_1 and adding the same multiple of a onto y_1. In other words, for any integer k we obtain a new solution $(x_1 + kb, y_1 - ka)$.[1] We can check that this is indeed a solution by computing

$$a(x_1 + kb) + b(y_1 - ka) = ax_1 + akb + by_1 - bka = ax_1 + by_1 = 1.$$

[1]Geometrically, we are starting from the known point (x_1, y_1) on the line $ax + by = 1$ and using the fact that the line has slope $-a/b$ to find new points $(x_1 + t, y_1 - (a/b)t)$. To get new points with integer coordinates, we need to let t be a multiple of b. Substituting $t = kb$ gives the new integer solution $(x_1 + kb, y_1 - ka)$.

$a = 5 \times b + 718$	$718 = a - 5b$
$b = 3 \times 718 + 193$	$193 = b - 3 \times 718$ $= b - 3 \times (a - 5b)$ $= -3a + 16b$
$718 = 3 \times 193 + 139$	$139 = 718 - 3 \times 193$ $= (a - 5b) - 3 \times (-3a + 16b)$ $= 10a - 53b$
$193 = 1 \times 139 + 54$	$54 = 193 - 139$ $= (-3a + 16b) - (10a - 53b)$ $= -13a + 69b$
$139 = 2 \times 54 + 31$	$31 = 139 - 2 \times 54$ $= (10a - 53b) - 2 \times (-13a + 69b)$ $= 36a - 191b$
$54 = 1 \times 31 + 23$	$23 = 54 - 31$ $= -13a + 69b - (36a - 191b)$ $= -49a + 260b$
$31 = 1 \times 23 + 8$	$8 = 31 - 23$ $= 36a - 191b - (-49a + 260b)$ $= 85a - 451b$
$23 = 2 \times 8 + 7$	$7 = 23 - 2 \times 8$ $= (-49a + 260b) - 2 \times (85a - 451b)$ $= -219a + 1162b$
$8 = 1 \times 7 + 1$	$1 = 8 - 7$ $= 85a - 451b - (-219a + 1162b)$ $= 304a - 1613b$
$7 = 7 \times 1 + 0$	

So, for example, if we start with the solution $(-1, 2)$ to $5x + 3y = 1$, we obtain new solutions $(-1 + 3k, 2 - 5k)$. Note that the integer k is allowed to be positive, negative, or zero. Putting in particular values of k gives the solutions

$$\ldots (-13, 22),\ (-10, 17),\ (-7, 12),\ (-4, 7),\ (-1, 2),$$
$$(2, -3),\ (5, -8),\ (8, -13),\ (11, -18) \ldots .$$

Still looking at the case that $\gcd(a, b) = 1$, we can show that this procedure gives all possible solutions. Suppose that we are given two solutions (x_1, y_1) and (x_2, y_2) to the equation $ax + by = 1$. In other words,

$$ax_1 + by_1 = 1 \qquad \text{and} \qquad ax_2 + by_2 = 1.$$

We are going to multiply the first equation by y_2, multiply the second equation by y_1, and subtract. This will eliminate b and, after a little bit of algebra, we are left with

$$ax_1y_2 - ax_2y_1 = y_2 - y_1.$$

Similarly, if we multiply the first equation by x_2, multiply the second equation by x_1, and subtract, we find that

$$bx_2y_1 - bx_1y_2 = x_2 - x_1.$$

So if we let $k = x_2y_1 - x_1y_2$, then we find that

$$x_2 = x_1 + kb \qquad \text{and} \qquad y_2 = y_1 - ka.$$

This means that the second solution (x_2, y_2) is obtained from the first solution (x_1, y_1) by adding a multiple of b onto x_1 and subtracting the same multiple of a from y_1. So every solution to $ax + by = 1$ can be obtained from the initial solution (x_1, y_1) by substituting different values of k into $(x_1 + kb, y_1 - ka)$.

What happens if $\gcd(a, b) > 1$? To make the formulas look a little bit simpler, we will let $g = \gcd(a, b)$. We know from the Euclidean algorithm method that there is at least one solution (x_1, y_1) to the equation

$$ax + by = g.$$

But g divides both a and b, so (x_1, y_1) is a solution to the simpler equation

$$\frac{a}{g}x + \frac{b}{g}y = 1.$$

Now our earlier work applies, so we know that every other solution can be obtained by substituting values for k in the formula

$$\left(x_1 + k \cdot \frac{b}{g},\ y_1 - k \cdot \frac{a}{g}\right).$$

This completes our description of the solutions to the equation $ax + by = g$, as summarized in the following theorem.

Theorem 6.1 (Linear Equation Theorem). *Let a and b be nonzero integers, and let $g = \gcd(a, b)$. The equation*

$$ax + by = g$$

always has a solution (x_1, y_1) in integers, and this solution can be found by the Euclidean algorithm method described earlier. Then every solution to the equation can be obtained by substituting integers k into the formula

$$\left(x_1 + k \cdot \frac{b}{g},\ y_1 - k \cdot \frac{a}{g}\right).$$

For example, we saw that the equation

$$60x + 22y = \gcd(60, 22) = 2$$

has the solution $x = -4$, $y = 11$. Then our Linear Equation Theorem says that every solution is obtained from the formula

$$(-4 + 11k, 11 - 30k) \qquad \text{with } k \text{ any integer.}$$

In particular, if we want a solution with x positive, then we can take $k = 1$, which gives the smallest such solution $(x, y) = (7, -19)$.

In this chapter we have shown that the equation

$$ax + by = \gcd(a, b)$$

always has a solution. This fact is extremely important for both theoretical and practical reasons, and we will be using it repeatedly in our number theoretic investigations. For example, we will need to solve the equation $ax + by = 1$ when we study cryptography in Chapter 18. And in the next chapter we will use this equation for our theoretical study of factorization of numbers into primes.

Exercises

6.1. (a) Find a solution in integers to the equation

$$12345x + 67890y = \gcd(12345, 67890).$$

(b) Find a solution in integers to the equation

$$54321x + 9876y = \gcd(54321, 9876).$$

6.2. Describe all integer solutions to each of the following equations.

(a) $105x + 121y = 1$

(b) $12345x + 67890y = \gcd(12345, 67890)$

(c) $54321x + 9876y = \gcd(54321, 9876)$

6.3. 💻 The method for solving $ax + by = \gcd(a, b)$ described in this chapter involves a considerable amount of manipulation and back substitution. This exercise describes an alternative way to compute x and y that is especially easy to implement on a computer.

(a) Show that the algorithm described in Figure 6.1 computes the greatest common divisor g of the positive integers a and b, together with a solution (x, y) in integers to the equation $ax + by = \gcd(a, b)$.

(b) Implement the algorithm on a computer using the computer language of your choice.

(c) Use your program to compute $g = \gcd(a, b)$ and integer solutions to $ax + by = g$ for the following pairs (a, b).
(i) $(19789, 23548)$ (ii) $(31875, 8387)$ (iii) $(22241739, 19848039)$

(d) What happens to your program if $b = 0$? Fix the program so that it deals with this case correctly.

(e) For later applications it is useful to have a solution with $x > 0$. Modify your program so that it always returns a solution with $x > 0$. [*Hint.* If (x, y) is a solution, then so is $(x + b, y - a)$.]

(1)	Set $x = 1$, $g = a$, $v = 0$, and $w = b$.
(2)	If $w = 0$ then set $y = (g - ax)/b$ and return the values (g, x, y).
(3)	Divide g by w with remainder, $g = qw + t$, with $0 \le t < w$.
(4)	Set $s = x - qv$.
(5)	Set $(x, g) = (v, w)$.
(6)	Set $(v, w) = (s, t)$.
(7)	Go to Step (2).

Figure 6.1: Efficient algorithm to solve $ax + by = \gcd(a, b)$

6.4. (a) Find integers x, y, and z that satisfy the equation

$$6x + 15y + 20z = 1.$$

(b) Under what conditions on a, b, c is it true that the equation

$$ax + by + cz = 1$$

has a solution? Describe a general method of finding a solution when one exists.

(c) Use your method from (b) to find a solution in integers to the equation

$$155x + 341y + 385z = 1.$$

6.5. Suppose that $\gcd(a, b) = 1$. Prove that for every integer c, the equation $ax + by = c$ has a solution in integers x and y (*Hint.* Find a solution to $au + by = 1$ and multiply by c.) Find a solution to $37x + 47y = 103$. Try to make x and y as small as possible.

6.6. Sometimes we are only interested in solutions to $ax + by = c$ using nonnegative values for x and y.

(a) Explain why the equation $3x + 5y = 4$ has no solutions with $x \ge 0$ and $y \ge 0$.

(b) Make a list of some of the numbers of the form $3x + 5y$ with $x \ge 0$ and $y \ge 0$. Make a conjecture as to which values are not possible. Then prove that your conjecture is correct.

(c) For each of the following values of (a, b), find the largest number that is not of the form $ax + by$ with $x \geq 0$ and $y \geq 0$.

(i) $(a, b) = (3, 7)$ (ii) $(a, b) = (5, 7)$ (iii) $(a, b) = (4, 11)$.

(d) Let $\gcd(a, b) = 1$. Using your results from (c), find a conjectural formula in terms of a and b for the largest number that is not of the form $ax + by$ with $x \geq 0$ and $y \geq 0$? Check your conjecture for at least two more values of (a, b).

(e) Prove that your conjectural formula in (d) is correct.

(f) Try to generalize this problem to sums of three terms $ax + by + cz$ with $x \geq 0$, $y \geq 0$ and $z \geq 0$. For example, what is the largest number that is not of the form $6x + 10y + 15z$ with nonnegative x, y, z?

Chapter 7

Factorization and the Fundamental Theorem of Arithmetic

A *prime number* is a number $p \geq 2$ whose only (positive) divisors are 1 and p. Numbers $m \geq 2$ that are not primes are called *composite numbers*. For example,

prime numbers	$2, 3, 5, 7, 11, 13, 17, 19, 23, 29, 31, \ldots$
composite numbers	$4, 6, 8, 9, 10, 12, 14, 15, 16, 18, 20, \ldots$

Prime numbers are characterized by the numbers by which they are divisible, that is, they are defined by the property that they are only divisible by 1 and by themselves. So it is not immediately clear that primes numbers should have special properties that involve the numbers that they divide. Thus the following fact concerning prime numbers is both nonobvious and important.

Claim 7.1. *Let p be a prime number, and suppose that p divides the product ab. Then either p divides a or p divides b (or p divides both a and b).*[1]

Verification. We are given that p divides the product ab. If p divides a, we are done, so we may as well assume that p does not divide a. Now consider what $\gcd(p, a)$ can be. It divides p, so it is either 1 or p. It also divides a, so it isn't p, since we have assumed that p does not divide a. Thus, $\gcd(p, a)$ must equal 1.

Now we use the Linear Equation Theorem (Chapter 6) with the numbers p and a. The Linear Equation Theorem says that we can find integers x and y that

[1]You may say that this claim is obvious if we just factor a and b into a product of primes. However, the fact that a number can be factored into a product of primes in exactly one way is itself a nonobvious fact. We will discuss this further later in this chapter.

solve the equation

$$px + ay = 1.$$

[Note that we are using the fact that $\gcd(p, a) = 1$.] Now multiply both sides of the equation by b. This gives

$$pbx + aby = b.$$

Certainly pbx is divisible by p, and also aby is divisible by p, since we know that p divides ab. It follows that p divides the sum

$$pbx + aby,$$

so p divides b. This completes the verification of the claim.[2] □

The claim says that if a prime divides a product ab, it must divide one of the factors. Notice that this is a special property of prime numbers; it is not true for composite numbers. For example, 6 divides the product $15 \cdot 14$, but 6 divides neither 15 nor 14. It is not hard to extend the claim to products with more than two factors.

Theorem 7.2 (Prime Divisibility Property). *Let p be a prime number, and suppose that p divides the product $a_1 a_2 \cdots a_r$. Then p divides at least one of the factors $a_1, a_2, \ldots, a_r$.*

Verification. If p divides a_1, we're done. If not, we apply the claim to the product

$$a_1(a_2 a_3 \cdots a_r)$$

to conclude that p must divide $a_2 a_3 \cdots a_r$. In other words, we are applying the claim with $a = a_1$ and $b = a_2 a_3 \cdots a_r$. We know that $p|ab$, so if $p \nmid a$, the claim says that p must divide b.

So now we know that p divides $a_2 a_3 \cdots a_r$. If p divides a_2, we're done. If not, we apply the claim to the product $a_2(a_3 \cdots a_r)$ to conclude that p must divide $a_3 \cdots a_r$. Continuing in this fashion, we must eventually find some a_i that is divisible by p. □

Later in this chapter we are going to use the Prime Divisibility Property to *prove* that every positive integer can be factored as a product of prime numbers in essentially one way. Unfortunately, this important fact is so familiar to most readers that they will question why it requires a proof. So before giving the proof,

[2] When we are verifying a claim or proving a statement, we use a little box □ to indicate that we have completed the verification.

I want to try to convince you that unique factorization into primes is far from being obvious. For this purpose, I invite you to leave the familiar behind and enter the[3]

Even Number World

(popularly known as the "$\mathbb{E}$-Zone")

Imagine yourself in a world where the only numbers that are known are the even numbers. So, in this world, the only numbers that exist are

$$\mathbb{E} = \{\ldots, -8, -6, -4, -2, 0, 2, 4, 6, 8, 10, \ldots\}.$$

Notice that in the $\mathbb{E}$-Zonewe can add, subtract, and multiply numbers just as usual, since the sum, difference, and product of even numbers is again an even number. We can also talk about divisibility. We say that a number m *$\mathbb{E}$-divides* a number n if there is a number k with $n = mk$. But remember that we're now in the $\mathbb{E}$-Zone, so the word "number" means an even number. For example, 6 $\mathbb{E}$-divides 12, since $12 = 6 \cdot 2$; but 6 does not $\mathbb{E}$-divide 18, since there is no (even) number k satisfying $18 = 6k$.

We can also talk about primes. We say that an (even) number p is an $\mathbb{E}$-prime if it is not divisible by any (even) numbers. (In the $\mathbb{E}$-Zone, a number is not divisible by itself!) For example, here are some $\mathbb{E}$-primes:

$$2,\ 6,\ 10,\ 14,\ 18,\ 22,\ 26,\ 30.$$

Recall the claim we proved above for ordinary numbers. We showed that if a prime p divides a product ab then either p divides a or p divides b. Now move to the $\mathbb{E}$-Zone and consider the $\mathbb{E}$-prime 6 and the numbers $a = 10$ and $b = 18$. The number 6 $\mathbb{E}$-divides $ab = 180$, since $180 = 6 \cdot 30$; but 6 $\mathbb{E}$-divides neither 10 nor 18. So our "obvious" claim is not true here in the $\mathbb{E}$-Zone!

There are other "self-evident facts" that are untrue in the $\mathbb{E}$-Zone. For example, consider the fact that every number can be factored as a product of primes in exactly one way. (Of course, rearranging the order of the factors is not considered a different factorization.) It's not hard to show, even in the $\mathbb{E}$-Zone, that every (even) number can be written as a product of $\mathbb{E}$-primes. But consider the following factorizations:

$$180 = 6 \cdot 30 = 10 \cdot 18.$$

Notice that all of the numbers 6, 30, 10, and 18 are $\mathbb{E}$-primes. This means that 180 can be written as a product of $\mathbb{E}$-primes in two fundamentally different ways! In fact, there is even a third way to write it as a product of $\mathbb{E}$-primes,

$$180 = 2 \cdot 90.$$

[3]Since this book is not a multimedia product, you'll have to use your imagination to supply the appropriate Twilight Zone music.

We are going to leave the $\mathbb{E}$-Zone now and return to the familiar world where odd and even numbers live together in peace and harmony. But we hope that our excursion into the $\mathbb{E}$-Zone has convinced you that facts that seem obvious require a healthy dose of skepticism. Especially, any "fact" that "must be true" because it is very familiar or because it is frequently proclaimed to be true is a fact that needs the most careful scrutiny.[4]

$\mathbb{E}$-Zone Border Crossing — Welcome Back Home

Everyone "knows" that a positive integer can be factored into a product of primes in exactly one way. But our visit to the $\mathbb{E}$-Zone provides convincing evidence that this obvious assertion requires a careful proof.

Theorem 7.3 (The Fundamental Theorem of Arithmetic). *Every integer $n \geq 2$ can be factored into a product of primes*

$$n = p_1 p_2 \cdots p_r$$

in exactly one way.

Before we commence the proof of the Fundamental Theorem of Arithmetic, a few comments are in order. First, if n itself is prime, then we just write $n = n$ and consider this to be a product consisting of a single number. Second, when we write $n = p_1 p_2 \cdots p_r$, we do not mean that $p_1, p_2, \ldots, p_r$ have to be different primes. For example, we would write $300 = 2 \cdot 2 \cdot 3 \cdot 5 \cdot 5$. Third, when we say that n can be written as a product in exactly one way, we do not consider rearrangement of the factors to be a new factorization. For example, $12 = 2 \cdot 2 \cdot 3$ and $12 = 2 \cdot 3 \cdot 2$ and $12 = 3 \cdot 2 \cdot 2$, but all these are treated as the same factorization.

Verification. The Fundamental Theorem of Arithmetic really contains two assertions.

Assertion 1. The number n can be factored into a product of primes in some way.

Assertion 2. There is only one such factorization (aside from rearranging the factors).

We begin with Assertion 1. We are going to give a proof by induction. Don't let this scare you, it just means that first we'll verify the assertion for $n = 2$, and

[4]The principle that well-known and frequently asserted "facts" should be carefully scrutinized also applies to endeavors far removed from mathematics. Politics and journalism come to mind, and the reader will undoubtedly be able to add many others to the list.

then for $n = 3$, and then for $n = 4$, and so on. We begin by observing that $2 = 2$ and $3 = 3$ and $4 = 2^2$, so each of these numbers can be written as a product of primes. This verifies Assertion 1 for $n = 2, 3, 4$. Now suppose that we've verified Assertion 1 for every n up to some number, call it N. This means we know that every number $n \leq N$ can be factored into a product of primes. Now we'll check that the same is true of $N + 1$.

There are two possibilities. First, $N + 1$ may already be prime, in which case it is its own factorization into primes. Second, $N + 1$ may be composite, which means that it can be factored as $N + 1 = n_1 n_2$ with $2 \leq n_1, n_2 \leq N$. But we know Assertion 1 is true for n_1 and n_2, since they are both less than or equal to N. This means that both n_1 and n_2 can be written as a product of primes, say

$$n_1 = p_1 p_2 \cdots p_r \qquad \text{and} \qquad n_2 = q_1 q_2 \cdots q_s.$$

Multiplying these two products together gives

$$N + 1 = n_1 n_2 = p_1 p_2 \cdots p_r q_1 q_2 \cdots q_s,$$

so $N + 1$ can be factored into a product of primes. This means that Assertion 1 is true for $N + 1$.

To recapitulate, we have shown that, if Assertion 1 is true for all numbers less than or equal to N, then it is also true for $N + 1$. But we have checked it is true for 2, 3, and 4, so taking $N = 4$, we see that it is also true for 5. But then we can take $N = 5$ to conclude that it is true for 6. Taking $N = 6$, we see that it is true for $N = 7$, and so on. Since we can continue this process indefinitely, it follows that Assertion 1 is true for every integer.

Next we tackle Assertion 2. It is possible to give an induction proof for this assertion, too, but we will proceed more directly. Suppose that we are able to factor n as a product of primes in two ways, say

$$n = p_1 p_2 p_3 p_4 \cdots p_r = q_1 q_2 q_3 q_4 \cdots q_s.$$

We need to check that the factorizations are the same, possibly after rearranging the order of the factors. We first observe that $p_1 | n$, so $p_1 | q_1 q_2 \cdots q_s$. The Prime Divisibility Property proved earlier in this chapter tells us that p_1 must divide (at least) one of the q_i's, so if we rearrange the q_i's, we can arrange matters so that $p_1 | q_1$. But q_1 is also a prime number, so its only divisors are 1 and q_1. Therefore, we must have $p_1 = q_1$.

Now we cancel p_1 (which is the same as q_1) from both sides of the equation. This gives the equation

$$p_2 p_3 p_4 \cdots p_r = q_2 q_3 q_4 \cdots q_s.$$

Briefly repeating the same argument, we note that p_2 divides the left-hand side of this equation, so p_2 divides the right-hand side, and hence by the Prime Divisibility Property, p_2 divides one of the q_i's. After rearranging the factors, we get $p_2|q_2$, and then the fact that q_2 is prime means that $p_2 = q_2$. This allows us to cancel p_2 (which equals q_2) to obtain the new equation

$$p_3 p_4 \cdots p_r = q_3 q_4 \cdots q_s.$$

We can continue in this fashion until either all the p_i's or all the q_i's are gone. But if all the p_i's are gone, then the left-hand side of the equation equals 1, so there cannot be any q_i's left, either. Similarly, if the q_i's are all gone, then the p_i's must all be gone. In other words, the number of p_i's must be the same as the number of q_i's. To recapitulate, we have shown that if

$$n = p_1 p_2 p_3 p_4 \cdots p_r = q_1 q_2 q_3 q_4 \cdots q_s,$$

where all the p_i's and q_i's are primes, then $r = s$, and we can rearrange the q_i's so that

$$p_1 = q_1 \quad \text{and} \quad p_2 = q_2 \quad \text{and} \quad p_3 = q_3 \quad \text{and} \quad \ldots \quad \text{and} \quad p_r = q_s.$$

This completes the verification that there is only one way to write n as a product of primes. □

The Fundamental Theorem of Arithmetic says that every integer $n \geq 2$ can be written as a product of prime numbers. Suppose we are given a particular integer n. As a practical matter, how can we write it as a product of primes? If n is fairly small (for example, $n = 180$) we can factor it by inspection,

$$180 = 2 \cdot 90 = 2 \cdot 2 \cdot 45 = 2 \cdot 2 \cdot 3 \cdot 15 = 2 \cdot 2 \cdot 3 \cdot 3 \cdot 5.$$

If n is larger (for example, $n = 9105293$) it may be more difficult to find a factorization. One method is to try dividing n by primes $2, 3, 5, 7, 11, \ldots$ until we find a divisor. For $n = 9105293$, we find after some work that the smallest prime dividing n is 37. We factor out the 37,

$$9105293 = 37 \cdot 246089,$$

and continue checking $37, 41, 43, \ldots$ to find a prime that divides 246089. We find that $43|246089$, since $246089 = 43 \cdot 5723$. And so on until we factor $5723 = 59 \cdot 97$, where we recognize that 59 and 97 are both primes. This gives the complete prime factorization

$$9105293 = 37 \cdot 43 \cdot 59 \cdot 97.$$

If n is not itself prime, then there must be a prime $p \leq \sqrt{n}$ that divides n. To see why this is true, we observe that, if p is the smallest prime that divides n, then $n = pm$ with $m \geq p$, and hence $n = pm \geq p^2$. Taking the square root of both sides yields $\sqrt{n} \geq p$. This gives the following foolproof method for writing any number n as a product of primes:

> To write n as a product of primes, try dividing it by every number (or just every prime number) $2, 3, \ldots$ that is less than or equal to $\sqrt{n}$. If you find no numbers that divide n, then n itself is prime. Otherwise, the first divisor that you find will be a prime p. Factor $n = pm$ and repeat the process with m.

This procedure, although fairly inefficient, works fine on a computer for numbers that are moderately large, say up to 10 digits. But how about a number like $n = 10^{128} + 1$? If n turns out to be prime, we won't find out until we've checked $\sqrt{n} \approx 10^{64}$ possible divisors. This is completely infeasible. If we could check 1,000,000,000 (that's one billion) possible divisors each second, it would still take approximately $3 \cdot 10^{48}$ years! This leads to the following two closely related questions:

Question 1. How can we tell if a given number n is prime or composite?

Question 2. If n is composite, how can we factor it into primes?

Although it might seem that these questions are the same, it turns out that Question 1 is much easier to answer than Question 2. We will later see how to write down large numbers that we know are composite, even though we will be unable to write down any of their factors. In a similar fashion, we will be able to find very large prime numbers p and q so that, if we were to send someone the value of the product $n = pq$, they would be unable to factor n to retrieve the numbers p and q. This curious fact, that it is very easy to multiply two numbers but very difficult to factor the product, lies at the heart of a remarkable application of number theory to the creation of very secure codes. We will describe these codes in Chapter 18.

Exercises

7.1. Suppose that $\gcd(a, b) = 1$, and suppose further that a divides the product bc. Show that a must divide c.

7.2. Suppose that $\gcd(a, b) = 1$, and suppose further that a divides c and that b divides c. Show that the product ab must divide c.

7.3. Give a proof by induction of each of the following formulas. [Notice that (a) is the formula that we proved in Chapter 1 using a geometric argument and that (c) is the first n terms of the geometric series.]

(a) $1+2+3+\cdot+n = \dfrac{n(n+1)}{2}$

(b) $1^2+2^2+3^2+\cdots+n^2 = \dfrac{n(n+1)(2n+1)}{6}$

(c) $1+a+a^2+a^3+\cdots+a^n = \dfrac{1-a^{n+1}}{1-a} \qquad (a \neq 1)$

(d) $\dfrac{1}{1\cdot 2}+\dfrac{1}{2\cdot 3}+\dfrac{1}{3\cdot 4}+\cdots+\dfrac{1}{(n-1)n} = \dfrac{n-1}{n}$

7.4. This exercise asks you to continue the investigation of the $\mathbb{E}$-Zone. Remember as you work that for the purposes of this exercise, odd numbers do not exist!

(a) Describe all $\mathbb{E}$-primes.

(b) Show that every even number can be factored as a product of $\mathbb{E}$-primes. (*Hint*. Mimic our proof of this fact for ordinary numbers.)

(c) We saw that 180 has three different factorizations as a product of $\mathbb{E}$-primes. Find the smallest number that has two different factorizations as a product of $\mathbb{E}$-primes. Is 180 the smallest number with three factorizations? Find the smallest number with four factorizations.

(d) The number 12 has only one factorization as a product of $\mathbb{E}$-primes: $12 = 2\cdot 6$. (As usual, we consider $2\cdot 6$ and $6\cdot 2$ to be the same factorization.) Describe all even numbers that have only one factorization as a product of $\mathbb{E}$-primes.

7.5. Welcome to $\mathbb{M}$-World, where the only numbers that exist are positive integers that leave a remainder of 1 when divided by 4. In other words, the only $\mathbb{M}$-numbers that exist are

$$\{1, 5, 9, 13, 17, 21, \ldots\}.$$

(Another description is that these are the numbers of the form $4t+1$ for $t = 0, 1, 2, \ldots$.) In the $\mathbb{M}$-World, we cannot add numbers, but we can multiply them, since if a and b both leave a remainder of 1 when divided by 4 then so does their product. (Do you see why this is true?)

We say that m $\mathbb{M}$-*divides* n if $n = mk$ for some $\mathbb{M}$-number k. And we say that n *is an* $\mathbb{M}$-*prime* if its only $\mathbb{M}$-divisors are 1 and itself. (Of course, we don't consider 1 itself to be an $\mathbb{M}$-prime.)

(a) Find the first six $\mathbb{M}$-primes.

(b) Find an $\mathbb{M}$-number n that has two *different* factorizations as a product of $\mathbb{M}$-primes.

7.6. In this exercise you are asked to write programs to factor a (positive) integer n into a product of primes. (If $n = 0$, be sure to return an error message instead of going into an infinite loop!) A convenient way to represent the factorization of n is as a $2\times r$ matrix. Thus, if

$$n = p_1^{k_1} p_2^{k_2} \cdots p_r^{k_r},$$

then store the factorization of n as the matrix

$$\begin{pmatrix} p_1 & p_2 & \cdots & p_r \\ k_1 & k_2 & \cdots & k_r \end{pmatrix}.$$

(If your computer doesn't allow dynamic storage allocation, you'll have to decide ahead of time how many factors to allow.)

(a) Write a program to factor n by trying each possible factor $d = 2, 3, 4, 5, 6, \ldots$. (This is an extremely inefficient method but will serve as a warm-up exercise.)

(b) Modify your program by storing the values of the first 100 (or more) primes and first removing these primes from n before looking for larger prime factors. You can speed up your program when trying larger d's as potential factors if you don't bother checking d's that are even, or divisible by 3, or by 5. You can also increase efficiency by using the fact that a number m is prime if it is not divisible by any number between 2 and $\sqrt{m}$. Use your program to find the complete factorization of all numbers between 1,000,000 and 1,000,030.

(c) Write a subroutine that prints the factorization of n in a nice format. Optimally, the exponents should appear as exponents; but if this is not possible, then print the factorization of (say) $n = 75460 = 2^2 \cdot 5 \cdot 7^3 \cdot 11$ as

```
2^2 * 5 * 7^3 * 11.
```

(To make the output easier to read, don't print exponents that equal 1.)

Chapter 8

Congruences

Divisibility is a powerful tool in the theory of numbers. We have seen this amply demonstrated in our work on Pythagorean triples, greatest common divisors, and factorization into primes. In this chapter we will discuss the theory of congruences. Congruences provide a convenient way to describe divisibility properties. In fact, they are so convenient and natural that they make the theory of divisibility very similar to the theory of equations.

We say that *a is congruent to b modulo m*, and we write

$$a \equiv b \pmod{m},$$

if m divides $a - b$. For example,

$$7 \equiv 2 \pmod{5} \qquad \text{and} \qquad 47 \equiv 35 \pmod{6},$$

since

$$5|(7-2) \qquad \text{and} \qquad 6|(47-35).$$

In particular, if a divided by m leaves a remainder of r, then a is congruent to r modulo m. Notice that the remainder satisfies $0 \le r < m$, so every integer is congruent, modulo m, to a number between 0 and $m - 1$.

The number m is called the *modulus* of the congruence. Congruences with the same modulus behave in many ways like ordinary equations. Thus, if

$$a_1 \equiv b_1 \pmod{m} \quad \text{and} \quad a_2 \equiv b_2 \pmod{m}, \quad \text{then}$$
$$a_1 \pm a_2 \equiv b_1 \pm b_2 \pmod{m} \quad \text{and} \quad a_1 a_2 \equiv b_1 b_2 \pmod{m}.$$

Warning. It is not always possible to divide congruences. In other words, if $ac \equiv bc \pmod{m}$, it need not be true that $a \equiv b \pmod{m}$. For example,

$15 \cdot 2 \equiv 20 \cdot 2 \pmod{10}$, but $15 \not\equiv 20 \pmod{10}$. Even more distressing, it is possible to have

$$uv \equiv 0 \pmod{m} \text{ with } u \not\equiv 0 \pmod{m} \text{ and } v \not\equiv 0 \pmod{m}.$$

Thus $6 \cdot 4 \equiv 0 \pmod{12}$, but $6 \not\equiv 0 \pmod{12}$ and $4 \not\equiv 0 \pmod{12}$. However, if $\gcd(c, m) = 1$, then it is okay to cancel c from the congruence $ac \equiv bc \pmod{m}$. You will be asked to verify this as an exercise.

Congruences with unknowns can be solved in the same way that equations are solved. For example, to solve the congruence

$$x + 12 \equiv 5 \pmod{8},$$

we subtract 12 from each side to get

$$x \equiv 5 - 12 \equiv -7 \pmod{8}.$$

This solution is fine, or we can use the equivalent solution $x \equiv 1 \pmod{8}$. Notice that -7 and 1 are the same modulo 8, since their difference is divisible by 8.

Here's another example. To solve

$$4x \equiv 3 \pmod{19},$$

we will multiply both sides by 5. This gives

$$20x \equiv 15 \pmod{19}.$$

But $20 \equiv 1 \pmod{19}$, so $20x \equiv x \pmod{19}$. Thus the solution is

$$x \equiv 15 \pmod{19}.$$

We can check our answer by substituting 15 into the original congruence. Is $4 \cdot 15 \equiv 3 \pmod{19}$? Yes, because $4 \cdot 15 - 3 = 57 = 3 \cdot 19$ is divisible by 19.

We solved this last congruence by a trick, but if all else fails, there's always the "climb every mountain" technique.[1] To solve a congruence modulo m, we can just try each value $0, 1, \ldots, m-1$ for each variable. For example, to solve the congruence

$$x^2 + 2x - 1 \equiv 0 \pmod{7},$$

we just try $x = 0, x = 1, \ldots, x = 6$. This leads to the two solutions $x \equiv 2 \pmod{7}$ and $x \equiv 3 \pmod{7}$. Of course, there are other solutions, such as $x \equiv 9 \pmod{7}$.

[1] Also known as the "ford every stream" technique for those who prefer wet feet to vertigo.

But 9 and 2 are not really different solutions, since they are the same modulo 7. So when we speak of "finding all the solutions to a congruence," we normally mean that we will find all incongruent solutions, that is, all solutions that are not congruent to one another.

We also observe that there are many congruences, such as $x^2 \equiv 3 \pmod{10}$, that have no solutions. This shouldn't be too surprising. After all, there are ordinary equations such as $x^2 = -1$ that have no (real) solutions.

Our final task in this chapter is to solve congruences that look like

$$ax \equiv c \pmod{m}.$$

Some congruences of this type have no solutions. For example, if

$$6x \equiv 15 \pmod{514}$$

were to have a solution, then 514 would have to divide $6x - 15$. But $6x - 15$ is always odd, so it cannot be divisible by the even number 514. Hence the congruence $6x \equiv 15 \pmod{514}$ has no solutions.

Before giving the general theory, let's try an example. We will solve the congruence

$$18x \equiv 8 \pmod{22}.$$

This means we need to find a value of x with 22 dividing $18x - 8$, so we have to find a value of x with $18x - 8 = 22y$ for some y. In other words, we need to solve the linear equation

$$18x - 22y = 8.$$

We know from Chapter 6 that we can solve the equation

$$18u - 22v = \gcd(18, 22) = 2,$$

and indeed we easily find the solution $u = 5$ and $v = 4$. But we really want the right-hand side to equal 8, so we multiply by 4 to get

$$18 \cdot (5 \cdot 4) - 22 \cdot (4 \cdot 4) = 8.$$

Thus, $18 \cdot 20 \equiv 8 \pmod{22}$, so $x \equiv 20 \pmod{22}$ is a solution to the original congruence. We will soon see that this congruence has two different solutions modulo 22; the other one turns out to be $x \equiv 9 \pmod{22}$.

Suppose now that we are asked to solve an arbitrary congruence of the form

$$ax \equiv c \pmod{m}.$$

We need to find an integer x so that m divides $ax - c$. The number m will divide the number $ax - c$ if we can find an integer y so that $ax - c = my$. Rearranging this last equation slightly, we see that $ax \equiv c \pmod{m}$ has a solution if, and only if, the linear equation $ax - my = c$ has a solution. This should look familiar; it is precisely the sort of problem we solved in Chapter 6.

To make our formulas a bit neater, we will let $g = \gcd(a, m)$. Our first observation is that every number of the form $ax - my$ is a multiple of g; so if g does not divide c, then $ax - my = c$ has no solutions and so $ax \equiv c \pmod{m}$ also has no solutions.

Next suppose that g does divide c. We know from the Linear Equation Theorem in Chapter 6 that there is always a solution to the equation

$$au + mv = g.$$

Suppose we find a solution $u = u_0$, $v = v_0$, either by trial and error or by using the Euclidean algorithm method described in Chapter 6. Since we are assuming that g divides c, we can multiply this equation by the integer c/g to obtain the equation

$$a\frac{cu_0}{g} + m\frac{cv_0}{g} = c.$$

This means that

$$x_0 \equiv \frac{cu_0}{g} \pmod{m} \quad \text{is a solution to the congruence} \quad ax \equiv c \pmod{m}.$$

Are there other solutions? Suppose that x_1 is some other solution to the congruence $ax \equiv c \pmod{m}$. Then $ax_1 \equiv ax_0 \pmod{m}$, so m divides $ax_1 - ax_0$. This implies that

$$\frac{m}{g} \quad \text{divides} \quad \frac{a(x_1 - x_0)}{g},$$

and we know that m/g and a/g have no common factors, so m/g must divide $x_1 - x_0$. In other words, there is some number k so that

$$x_1 = x_0 + k \cdot \frac{m}{g}.$$

But any two solutions that differ by a multiple of m are considered to be the same, so there will be exactly g different solutions that are obtained by taking $k = 0, 1, \ldots, g - 1$.

This completes our analysis of the congruence $ax \equiv c \pmod{m}$. We summarize our findings in the following statement.

Theorem 8.1 (Linear Congruence Theorem). *Let a, c, and m be integers with $m \geq 1$, and let $g = \gcd(a, m)$.*

(a) *If $g \nmid c$, then the congruence $ax \equiv c \pmod{m}$ has no solutions.*

(b) *If $g|c$, then the congruence $ax \equiv c \pmod{m}$ has exactly g incongruent solutions. To find the solutions, first find a solution (u_0, v_0) to the linear equation*

$$au + mv = g.$$

(A method for solving this equation is described in Chapter 6.) Then $x_0 = cu_0/g$ is a solution to $ax \equiv c \pmod{m}$, and a complete set of incongruent solutions is given by

$$x \equiv x_0 + k \cdot \frac{m}{g} \pmod{m} \quad \text{for } k = 0, 1, 2, \ldots, g - 1.$$

For example, the congruence

$$943x \equiv 381 \pmod{2576}$$

has no solutions, since $\gcd(943, 2576) = 23$ does not divide 381. On the other hand, the congruence

$$893x \equiv 266 \pmod{2432}$$

has 19 solutions, since $\gcd(893, 2432) = 19$ does divide 266. Notice that we are able to determine the number of solutions without having computed any of them. To actually find the solutions, we first solve

$$893u - 2432v = 19.$$

Using the methods from Chapter 6, we find the solution $(u, v) = (79, 29)$. Multiplying by $266/19 = 14$ gives the solution

$$(x, y) = (1106, 406) \qquad \text{to the equation} \qquad 893x - 2432y = 266.$$

Finally, the complete set of solutions to

$$893x \equiv 266 \pmod{2432}$$

is obtained by starting with $x \equiv 1106 \pmod{2432}$ and adding multiples of the quantity $2432/19 = 128$. (Don't forget that if the numbers go above 2432 we are allowed to subtract 2432.) The 19 incongruent solutions are

$$1106, 1234, 1362, 1490, 1618, 1746, 1874, 2002, 2130, 2258,$$
$$2386, 82, 210, 338, 466, 594, 722, 850, 978.$$

Important Note. The most important case of the Linear Congruence Theorem is when $\gcd(a, m) = 1$. In this case, it says that the congruence

$$ax \equiv c \pmod{m} \qquad (*)$$

has exactly one solution. We might even write the solution as a fraction

$$x \equiv \frac{c}{a} \pmod{m},$$

but if we do, then we must remember that the symbol "$\frac{c}{a} \pmod{m}$" is really only a convenient shorthand for the solution to the congruence $(*)$.

Exercises

8.1. Suppose that $a_1 \equiv b_1 \pmod{m}$ and $a_2 \equiv b_2 \pmod{m}$.
(a) Verify that $a_1 + a_2 \equiv b_1 + b_2 \pmod{m}$ and that $a_1 - a_2 \equiv b_1 - b_2 \pmod{m}$.
(b) Verify that $a_1 a_2 \equiv b_1 b_2 \pmod{m}$.

8.2. Suppose that

$$ac \equiv bc \pmod{m}$$

and also assume that $\gcd(c, m) = 1$. Prove that $a \equiv b \pmod{m}$.

8.3. Find all incongruent solutions to each of the following congruences.
(a) $7x \equiv 3 \pmod{15}$ (b) $6x \equiv 5 \pmod{15}$
(c) $x^2 \equiv 1 \pmod{8}$ (d) $x^2 \equiv 2 \pmod{7}$
(e) $x^2 \equiv 3 \pmod{7}$

8.4. Prove that the following divisibility tests work.
(a) The number a is divisible by 4 if and only if its last two digits are divisible by 4.
(b) The number a is divisible by 8 if and only if its last three digits are divisible by 8.
(c) The number a is divisible by 3 if and only if the sum of its digits is divisible by 3.
(d) The number a is divisible by 9 if and only if the sum of its digits is divisible by 9.
(e) The number a is divisible by 11 if and only if the alternating sum of the digits of a is divisible by 11. (If the digits of a are $a_1 a_2 a_3 \ldots a_{d-1} a_d$, the alternating sum means to take $a_1 - a_2 + a_3 - \cdots$ with alternating plus and minus signs.)
[*Hint*. For (a), reduce modulo 100, and similarly for (b). For (c), (d), and (e), write a as a sum of multiples of powers of 10 and reduce modulo 3, 9, and 11.]

8.5. Find all incongruent solutions to each of the following linear congruences.
(a) $8x \equiv 6 \pmod{14}$
(b) $66x \equiv 100 \pmod{121}$
(c) $21x \equiv 14 \pmod{91}$

8.6. Determine the number of incongruent solutions for each of the following congruences. You need not write down the actual solutions.

(a) $72x \equiv 47 \pmod{200}$

(b) $4183x \equiv 5781 \pmod{15087}$

(c) $1537x \equiv 2863 \pmod{6731}$

8.7. Write a program that solves the congruence

$$ax \equiv c \pmod{m}.$$

[If $\gcd(a, m)$ does not divide c, return an error message and the value of $\gcd(a, m)$.]

Chapter 9

Congruences, Powers, and Fermat's Little Theorem

Take a number a and consider its powers $a, a^2, a^3, \ldots$ modulo m. Is there any pattern to these powers? We will start by looking at a prime modulus $m = p$, since the pattern is easier to spot. This is a common situation in the theory of numbers, especially when working with congruences. So whenever you're faced with discovering a congruence pattern, it's usually a good idea to begin with a prime modulus.

For each of the primes $p = 3$, $p = 5$, and $p = 7$, we have listed integers $a = 0, 1, 2, \ldots$ and some of their powers modulo p. Before reading further, you should stop, examine these tables, and try to formulate some conjectural patterns. Then test your conjectures by creating a similar table for $p = 11$ and seeing if your patterns are still true.

a	a^2	a^3	a^4
0	0	0	0
1	1	1	1
2	1	2	1

a^k modulo 3

a	a^2	a^3	a^4	a^5	a^6
0	0	0	0	0	0
1	1	1	1	1	1
2	4	3	1	2	4
3	4	2	1	3	4
4	1	4	1	4	1

a^k modulo 5

a	a^2	a^3	a^4	a^5	a^6	a^7	a^8
0	0	0	0	0	0	0	0
1	1	1	1	1	1	1	1
2	4	1	2	4	1	2	4
3	2	6	4	5	1	3	2
4	2	1	4	2	1	4	2
5	4	6	2	3	1	5	4
6	1	6	1	6	1	6	1

a^k modulo 7

Many interesting patterns are visible in these tables. The one that we will be concerned with in this chapter can be seen in the columns

$$a^2 \pmod{3}, \qquad a^4 \pmod{5}, \qquad \text{and} \qquad a^6 \pmod{7}.$$

Every entry in these columns, aside from the top one, is equal to 1. Does this pattern continue to hold for larger primes? You can check the table you made for $p = 11$, and you will find that

$$1^{10} \equiv 1 \pmod{11}, \quad 2^{10} \equiv 1 \pmod{11}, \quad 3^{10} \equiv 1 \pmod{11} \ldots$$
$$9^{10} \equiv 1 \pmod{11}, \quad \text{and} \quad 10^{10} \equiv 1 \pmod{11}.$$

This leads us to make the following conjecture:

$$a^{p-1} \equiv 1 \pmod{p} \qquad \text{for every integer } 1 \le a < p.$$

Of course, we don't really need to restrict a to be between 1 and $p-1$. If a_1 and a_2 differ by a multiple of p then their powers will be the same modulo p. So the real condition on a is that it not be a multiple of p. This result was first stated by Pierre de Fermat in a letter to Frénicle de Bessy dated 1640, but Fermat gave no indication of his proof. The first known proof appears to be due to Gottfried Leibniz.[1]

Theorem 9.1 (Fermat's Little Theorem). *Let p be a prime number, and let a be any number with $a \not\equiv 0 \pmod{p}$. Then*

$$a^{p-1} \equiv 1 \pmod{p}.$$

Before giving the proof of Fermat's Little Theorem, we want to indicate its power and show how it can be used to simplify computations. As a particular example, consider the congruence

$$6^{22} \equiv 1 \pmod{23}.$$

This says that the number $6^{22} - 1$ is a multiple of 23. If we wanted to check this fact without using Fermat's Little Theorem, we would have to multiply out 6^{22}, subtract 1, and divide by 23. Here's what we get:

$$6^{22} - 1 = 23 \cdot 5722682775750745.$$

[1]Gottfried Leibniz (1646–1716) is best known as one of the discoverers of the calculus. He and Isaac Newton worked out the main theorems of the calculus independently and at about the same time. The German and English mathematical communities spent the next two centuries arguing over who deserved priority. The current consensus is that both Leibniz and Newton should be given joint credit as the (independent) discoverers of the calculus.

Similarly, in order to verify directly that $73^{100} \equiv 1 \pmod{101}$, we would have to compute $73^{100} - 1$. Unfortunately, $73^{100} - 1$ has 187 digits! And notice that this example only uses $p = 101$, which is a comparatively small prime. Fermat's Little Theorem thus describes a very surprising fact about extremely large numbers.

We can use Fermat's Little Theorem to simplify computations. For example, in order to compute $2^{35} \pmod 7$, we can use the fact that $2^6 \equiv 1 \pmod 7$. So we write $35 = 6 \cdot 5 + 5$ and use the law of exponents to compute

$$2^{35} = 2^{6 \cdot 5+5} = (2^6)^5 \cdot 2^5 \equiv 1^5 \cdot 2^5 \equiv 32 \equiv 4 \pmod 7.$$

Similarly, suppose that we want to solve the congruence $x^{103} \equiv 4 \pmod{11}$. Certainly, $x \not\equiv 0 \pmod{11}$, so Fermat's Little Theorem tells us that

$$x^{10} \equiv 1 \pmod{11}.$$

Raising both sides to the tenth power gives $x^{100} \equiv 1 \pmod{11}$, and then multiplying by x^3 gives $x^{103} \equiv x^3 \pmod{11}$. So, to solve the original congruence, we just need to solve $x^3 \equiv 4 \pmod{11}$. This can be solved by trying successively $x = 1$, $x = 2, \ldots$. Thus,

$x \pmod{11}$	0	1	2	3	4	5	6	7	8	9	10
$x^3 \pmod{11}$	0	1	8	5	9	4	7	2	6	3	10

So the congruence $x^{103} \equiv 4 \pmod{11}$ has the solution $x \equiv 5 \pmod{11}$.

We are now ready to prove Fermat's Little Theorem. In order to illustrate the method of proof, we will first prove that $3^6 \equiv 1 \pmod 7$. Of course, there is no need to give a fancy proof of this fact, since $3^6 - 1 = 728 = 7 \cdot 104$. Nevertheless, when attempting to understand a proof or when attempting to construct a proof, it is often worthwhile using specific numbers. Of course, the idea is to devise a proof that doesn't really use the fact that we are considering specific numbers and then hope that the proof can be made to work in general.

To prove that $3^6 \equiv 1 \pmod 7$, we start with the numbers

$$1, 2, 3, 4, 5, 6,$$

multiply each of them by 3, and reduce modulo 7. The results are listed in the following table:

$x \pmod 7$	1	2	3	4	5	6
$3x \pmod 7$	3	6	2	5	1	4

Notice that each of the numbers $1, 2, 3, 4, 5, 6$ reappears exactly once in the second row. So if we multiply together all the numbers in the second row, we get the same result as multiplying together all the numbers in the first row. Of course, we must work modulo 7. Thus,

$$\underbrace{(3 \cdot 1)(3 \cdot 2)(3 \cdot 3)(3 \cdot 4)(3 \cdot 5)(3 \cdot 6)}_{\text{numbers in second row}} \equiv \underbrace{1 \cdot 2 \cdot 3 \cdot 4 \cdot 5 \cdot 6}_{\text{numbers in first row}} \pmod{7}.$$

To save space, we use the standard symbol $n!$ for the number n *factorial*, which is the product of $1, 2, \ldots, n$. In other words,

$$n! = 1 \cdot 2 \cdot 3 \cdots (n-1) \cdot n.$$

Factoring out the six factors of 3 on the left-hand side of our congruence gives

$$3^6 \cdot 6! \equiv 6! \pmod{7}.$$

Notice that $6!$ is relatively prime to 7, so we can cancel the $6!$ from both sides. This gives $3^6 \equiv 1 \pmod{7}$, which is exactly Fermat's Little Theorem.

We are now ready to prove Fermat's Little Theorem in general. The key observation in our proof for $3^6 \pmod{7}$ was that multiplication by 3 rearranged the numbers $1, 2, 3, 4, 5, 6 \pmod{7}$. So first we are going to verify the following claim:

Claim 9.2. *Let p be a prime number and let a be any number with $a \not\equiv 0 \pmod{p}$. Then the numbers*

$$a, 2a, 3a, \ldots, (p-1)a \pmod{p}$$

are the same as the numbers

$$1, 2, 3, \ldots, (p-1) \pmod{p},$$

although they may be in a different order.

Verification. The list $a, 2a, 3a, \ldots, (p-1)a$ contains $p-1$ numbers, and clearly none of them are divisible by p. Suppose that we take two numbers ja and ka in this list, and suppose that they happen to be congruent,

$$ja \equiv ka \pmod{p}.$$

Then $p | (j-k)a$, so $p | (j-k)$, since we are assuming that p does not divide a. Notice that we are using the Prime Divisibility Property proved in Chapter 7, which says that if a prime divides a product then it divides one of the factors. On the other hand, we know that $1 \le j, k \le p-1$, so $|j-k| < p-1$. There is only one number with absolute value less than $p-1$ that is divisible by p and that number is zero. Hence, $j = k$. This shows that different multiples in the list $a, 2a, 3a, \ldots, (p-1)a$ are distinct modulo p.

So we now know that the list $a, 2a, 3a, \ldots, (p-1)a$ contains $p-1$ distinct nonzero values modulo p. But there are only $p-1$ distinct nonzero values modulo p, that is, the numbers $1, 2, 3, \ldots, (p-1)$. Hence, the list $a, 2a, 3a, \ldots, (p-1)a$ and the list $1, 2, 3, \ldots, (p-1)$ must contain the same numbers modulo p, although the numbers may appear in a different order. This finishes the verification of the claim.

Using the claim, it is easy to finish the proof of Fermat's Little Theorem. The claim says that the lists of numbers

$$a, 2a, 3a, \ldots, (p-1)a \pmod{p} \quad \text{and} \quad 1, 2, 3, \ldots, (p-1) \pmod{p}$$

are the same, so the product of the numbers in the first list is equal to the product of the numbers in the second list:

$$a \cdot (2a) \cdot (3a) \cdots ((p-1)a) \equiv 1 \cdot 2 \cdot 3 \cdots (p-1) \pmod{p}.$$

Next we factor our $p-1$ copies of a from the left-hand side to obtain

$$a^{p-1} \cdot (p-1)! \equiv (p-1)! \pmod{p}.$$

Finally, we observe that $(p-1)!$ is relatively prime to p, so we may cancel it from both sides to obtain Fermat's Little Theorem,

$$a^{p-1} \equiv 1 \pmod{p}. \qquad \square$$

Fermat's Little Theorem can be used to show that a number is not a prime without actually factoring it. For example, it turns out that

$$2^{1234566} \equiv 899557 \pmod{1234567}.$$

This means that 1234567 cannot be a prime, since if it were, Fermat's Little Theorem would tell us that $2^{1234566}$ must be congruent to 1 modulo 1234567. [If you're wondering how we computed $2^{1234566} \pmod{1234567}$, don't fret; we'll describe how to do it in Chapter 16.] It turns out that $1234567 = 127 \cdot 9721$, so in this case we can actually find a factor. But consider the number

$$m = 10^{100} + 37.$$

When we compute $2^{m-1} \pmod{m}$, we get

$$\begin{aligned} 2^{m-1} \equiv 3626360327545861062487760199633583910&8 \\ 368732530191513801283208240911248594&63 \\ 5794590597300702318443&97 \pmod{m}. \end{aligned}$$

Again we deduce from Fermat's Little Theorem that $10^{100} + 37$ is not prime, but it is not at all clear how to find a factor. A quick check on a desktop computer reveals no prime factors less than 200,000. It is somewhat surprising that we can easily write down numbers that we know are composite, yet for which we are unable to find any factors.

Exercises

9.1. Use Fermat's Little Theorem to perform the following tasks.

(a) Find a number $0 \le a < 73$ with $a \equiv 9^{794} \pmod{73}$.

(b) Solve $x^{86} \equiv 6 \pmod{29}$.

(c) Solve $x^{39} \equiv 3 \pmod{13}$.

9.2. The quantity $(p-1)! \pmod{p}$ appeared in our proof of Fermat's Little Theorem, although we didn't need to know its value.

(a) Compute $(p-1)! \pmod{p}$ for some small values of p, find a pattern, and make a conjecture.

(b) Prove that your conjecture is correct. [Try to discover why $(p-1)! \pmod{p}$ has the value it does for small values of p, and then generalize your observation to prove the formula for all values of p.]

9.3. Exercise 9.2 asked you to determine the value of $(p-1)! \pmod{p}$ when p is a prime number.

(a) Compute the value of $(m-1)! \pmod{m}$ for some small values of m that are not prime. Do you find the same pattern as you found for primes?

(b) If you know the value of $(n-1)! \pmod{n}$, how can you use the value to definitely distinguish whether n is prime or composite?

9.4. If p is a prime number and if $a \not\equiv 0 \pmod{p}$, then Fermat's Little Theorem tells us that $a^{p-1} \equiv 1 \pmod{p}$.

(a) The congruence $7^{1734250} \equiv 1660565 \pmod{1734251}$ is true. Can you conclude that 1734251 is a composite number?

(b) The congruence $129^{64026} \equiv 15179 \pmod{64027}$ is true. Can you conclude that 64027 is a composite number?

(c) The congruence $2^{52632} \equiv 1 \pmod{52633}$ is true. Can you conclude that 52633 is a prime number?

Chapter 10

Congruences, Powers, and Euler's Formula

In the previous chapter we proved Fermat's Little Theorem: If p is a prime and $p \nmid a$, then $a^{p-1} \equiv 1 \pmod{p}$. This formula is certainly not true if we replace p by a composite number. For example, $5^5 \equiv 5 \pmod{6}$ and $2^8 \equiv 4 \pmod{9}$. So we ask whether there is some power, depending on the modulus m, so that

$$a \equiv 1 \pmod{m}.$$

Our first observation is that this is impossible if $\gcd(a, m) > 1$. To see why, suppose that $a^k \equiv 1 \pmod{m}$. Then $a^k = 1 + my$ for some integer y, so $\gcd(a, m)$ divides $a^k - my = 1$. In other words, if some power of a is congruent to 1 modulo m, then we must have $\gcd(a, m) = 1$. This suggests that we look at the set of numbers that are relatively prime to m,

$$\{a : 1 \le a \le m \text{ and } \gcd(a, m) = 1\}.$$

For example,

m	$\{a : 1 \le a \le m \text{ and } \gcd(a, m) = 1\}$
1	$\{1\}$
2	$\{1\}$
3	$\{1, 2\}$
4	$\{1, 3\}$
5	$\{1, 2, 3, 4\}$
6	$\{1, 5\}$
7	$\{1, 2, 3, 4, 5, 6\}$
8	$\{1, 3, 5, 7\}$
9	$\{1, 2, 4, 5, 7, 8\}$
10	$\{1, 3, 7, 9\}$

The number of integers between 0 and m that are relatively prime to m is an important quantity, so we give this quantity a name:

$$\phi(m) = \#\{a : 1 \le a \le m \text{ and } \gcd(a, m) = 1\}.$$

The function ϕ is called *Euler's phi function*. From the preceding table, we can read off the value of $\phi(m)$ for $1 \le m \le 10$. Thus

m	1	2	3	4	5	6	7	8	9	10
$\phi(m)$	1	1	2	2	4	2	6	4	6	4

Notice that if p is a prime number then every integer $1 \le a < p$ is relatively prime to p. So for prime numbers we have the formula

$$\phi(p) = p - 1.$$

We are going to try to mimic our proof of Fermat's Little Theorem. Suppose, for example, that we want to find a power of 7 that is congruent to 1 modulo 10. Rather than taking all the numbers $1 \le a < 10$, we will just take the numbers that are relatively prime to 10. They are

$$1,\ 3,\ 7,\ 9 \pmod{10}.$$

If we multiply each of them by 7, we get

$$7 \cdot 1 \equiv 7 \pmod{10}, \qquad 7 \cdot 3 \equiv 1 \pmod{10},$$
$$7 \cdot 7 \equiv 9 \pmod{10}, \qquad 7 \cdot 9 \equiv 3 \pmod{10}.$$

Notice that we get back the same numbers, but rearranged. So if we multiply them together, we get the same product,

$$(7 \cdot 1)(7 \cdot 3)(7 \cdot 7)(7 \cdot 9) \equiv 1 \cdot 3 \cdot 7 \cdot 9 \pmod{10}$$
$$7^4(1 \cdot 3 \cdot 7 \cdot 9) \equiv 1 \cdot 3 \cdot 7 \cdot 9 \pmod{10}.$$

Now we can cancel $1 \cdot 3 \cdot 7 \cdot 9$ to get $7^4 \equiv 1 \pmod{10}$.

Where does the exponent 4 come from? It's equal to the number of integers between 0 and 10 that are relatively prime to 10; that is, the exponent is 4 because $\phi(10) = 4$. This suggests the truth of the following formula.

Theorem 10.1 (Euler's Formula). *If* $\gcd(a, m) = 1$, *then*

$$a^{\phi(m)} \equiv 1 \pmod{m}.$$

Proof. Now that we have identified the correct set of numbers to consider, the proof of Euler's formula is almost identical to the proof of Fermat's Little Theorem. So we let

$$1 \le b_1 < b_2 < \cdots < b_{\phi(m)} < m$$

be the $\phi(m)$ numbers between 0 and m that are relatively prime to m.

Claim 10.2. *If* $\gcd(a, m) = 1$, *then the numbers*

$$b_1 a,\ b_2 a,\ b_3 a,\ \ldots,\ b_{\phi(m)} a \pmod{m}$$

are the same as the numbers

$$b_1,\ b_2,\ b_3,\ \ldots,\ b_{\phi(m)} \pmod{m},$$

although they may be in a different order.

Verification of the claim. We note that if b is relatively prime to m, then ab is also relatively prime to m. Hence, each of the numbers in the list

$$b_1 a, b_2 a, b_3 a, \ldots, b_{\phi(m)} a \pmod{m}$$

is congruent to one number in the list

$$b_1, b_2, b_3, \ldots, b_{\phi(m)} \pmod{m}.$$

Furthermore, there are $\phi(m)$ numbers in each list. So if we can show that the numbers in the first list are distinct modulo m, it will follow that the two lists are the same (after rearranging).

Suppose that we take two numbers $b_j a$ and $b_k a$ from the first list, and suppose that they are congruent,

$$b_j a \equiv b_k a \pmod{m}.$$

Then $m|(b_j - b_k)a$. But m and a are relatively prime, so we find that $m|b_j - b_k$. On the other hand, b_j and b_k are between 1 and m, which implies $|b_j - b_k| \le m - 1$. There is only one number with absolute value strictly less than m that is divisible by m and that number is zero. Hence, $b_j = b_k$. This shows that the numbers in the list

$$b_1 a, b_2 a, b_3 a, \ldots, b_{\phi(m)} a \pmod{m}$$

are all distinct modulo m, which completes the verification that the claim is true.

Using the claim, we can easily finish the proof of Euler's formula. The claim says that the lists of numbers

$$b_1 a, b_2 a, b_3 a, \ldots, b_{\phi(m)} a \pmod{m}$$

and

$$b_1, b_2, b_3, \ldots, b_{\phi(m)} \pmod{m}$$

are the same, so the product of the numbers in the first list is equal to the product of the numbers in the second list:

$$(b_1 a) \cdot (b_2 a) \cdot (b_3 a) \cdots (b_{\phi(m)} a) \equiv b_1 \cdot b_2 \cdot b_3 \cdots b_{\phi(m)} \pmod{m}.$$

We can factor out $\phi(m)$ copies of a from the left-hand side to obtain

$$a^{\phi(m)} B \equiv B \pmod{m}, \qquad \text{where } B = b_1 b_2 b_3 \cdots b_{\phi(m)}.$$

Finally, we observe that B is relatively prime to m, since each of the b_i's is relatively prime to m. This means we may cancel B from both sides to obtain Euler's formula

$$a^{\phi(m)} \equiv 1 \pmod{m}. \qquad \square$$

Exercises

10.1. Let $b_1 < b_2 < \cdots < b_{\phi(m)}$ be the integers between 1 and m that are relatively prime to m (including 1), and let $B = b_1 b_2 b_3 \cdots b_{\phi(m)}$ be their product. The quantity B came up during the proof of Euler's formula.

(a) Show that either $B \equiv 1 \pmod{m}$ or $B \equiv -1 \pmod{m}$.

(b) Compute B for some small values of m and try to find a pattern for when it is equal to $+1 \pmod{m}$ and when it is equal to $-1 \pmod{m}$.

10.2. If $\phi(m) = 1000$, find a small number a (say less than 2000) that is not divisible by 7 and satisfies the congruence $a \equiv 7^{3003} \pmod{m}$. You should be able to do this problem without using a calculator.

10.3. A composite number m is called a *Carmichael number* if the congruence $a^{m-1} \equiv 1 \pmod{m}$ is true for every number a with $\gcd(a, m) = 1$.

(a) Verify that $m = 561 = 3 \cdot 11 \cdot 17$ is a Carmichael number. [*Hint*. It is not necessary to actually compute $a^{m-1} \pmod{m}$ for all 320 values of a. Instead, use Fermat's Little Theorem to check that $a^{m-1} \equiv 1 \pmod{p}$ for each prime p dividing m, and then explain why this implies that $a^{m-1} \equiv 1 \pmod{m}$.]

(b) Try to find another Carmichael number. Do you think that there are infinitely many of them?

Chapter 11

Euler's Phi Function and the Chinese Remainder Theorem

Euler's formula

$$a^{\phi(m)} \equiv 1 \pmod{m}$$

is a beautiful and powerful result, but it won't be of much use to us unless we can find an efficient way to compute the value of $\phi(m)$. Clearly, we don't want to list all the numbers from 1 to $m-1$ and check each to see if it is relatively prime to m. This would be very time consuming if $m \approx 1000$, for example, and it would be impossible for $m \approx 10^{100}$. As we observed in the last chapter, one case where $\phi(m)$ is easy to compute is when $m = p$ is a prime, since then every integer $1 \le a \le p-1$ is relatively prime to m. Thus, $\phi(p) = p - 1$.

We can easily derive a similar formula for $\phi(p^k)$ when $m = p^k$ is a power of a prime. Rather than trying to count the numbers between 1 and p^k that are relatively prime to p^k, we will instead start with all numbers $1 \le a \le p^k$, and then we will discard the ones that are not relatively prime to p^k.

When is a number a not relatively prime to p^k? The only factors of p^k are powers of p, so a is not relatively prime to p^k exactly when it is divisible by p. In other words,

$$\phi(p^k) = p^k - \#\{a : 1 \le a \le p^k \text{ and } p|a\}.$$

So we have to count how many integers between 1 and p^k are divisible by p. That's easy, they are the multiples of p,

$$p, 2p, 3p, 4p, \ldots (p^{k-1}-2)p, (p^{k-1}-1)p, p^k.$$

There are p^{k-1} of them, which gives us the formula

$$\phi(p^k) = p^k - p^{k-1}.$$

For example,

$$\phi(2401) = \phi(7^4) = 7^4 - 7^3 = 2058.$$

This means that there are 2058 integers between 1 and 2401 that are relatively prime to 2401.

We now know how to compute $\phi(m)$ when m is a power of a prime. Next suppose that m is the product of two primes powers, $m = p^j q^k$. To formulate a conjecture, we compute $\phi(p^j q^k)$ for some small values and compare it with the values of $\phi(p^j)$ and $\phi(q^k)$.

p^j	q^k	$p^j q^k$	$\phi(p^j)$	$\phi(q^k)$	$\phi(p^j q^k)$
2	3	6	1	2	2
4	5	20	2	4	8
3	7	21	2	6	12
8	9	72	4	6	24
9	25	225	6	20	120

This table suggests that $\phi(p^j q^k) = \phi(p^j)\phi(q^k)$. We can also try some examples with numbers that are not prime powers, such as

$$\phi(14) = 6, \qquad \phi(15) = 8, \qquad \phi(210) = \phi(14 \cdot 15) = 48.$$

all this leads us to guess that the following assertion is true:

If $\gcd(m, n) = 1$, then $\phi(mn) = \phi(m)\phi(n)$.

Before trying to prove this multiplication formula, we show how it can be used to easily compute $\phi(m)$ for any m or, more precisely, for any m that you are able to factor as a product of primes.

Suppose that we are given a number m, and suppose that we have factored m as a product of primes, say

$$m = p_1^{k_1} \cdot p_2^{k_2} \cdots p_r^{k_r},$$

where $p_1, p_2, \ldots, p_r$ are all different. First we use the multiplication formula to compute

$$\phi(m) = \phi(p_1^{k_1}) \cdot \phi(p_2^{k_2}) \cdots \phi(p_r^{k_r}).$$

Then we use the prime power formula $\phi(p^k) = p^k - p^{k-1}$ to obtain

$$\phi(m) = (p_1^{k_1} - p_1^{k_1-1}) \cdot (p_2^{k_2} - p_2^{k_2-1}) \cdots (p_r^{k_r} - p_r^{k_r-1}).$$

This formula may look complicated, but the procedure to compute $\phi(m)$ is really very simple. For example,

$$\begin{aligned}\phi(1512) &= \phi(2^3 \cdot 3^3 \cdot 7) = \phi(2^3) \cdot \phi(3^3) \cdot \phi(7) \\ &= (2^3 - 2^2) \cdot (3^3 - 3^2) \cdot (7 - 1) = 4 \cdot 18 \cdot 6 = 432.\end{aligned}$$

So there are 432 numbers between 1 and 1512 that are relatively prime to 1512.

We are now ready to prove the multiplication formula for Euler's phi function. We also restate the formula for prime powers so as to have both formulas conveniently listed together.

Theorem 11.1 (Phi Function Formulas). (a) *If p is a prime and* $k \geq 1$, *then*

$$\phi(p^k) = p^k - p^{k-1}.$$

(b) *If* $\gcd(m, n) = 1$, *then* $\phi(mn) = \phi(m)\phi(n)$.

Proof. We verified the prime power formula (a) earlier in this chapter, so we need to check the product formula (b). We will do this by using one of the most powerful tools available in number theory:

COUNTING

Briefly, we are going to find a set that contains $\phi(mn)$ elements, and we are going to find a second set that contains $\phi(m)\phi(n)$ elements. Then we will show that the two sets contain the same number of elements.

The first set is

$$\{a : 1 \leq a \leq mn \text{ and } \gcd(a, mn) = 1\}.$$

It is clear that this set contains $\phi(mn)$ elements, since that's just the definition of $\phi(mn)$. The second set is

$$\begin{aligned}\{(b, c) : 1 \leq b \leq m \quad \text{and} \quad \gcd(b, m) = 1 \quad \text{and} \quad 1 \leq c \leq n \\ \text{and} \quad \gcd(c, n) = 1\}.\end{aligned}$$

How many pairs (b, c) are in this second set? Well, there are $\phi(m)$ choices for b, since that's the definition of $\phi(m)$, and there are $\phi(n)$ choices for c, since that's the definition of $\phi(n)$. So there are $\phi(m)$ choices for the first coordinate b and $\phi(n)$ choices for the second coordinate c; so there are a total of $\phi(m)\phi(n)$ choices for the pair (b, c).

For example, suppose that we take $m = 4$ and $n = 5$. Then the first set consists of the numbers

$$\{1, 3, 7, 9, 11, 13, 17, 19\}$$

that are relatively prime to 20. The second set consists of the pairs

$$\{(1,1),\ (1,2),\ (1,3),\ (1,4),\ (3,1),\ (3,2),\ (3,3),\ (3,4)\}$$

where the first number in each pair is relatively prime to 4 and the second number in each pair is relatively prime to 5.

Going back to the general case, we are going to take each element in the first set and assign it to a pair in the second set in the following way:

$$\begin{array}{ccc} \left\{ a : \begin{array}{c} 1 \le a \le mn \\ \gcd(a, mn) = 1 \end{array} \right\} & \longrightarrow & \left\{ (b,c) : \begin{array}{ll} 1 \le b \le m, & \gcd(b,m) = 1 \\ 1 \le c \le n, & \gcd(c,n) = 1 \end{array} \right\} \\ a \bmod mn & \longmapsto & (a \bmod m, a \bmod n) \end{array}$$

What this means is that we take the integer a in the first set and send it to the pair (b, c) with

$$a \equiv b \pmod{m} \qquad \text{and} \qquad a \equiv c \pmod{n}.$$

This is probably clearer if we look again at our example with $m = 4$ and $n = 5$. Then, for example, the number 13 in the first set gets sent to the pair $(1, 3)$ in the second set, since $13 \equiv 1 \pmod 4$ and $13 \equiv 3 \pmod 5$. We do the same for each of the other numbers in the first set.

$$\{1,3,7,9,11,13,17,19\} \longrightarrow \begin{array}{l} \{(1,1),\ (1,2),\ (1,3),\ (1,4), \\ \qquad (3,1),\ (3,2),\ (3,3),\ (3,4)\} \end{array}$$

$$\begin{array}{ll} 1 \longmapsto (1,1) & 11 \longmapsto (3,1) \\ 3 \longmapsto (3,3) & 13 \longmapsto (1,3) \\ 7 \longmapsto (3,2) & 17 \longmapsto (1,2) \\ 9 \longmapsto (1,4) & 19 \longmapsto (3,4) \end{array}$$

In this example, you can see that each pair in the second set is matched with exactly one number in the first set. This means that the two sets have the same number of elements. We want to check that the same matching occurs in general.

We need to check that the following two statements are correct:

1. Different numbers in the first set get sent to different pairs in the second set.

2. Every pair in the second set is hit by some number in the first set.

Once we verify these two statements, we will know that the two sets have the same number of elements. But we know that the first set has $\phi(mn)$ elements and the second set has $\phi(m)\phi(n)$ elements. So in order to finish the proof that $\phi(mn) = \phi(m)\phi(n)$, we just need to verify (1) and (2).

To check (1), we take two numbers a_1 and a_2 in the first set, and we suppose that they have the same image in the second set. This means that

$$a_1 \equiv a_2 \pmod{m} \qquad \text{and} \qquad a_1 \equiv a_2 \pmod{n}.$$

Thus, $a_1 - a_2$ is divisible by both m and n. However, m and n are relatively prime, so $a_1 - a_2$ must be divisible by the product mn. In other words,

$$a_1 \equiv a_2 \pmod{mn},$$

which shows that a_1 and a_2 are the same element in the first set. This completes our verification of statement (1).

To check statement (2), we need to show that for any given values of b and c we can find at least one integer a satisfying

$$a \equiv b \pmod{m} \qquad \text{and} \qquad a \equiv c \pmod{n}.$$

The fact that these simultaneous congruences have a solution is of sufficient importance to warrant having its own name.

Theorem 11.2 (Chinese Remainder Theorem). *Let m and n be integers with $\gcd(m, n) = 1$, and let b and c be any integers. Then the simultaneous congruences*

$$x \equiv b \pmod{m} \qquad \textit{and} \qquad x \equiv c \pmod{n}$$

have exactly one solution with $0 \le x < mn$.

Proof. Let's start, as usual, with an example. Suppose we want to solve

$$x \equiv 8 \pmod{11} \qquad \text{and} \qquad x \equiv 3 \pmod{19}.$$

The solution to the first congruence consists of all numbers that have the form $x = 11y + 8$. We substitute this into the second congruence, simplify, and try to solve. Thus,

$$11y + 8 \equiv 3 \pmod{19}$$
$$11y \equiv 14 \pmod{19}.$$

We know how to solve linear congruences of this sort (see the Linear Congruence Theorem in Chapter 8). The solution is $y_1 \equiv 3 \pmod{19}$, and then we can find the solution to the original congruences using $x_1 = 11y_1 + 8 = 11 \cdot 3 + 8 = 41$. Finally, we should check our answer: $(41 - 8)/11 = 3$ and $(41 - 3)/19 = 2$. ✓

For the general case, we again begin by solving the first congruence $x \equiv b \pmod{m}$. The solution consists of all numbers of the form $x = my + b$. We substitute this into the second congruence, which yields

$$my \equiv c - b \pmod{n}.$$

We are given that $\gcd(m, n) = 1$, so the Linear Congruence Theorem of Chapter 8 tells us that there is exactly one solution y_1 with $0 \le y_1 < n$. Then the solution to the original pair of congruences is given by

$$x_1 = my_1 + b;$$

and this will be the only solution x_1 with $0 \le x_1 < mn$, since there is only one y_1 between 0 and n, and we multiplied y_1 by m to get x_1. This completes our verification of the Chinese Remainder Theorem and, with it, our proof of the formula $\phi(mn) = \phi(m)\phi(n)$. □

Historical Interlude. The first recorded instance of the Chinese Remainder Theorem appears in a Chinese mathematical work from the late third or early fourth century. Somewhat surprisingly, it deals with the harder problem of three simultaneous congruences.

> "We have a number of things, but we do not know exactly how many. If we count them by threes, we have two left over. If we count them by fives, we have three left over. If we count them by sevens, we have two left over. How many things are there?"
> *Sun Tzu Suan Ching* (Master Sun's Mathematical Manual)
> Circa AD 300, volume 3, problem 26.

Exercises

11.1. (a) Find the value of $\phi(97)$.
(b) Find the value of $\phi(8800)$.

11.2. (a) If $m \ge 3$, explain why $\phi(m)$ is always even.
(b) $\phi(m)$ is "usually" divisible by 4. Describe all the m's for which $\phi(m)$ is not divisible by 4.

11.3. Suppose that $p_1, p_2, \ldots, p_r$ are the distinct primes that divide m. Show that the following formula for $\phi(m)$ is correct.

$$\phi(m) = m\left(1 - \frac{1}{p_1}\right)\left(1 - \frac{1}{p_2}\right)\cdots\left(1 - \frac{1}{p_r}\right).$$

Use this formula to compute $\phi(1000000)$.

11.4. Write a program to compute $\phi(n)$, the value of Euler's phi function. You should compute $\phi(n)$ by using a factorization of n into primes, not by finding all the a's between 1 and n that are relatively prime to n.

11.5. For each part, find an x that solves the given simultaneous congruences.

(a) $x \equiv 3 \pmod{7}$ and $x \equiv 5 \pmod{9}$

(b) $x \equiv 3 \pmod{37}$ and $x \equiv 1 \pmod{87}$

(c) $x \equiv 5 \pmod{7}$ and $x \equiv 2 \pmod{12}$ and $x \equiv 8 \pmod{13}$

11.6. Solve the 1700-year-old Chinese remainder problem from the *Sun Tzu Suan Ching* stated on page 75.

11.7. A farmer is on the way to market to sell eggs when a meteorite hits his truck and destroys all of his produce. In order to file an insurance claim, he needs to know how many eggs were broken. He knows that when he counted the eggs by 2's, there was 1 left over, when he counted them by 3's, there was 1 left over, when he counted them by 4's, there was 1 left over, when he counted them by 5's, there was 1 left over, and when he counted them by 6's, there was 1 left over, but when he counted them by 7's, there were none left over. What is the smallest number of eggs that were in the truck?

11.8. Write a program that takes as input four integers (b, m, c, n) with $\gcd(m, n) = 1$ and computes an integer x with $0 \le x < mn$ satisfying

$$x \equiv b \pmod{m} \qquad \text{and} \qquad x \equiv c \pmod{n}.$$

11.9. In this exercise you will prove a version of the Chinese Remainder Theorem for three congruences. Let m_1, m_2, m_3 be positive integers so that each pair is relatively prime. That is,

$$\gcd(m_1, m_2) = 1 \quad \text{and} \quad \gcd(m_1, m_3) = 1 \quad \text{and} \quad \gcd(m_2, m_3) = 1.$$

Let a_1, a_2, a_3 be any three integers. Show that there is exactly one integer x in the interval $0 \le x < m_1 m_2 m_3$ that simultaneously solves the three congruences

$$x \equiv a_1 \pmod{m_1}, \qquad x \equiv a_2 \pmod{m_2}, \qquad x \equiv a_3 \pmod{m_3}.$$

Can you figure out how to generalize this problem to deal with lots of congruences

$$x \equiv a_1 \pmod{m_1}, \qquad x \equiv a_2 \pmod{m_2}, \ldots, \qquad x \equiv a_r \pmod{m_r}?$$

In particular, what conditions do the moduli $m_1, m_2, \ldots, m_r$ need to satisfy?

11.10. What can you say about n if the value of $\phi(n)$ is a prime number? What if it is the square of a prime number?

11.11. Find at least five different numbers n with $\phi(n) = 160$. How many more can you find?

11.12. Find all values of n that solve each of the following equations.

(a) $\phi(n) = n/2$ (b) $\phi(n) = n/3$ (c) $\phi(n) = n/6$

(*Hint.* The formula in Exercise 11.3 might be useful.)

11.13. (a) For each integer $2 \le a \le 10$, find the last four digits of a^{1000}.

(b) Based on your experiments in (a) and further experiments if necessary, give a simple criterion that allows you to predict the last four digits of a^{1000} from the value of a.

(c) Prove that your criterion in (a) is correct.

Chapter 12

Prime Numbers

Prime numbers are the basic building blocks of number theory. That's what the Fundamental Theorem of Arithmetic, discussed in Chapter 7, tells us. Every number is built up in a unique fashion by multiplying together prime numbers. There are analogous situations in other areas of science, and without exception the discovery and description of the building blocks has had a profound effect on its discipline. For example, the field of chemistry was revolutionized by the discovery that every chemical is formed from a few basic elements and by Mendeleev cataloging these elements into families whose properties recur periodically. We will do something similar below when we split the set of prime numbers into various subsets, for example, into the set congruent to 1 modulo 4 and the set congruent to 3 modulo 4. Similarly, a tremendous advance in physics occurred when scientists discovered that the atoms comprising every element are made up of three basic particles, protons, neutrons, and electrons,[1] and that the number of each determines the chemical and physical attributes of the atom. For example, an atom made up of 92 protons and only 143 neutrons has properties that clearly distinguish it from its cousin with three additional neutrons.

The fact that prime numbers are basic building blocks is sufficient reason to study their properties. Of course, this doesn't imply that those properties will be interesting. Studying how to conjugate irregular verbs is important when learning a language, but that doesn't make it very appealing. Luckily, the more one studies prime numbers, the more interesting they become, and the more beautiful and surprising become the relationships that one discovers. In this brief chapter we will only have time to mention a few of the many remarkable properties of prime numbers.

---

[1]This description of an atom is a simplification, but it is a fairly accurate portrayal of the original atomic theories advanced in the early part of the twentieth century.

To begin with, let's list the first few primes:

$$2, 3, 5, 7, 11, 13, 17, 19, 23, 29, 31, 37, 41, 43, 47, 53, 59, 61, \ldots.$$

What can we glean from this list? First, it looks like 2 is the only even prime. This is true, of course. If n is even and larger than 2 then it factors as $n = 2 \cdot (n/2)$. This makes 2 somewhat unusual among the set of primes, so people have been known to say that

"2 is the oddest prime!"[2]

A more important observation from our list of primes is signified by the ellipsis (three dots) appended at the end. This means that the list is not complete. For example, 67 and 71 are the next two primes. However, the real issue is whether the list ends or whether it continues indefinitely. In other words, are there infinitely many prime numbers? The answer is yes. We now give a beautiful proof that appeared in Euclid's *Elements* more than 2000 years ago.

Theorem 12.1 (Infinitely Many Primes Theorem). *There are infinitely many prime numbers.*

Euclid's Proof. Suppose that you have already compiled a (finite) list of primes. I am going to show you how to find a new prime that isn't in your list. Since you can then add the new prime to the list and repeat the process, this will show that there must be infinitely many primes.

So suppose we start with some list of primes $p_1, p_2, \ldots, p_r$. We multiply them together and add 1, which gives the number

$$A = p_1 p_2 \cdots p_r + 1.$$

If A itself is prime, we're done, since A is too large to be in the original list. But even if A is not prime, it will certainly be divisible by some prime, since every number can be written as a product of primes. Let q be some prime dividing A, for example, the smallest one. I claim that q is not in the original list, so it will be the desired new prime.

Why isn't q in the original list? We know that q divides A, so

$$q \text{ divides } p_1 p_2 \cdots p_r + 1.$$

If q were to equal one of the p_i's, then it would have to divide 1, which is not possible. This means that q is a new prime that may be added to our list. Repeating

[2]Naturally, I would never even consider repeating such a weak joke! Notice that this is one of those jokes that is language specific. For example, it doesn't work in French, since an odd number is *impair*, while an odd person or event is *étrange* or *bizarre*.

this process, we can create a list of primes that is as long as we want. This shows that there must be infinitely many prime numbers. □

Euclid's proof is very clever and beautiful. We will illustrate the ideas in Euclid's proof by using them to create a list of primes. We start with a list consisting of the single prime $\{2\}$. Following Euclid, we compute $A = 2 + 1 = 3$. This A is already prime, so we append it to our list. Now we have two primes, $\{2, 3\}$. Again using Euclid's argument, we compute $A = 2 \cdot 3 + 1 = 7$, and again A is prime and can be added to the list. This gives three primes, $\{2, 3, 7\}$. Repeating the argument gives $A = 2 \cdot 3 \cdot 7 + 1 = 43$, another prime! So now our list has four primes, $\{2, 3, 7, 43\}$. Into the breach once more, we compute $A = 2 \cdot 3 \cdot 7 \cdot 43 + 1 = 1807$. This time, A is not prime, it factors as $A = 13 \cdot 139$. We add 13 to our list, which now reads $\{2, 3, 7, 43, 13\}$. One more time, we compute $A = 2 \cdot 3 \cdot 7 \cdot 43 \cdot 13 + 1 = 23479$. This A also factors, $A = 53 \cdot 443$. This gives the list $\{2, 3, 7, 43, 13, 53\}$, and we will stop here. But in principle we could continue this process to produce a list of primes of any specified length.

We now know that the list of primes continues without end, and we also observed that 2 is the only even prime. Every odd number is congruent to either 1 or 3 modulo 4, so we might ask which primes are congruent to 1 modulo 4 and which are congruent to 3 modulo 4. This separates the set of (odd) primes into two families, just as the periodic table separates the elements into families having similar properties. In the following list, we have boxed the primes congruent to 1 modulo 4:

$$3, \boxed{5}, 7, 11, \boxed{13}, \boxed{17}, 19, 23, \boxed{29}, 31, \boxed{37}, \boxed{41}, 43, 47, \boxed{53}, 59,$$
$$\boxed{61}, 67, 71, \boxed{73}, 79, 83, \boxed{89}, \boxed{97}, \boxed{101}, \ldots.$$

There doesn't seem to be any obvious pattern, although there do seem to be plenty of primes of each kind. Here's a longer list.

$p \equiv 1 \pmod 4$	$5, 13, 17, 29, 37, 41, 53, 61, 73, 89, 97, 101, 109,$ $113, 137, 149, 157, 173, 181, 193, 197, \ldots$
$p \equiv 3 \pmod 4$	$3, 7, 11, 19, 23, 31, 43, 47, 59, 67, 71, 79, 83, 103,$ $107, 127, 131, 139, 151, 163, 167, 179, \ldots$

Is it possible that one of the lines in this list eventually stops, or are there infinitely many primes in each family? It turns out that each line continues indefinitely. We will use a variation of Euclid's proof to show that there are infinitely many primes congruent to 3 modulo 4. In Chapter 24 we use a slightly different argument to deal with the 1 modulo 4 primes.

Theorem 12.2 (Primes 3 (Mod 4) Theorem). *There are infinitely many primes that are congruent to 3 modulo 4.*

Proof. We suppose that we have already compiled a (finite) list of primes, all of which are congruent to 3 modulo 4. Our goal is to make the list longer by finding a new 3 modulo 4 prime. Repeating this process gives a list of any desired length, thereby proving that there are infinitely many primes congruent to 3 modulo 4.

Suppose that our initial list of primes congruent to 3 modulo 4 is

$$3, p_1, p_2, \ldots, p_r.$$

Consider the number

$$A = 4p_1p_2 \cdots p_r + 3.$$

(Notice that we don't include the prime 3 in the product.) We know that A can be factored into a product of primes, say

$$A = q_1q_2 \cdots q_s.$$

I claim that among the primes $q_1, q_2, \ldots, q_s$ at least one of them must be congruent to 3 modulo 4. This is the key step in the proof. Why is it true? Well, if not, then $q_1, q_2, \ldots, q_s$ would all be congruent to 1 modulo 4, in which case their product A would be congruent to 1 modulo 4. But you can see from its definition that A is clearly congruent to 3 modulo 4. Hence, at least one of $q_1, q_2, \ldots, q_s$ must be congruent to 3 modulo 4, say $q_i \equiv 3 \pmod 4$.

My second claim is that q_i is not in the original list. Why not? Well, we know that q_i divides A, while it is clear from the definition of A that none of $3, p_1, p_2, \ldots, p_r$ divides A. Thus, q_i is not in our original list, so we may add it to the list and repeat the process. In this way we can create as long a list as we want, which shows that there must be infinitely many primes congruent to 3 modulo 4. □

We can use the ideas in the proof of the Primes 3 (Mod 4) Theorem to create a list of primes congruent to 3 modulo 4. We need to start with a list containing at least one such prime, and remember that 3 is not allowed in our list. So we start with the list consisting of the single prime $\{7\}$. We compute $A = 4 \cdot 7 + 3 = 31$. This A is itself prime, so it is a new $3 \pmod 4$ prime to add to our list. The list now reads $\{7, 31\}$, so we compute $A = 4 \cdot 7 \cdot 31 + 3 = 871$. This A is not prime; it factors as $A = 13 \cdot 67$. The proof of the theorem tells us that at least one of the prime factors will be congruent to 3 modulo 4. In this case, the prime 67 is $3 \pmod 4$, so we add it to our list. Next we take $\{7, 31, 67\}$, compute $A = 4 \cdot 7 \cdot 31 \cdot 67 + 3 = 58159$, and factor it as $A = 19 \cdot 3061$. This time it is the first factor 19 that is 3 $(\mathrm{mod}\ 4)$, so our list becomes $\{7, 31, 67, 19\}$. We will repeat the process one more time. So

$$A = 4 \cdot 7 \cdot 31 \cdot 67 \cdot 19 + 3 = 1104967 = 179 \cdot 6173,$$

which gives the prime 179 to add to the list, $\{7, 31, 67, 19, 179\}$.

Why won't the same idea work for 1 $(\mathrm{mod}\ 4)$ primes? This is not an idle question; it's almost as important to understand the limitations of an argument as it is to understand why the argument is valid. So suppose we try to create a list of 1 $(\mathrm{mod}\ 4)$ primes. If we start with the list $\{p_1, p_2, \ldots, p_r\}$, we can compute the number $A = 4p_1p_2 \cdots p_r + 1$, factor it, and try to find a prime factor that is a new 1 $(\mathrm{mod}\ 4)$ prime. What happens if we start with the list $\{5\}$? We compute $A = 4 \cdot 5 + 1 = 21 = 3 \cdot 7$, and neither of the factors 3 or 7 is a 1 $(\mathrm{mod}\ 4)$ number. So we're stuck. The problem is that it is possible to multiply two 3 $(\mathrm{mod}\ 4)$ numbers, such as 3 and 7, and end up with a 1 $(\mathrm{mod}\ 4)$ number like $A = 21$. In general, we cannot use the fact that $A \equiv 1 \pmod 4$ to deduce that some prime factor of A is 1 $(\mathrm{mod}\ 4)$, and that's why this proof won't work for primes congruent to 1 modulo 4.

There is no particular reason to consider only congruences modulo 4. For example, every number is congruent to either 0, 1, 2, 3, or 4 modulo 5; and except for 5 itself, every prime number is congruent to one of 1, 2, 3, or 4 modulo 5. (Why?) So we can break up the set of prime numbers into four families, depending on their congruence class modulo 5. Here's a list of the first few numbers in each family:

$p \equiv 1 \pmod 5$	$11, 31, 41, 61, 71, 101, 131, 151, 181, 191, 211, 241$
$p \equiv 2 \pmod 5$	$2, 7, 17, 37, 47, 67, 97, 107, 127, 137, 157, 167, 197$
$p \equiv 3 \pmod 5$	$3, 13, 23, 43, 53, 73, 83, 103, 113, 163, 173, 193, 223$
$p \equiv 4 \pmod 5$	$19, 29, 59, 79, 89, 109, 139, 149, 179, 199, 229, 239$

Again there seem to be lots of primes in each family, so we might guess that each contains infinitely many prime numbers.

In general, if we fix a modulus m and a number a, when might we expect there to be infinitely many primes congruent to a modulo m? There is one situation in which this cannot happen, that is if a and m have a common factor. For example, suppose that p is a prime and that $p \equiv 35 \pmod{77}$. This means that $p = 35 + 77y = 7(5 + 11y)$, so the only possibility is $p = 7$, and even $p = 7$ doesn't work. Generally, if p is a prime satisfying $p \equiv a \pmod m$, then $\gcd(a, m)$ divides p. So either $\gcd(a, m) = 1$ or else $\gcd(a, m) = p$, which means there is at most one possibility for p. Thus, it is really only interesting to ask about primes congruent to a modulo m if we assume that $\gcd(a, m) = 1$. A famous theorem of Dirichlet from 1837 says that with this assumption there are always infinitely many primes congruent to a modulo m.

Theorem 12.3 (Dirichlet's Theorem on Primes in Arithmetic Progressions[3]). *Let a and m be integers with $\gcd(a, m) = 1$. Then there are infinitely many primes that are congruent to a modulo m. That is, there are infinitely many prime numbers p satisfying*

$$p \equiv a \pmod{m}.$$

Earlier in this chapter we proved Dirichlet's Theorem for $(a, m) = (3, 4)$, and Exercise 12.2 asks you to do $(a, m) = (5, 6)$. In Chapter 24, we will deal with $(a, m) = (1, 4)$. Unfortunately, the proof of Dirichlet's Theorem for all (a, m) is quite complicated, so we will not be able to give it in this book. The proof uses advanced methods from calculus and, in fact, calculus with complex numbers!

Exercises

12.1. Start with the list consisting of the single prime $\{5\}$ and use the ideas in Euclid's proof that there are infinitely many primes to create a list of primes until the numbers get too large for you to easily factor. (You should be able to factor any number less than 1000.)

12.2. (a) Show that there are infinitely many primes that are congruent to 5 modulo 6. (*Hint*. Use $A = 6p_1p_2\cdots p_r + 5$.)

(b) Try to use the same idea (with $A = 5p_1p_2\cdots p_r + 4$) to show that there are infinitely many primes congruent to 4 modulo 5. What goes wrong? In particular, what happens if you start with $\{19\}$ and try to make a longer list?

12.3. Let p be an odd prime number. Write the quantity

$$1 + \frac{1}{2} + \frac{1}{3} + \frac{1}{4} + \cdots + \frac{1}{p-1}$$

as a fraction A_p/B_p in lowest terms.

(a) Find the value of $A_p \pmod{p}$ and prove that your answer is correct.

(b) Make a conjecture for the value of $A_p \pmod{p^2}$.

(c) Prove your conjecture in (b). (This is quite difficult.)

[3] An arithmetic progression is a list of numbers with a common difference. For example, 2, 7, 12, 17, 22, ... is an arithmetic progression with common difference 5. The numbers congruent to a modulo m form an arithmetic progression with common difference m, which explains the name of Dirichlet's Theorem.

12.4. Let m be a positive integer, let $a_1, a_2, \ldots, a_{\phi(m)}$ be the integers between 1 and m that are relatively prime to m, and write the quantity

$$\frac{1}{a_1} + \frac{1}{a_2} + \frac{1}{a_3} + \cdots + \frac{1}{a_{\phi(m)}}$$

as a fraction A_m/B_m in lowest terms.

(a) Find the value of $A_m \pmod{m}$ and prove that your answer is correct.
(b) Make a conjecture for the value of $A_m \pmod{m^2}$.
(c) Prove your conjecture in (b). [This is quite difficult.]

12.5. Recall that the number n *factorial*, which is written $n!$, is equal to the product

$$n! = 1 \cdot 2 \cdot 3 \cdots (n-1) \cdot n.$$

(a) Find the highest power of 2 dividing each of the numbers $1!, 2!, 3!, \ldots, 10!$.
(b) Formulate a rule that gives the highest power of 2 dividing $n!$. Use your rule to compute the highest power of 2 dividing 100! and 1000!.
(c) Prove that your rule in (b) is correct.
(d) Repeat (a), (b), and (c), but this time for the largest power of 3 dividing $n!$.
(e) Try to formulate a general rule for the highest power of a prime p that divides $n!$. Use your rule to find the highest power of 7 dividing 1000! and the highest power of 11 dividing 5000!.
(f) Using your rule from (e) or some other method, prove that if p is prime and if p^m divides $n!$ then $m < n/(p-1)$. (This inequality is very important in many areas of advanced number theory.)

12.6. (a) Find a prime p satisfying $p \equiv 1338 \pmod{1115}$. Are there infinitely many such primes?
(b) Find a prime p satisfying $p \equiv 1438 \pmod{1115}$. Are there infinitely many such primes?

Chapter 13

Counting Primes

How many prime numbers are there? We have already given the answer that there are infinitely many. Of course, there are also infinitely many composite numbers. Which are there more of, primes or composites? Despite the fact that there are infinitely many of each, we can compare them by using a counting function.

First, let's start with an easier question that will illustrate the underlying idea. Our intuition says that approximately half of all numbers are even. We can put this intuition onto firmer ground by looking at the even number counting function:

$$E(x) = \#\{\text{even numbers } n \text{ with } 1 \le n \le x\}.$$

This function counts how many even numbers there are less than or equal to x. For example,

$$\begin{gathered} E(3) = 1, \quad E(4) = 2, \quad E(5) = 2, \quad \dots \\ E(100) = 50, \quad E(101) = 50, \dots . \end{gathered}$$

To study what fraction of all numbers are even, we should look at the ratio $E(x)/x$. Thus,

$$\begin{gathered} \frac{E(3)}{3} = \frac{1}{3}, \quad \frac{E(4)}{4} = \frac{1}{2}, \quad \frac{E(5)}{5} = \frac{2}{5}, \quad \dots \\ \frac{E(100)}{100} = \frac{1}{2}, \quad \frac{E(101)}{101} = \frac{50}{101}, \dots . \end{gathered}$$

It is certainly not true that the ratio $E(x)/x$ is always equal to $\frac{1}{2}$, but it is true that when x is large $E(x)/x$ will be close to $\frac{1}{2}$. If you have taken a little bit of calculus, you will recognize that we are trying to say that

$$\lim_{x \to \infty} \frac{E(x)}{x} = \frac{1}{2}.$$

This statement[1] just means that as x gets larger and larger the distance between $E(x)/x$ and $\frac{1}{2}$ gets closer and closer to 0.

Now let's do the same thing for prime numbers. The counting function for prime numbers is called $\pi(x)$, where "π" is an abbreviation for "prime." (This use of the Greek letter π has nothing to do with the number $3.14159\ldots$.) Thus

$$\pi(x) = \#\{\text{primes } p \text{ with } p \le x\}$$

For example, $\pi(10) = 4$, since the primes less than 10 are 2, 3, 5, and 7. Similarly, the primes less than 60 are

$$2, 3, 5, 7, 11, 13, 17, 19, 23, 29, 31, 37, 41, 43, 47, 53, 59,$$

so $\pi(60) = 17$. Here's a short table giving the values of $\pi(x)$ and the ratio $\pi(x)/x$.

x	10	25	50	100	200	500	1000	5000
$\pi(x)$	4	9	15	25	46	95	168	669
$\pi(x)/x$	0.400	0.360	0.300	0.250	0.230	0.190	0.168	0.134

It certainly looks like the ratio $\pi(x)/x$ is getting smaller and smaller as x gets larger. Assuming that this pattern continues, we would be justified in saying that "most numbers are not prime." This raises the further question of just how rapidly $\pi(x)/x$ decreases. The answer is provided by the following celebrated result, which is one of the pinnacles of nineteenth-century number theory.

Theorem 13.1 (The Prime Number Theorem). *When x is large, the number of primes less than x is approximately equal to $x/\ln(x)$. In other words,*

$$\lim_{x\to\infty} \frac{\pi(x)}{x/\ln(x)} = 1.$$

The quantity $\ln(x)$, which is called the natural logarithm of x, is the logarithm of x to the base $e = 2.7182818\ldots$.[2] Here is a table that compares the values

[1]This mathematical statement is read "the limit, as x goes to infinity, of $E(x)/x$, is equal to $1/2$."

[2]If you are not familiar with natural logarithms, you can just think of $\ln(x)$ as being approximately equal to $2.30259\log(x)$, where $\log(x)$ is the usual logarithm to the base 10. The natural logarithm is so important in mathematics and science that most scientific calculators have a special button to compute it. The natural logarithm appears "naturally" in problems involving compound growth, such as population growth, interest payments, and decay of radioactive materials. It is a wonderful fact that this widely applicable function also appears in the purely mathematical problem of counting prime numbers.

of $\pi(x)$ and $x/\ln(x)$.

x	10	100	1000	10^4	10^6	10^9
$\pi(x)$	4	25	168	1229	78498	50847534
$x/\ln(x)$	4.34	21.71	144.76	1085.74	72382.41	48254942.43
$\pi(x)/(x/\ln(x))$	0.921	1.151	1.161	1.132	1.084	1.054

By examining similar, but shorter, tables around 1800, Karl Friedrich Gauss and Adrien-Marie Legendre independently were led to conjecture that the Prime Number Theorem should be true. Almost a century passed before a proof was found. In 1896 Jacques Hadamard and Ch. de la Vallée Poussin each managed to prove the Prime Number Theorem. Just as with Dirichlet's Theorem, the proof uses methods from complex analysis (i.e., calculus with complex numbers). More recently, in 1948, Paul Erdös and Atle Selberg found an "elementary" proof of the Prime Number Theorem. Their proof is elementary in the sense that it does not require methods from complex analysis, but it is by no means easy, so we are not be able to present it here.

It is somewhat surprising that to prove theorems about whole numbers, such as Dirichlet's Theorem and the Prime Number Theorem, mathematicians have to use tools from calculus. An entire branch of mathematics called Analytic Number Theory is devoted to proving theorems in number theory using calculus methods.

There are many famous unsolved problems involving prime numbers. We conclude this chapter by describing three such problems with a little bit of their history.

Conjecture 13.2 (Goldbach's Conjecture). *Every even number $n \geq 4$ is a sum of two primes.*

Goldbach proposed this conjecture to Euler in a letter dated June 7, 1742. It is not hard to check that Goldbach's Conjecture is true for the first few even numbers. Thus,

$$4 = 2+2, \quad 6 = 3+3, \quad 8 = 3+5, \quad 10 = 3+7, \quad 12 = 5+7,$$
$$14 = 3+11, \quad 16 = 3+13, \quad 18 = 5+13, \quad 20 = 7+13 \ldots.$$

This verifies Goldbach's Conjecture for all even numbers up to 20. Using computers, Goldbach's conjecture has been checked for all even numbers up to $2 \cdot 10^{10}$. Even better, mathematicians have been able to prove results that are similar to Goldbach's Conjecture. These suggest that Goldbach's Conjecture is also true. One such theorem was proved by I.M. Vinogradov in 1937. He showed that every (sufficiently large) odd number n is a sum of three primes. A second theorem, proved by Chen Jing-run in 1966, says that every (sufficiently large) even number is a sum of two numbers $p + a$, where p is a prime number and a is either prime or a product of two primes.

Conjecture 13.3 (The Twin Primes Conjecture). *There are infinitely many prime numbers p so that $p + 2$ is also prime.*

The list of prime numbers is quite irregular, and there are often very large gaps between consecutive primes. For example, there are 111 composite numbers following the prime 370,261. On the other hand, there seem to be quite a few instances in which a prime p is followed almost immediately by another prime $p + 2$. (Of course, $p + 1$ cannot be prime, since it is even.) These pairs are called *twin primes*, and the Twin Primes Conjecture says that the list of twin primes should never end. The first few twin primes are

$$\begin{gathered}(3,5),\ (5,7),\ (11,13),\ (17,19),\ (29,31),\ (41,43),\ (59,61),\ (71,73),\\ (101,103),\ (107,109),\ (137,139),\ (149,151),\ (179,181),\ (191,193),\\ (197,199),\ (227,229),\ (239,241),\ (269,271),\ (281,283),\ (311,313).\end{gathered}$$

Just as with Goldbach's Conjecture, people have used computers to compile long lists of twin primes, including, for example, the tremendous pair consisting of

$$242206083 \cdot 2^{38880} - 1 \qquad \text{and} \qquad 242206083 \cdot 2^{38880} + 1.$$

As further evidence for the validity of the conjecture, Chen Jing-run proved in 1966 that there are infinitely many primes p so that $p + 2$ is either a prime or a product of two primes.

Conjecture 13.4 (The $N^2 + 1$ Conjecture). *There are infinitely many primes of the form $N^2 + 1$.*

If N is odd, then $N^2 + 1$ is even, so it cannot be prime (unless $N = 1$). However, if N is even, then $N^2 + 1$ seems frequently to be prime. The $N^2 + 1$ Conjecture says that this should happen infinitely often. The first few primes of this form are

$$\begin{gathered}2^2 + 1 = 5, \quad 4^2 + 1 = 17, \quad 6^2 + 1 = 37, \quad 10^2 + 1 = 101,\\ 14^2 + 1 = 197, \quad 16^2 + 1 = 257, \quad 20^2 + 1 = 401, \quad 24^2 + 1 = 577,\\ 26^2 + 1 = 677, \quad 36^2 + 1 = 1297, \quad 40^2 + 1 = 1601.\end{gathered}$$

The best result currently known was proved by Hendrik Iwaniec in 1978. He showed that there are infinitely many values of N for which $N^2 + 1$ is either prime or a product of two primes.

Although no one knows if there are infinitely many twin primes or infinitely many primes of the form N^2+1, mathematicians have guessed what their counting functions should look like. Let

$$T(x) = \#\{\text{primes } p \le x \text{ such that } p+2 \text{ is also prime}\},$$
$$S(x) = \#\{\text{primes } p \le x \text{ such that } p \text{ has the form } N^2+1\}.$$

Then it is conjectured that

$$\lim_{x\to\infty} \frac{T(x)}{x/(\ln x)^2} = C \qquad \text{and} \qquad \lim_{x\to\infty} \frac{S(x)}{\sqrt{x}/\ln x} = C'.$$

The numbers C and C' are a bit complicated to describe precisely. For example, C is approximately equal to 0.66016.

Exercises

13.1. (a) Explain why the statement "one-fifth of all numbers are congruent to 2 modulo 5" makes sense by using the counting function

$$F(x) = \#\{\text{positive numbers } n \le x \text{ satisfying } n \equiv 2 \pmod 5\}.$$

(b) Explain why the statement "most numbers are not squares" makes sense by using the counting function

$$S(x) = \#\{\text{square numbers less than } x\}.$$

Find a simple function of x that is approximately equal to $S(x)$ when x is large.

13.2. (a) Check that every even number between 70 and 100 is a sum of two primes.
(b) How many different ways can 70 be written as a sum of two primes $70 = p + q$ with $p \le q$? Same question for 90? Same question for 98?

13.3. The number $n!$ (n factorial) is the product of all numbers from 1 to n. For example, $4! = 1\cdot 2\cdot 3\cdot 4 = 24$ and $7! = 1\cdot 2\cdot 3\cdot 4\cdot 5\cdot 6\cdot 7 = 5040$. If $n \ge 2$, show that all the numbers

$$n!+2, \quad n!+3, \quad n!+4, \quad \dots, \quad n!+(n-1), \quad n!+n$$

are composite numbers.

13.4. (a) Do you think there are infinitely many primes of the form N^2+2?
(b) Do you think there are infinitely many primes of the form N^2-2?
(c) Do you think there are infinitely many primes of the form N^2+3N+2?
(d) Do you think there are infinitely many primes of the form N^2+2N+2?

13.5. The Prime Number Theorem says that the number of primes smaller than x is approximately $x/\ln(x)$. This exercise asks you to explain why certain statements are plausible. So do not try to write down formal mathematical proofs. Instead, explain as convincingly as you can in words why the Prime Number Theorem makes each of the following statements reasonable.

(a) If you choose a random integer between 1 and x, then the probability that you chose a prime number is approximately $1/\ln(x)$.

(b) If you choose two random integers between 1 and x, then the probability that both of them are prime numbers is approximately $1/(\ln x)^2$.

(c) The number of twin primes between 1 and x should be approximately $x/(\ln x)^2$. [Notice that this explains the conjectured limit formula for the twin prime counting function $T(x)$.]

13.6. (This exercise is for people who have taken some calculus.) The Prime Number Theorem says that the counting function for primes, $\pi(x)$, is approximately equal to $x/\ln(x)$ when x is large. It turns out that $\pi(x)$ is even closer to the value of the definite integral $\int_2^x dt/\ln(t)$.

(a) Show that

$$\lim_{x\to\infty} \left(\int_2^x \frac{dt}{\ln(t)}\right) \Big/ \left(\frac{x}{\ln(x)}\right) = 1.$$

This means that $\int_2^x dt/\ln(t)$ and $x/\ln(x)$ are approximately the same when x is large. (*Hint*. Use L'Hôpital's rule and the Second Fundamental Theorem of Calculus.)

(b) It can be shown that

$$\int \frac{dt}{\ln(t)} = \ln(\ln(t)) + \ln(t) + \frac{(\ln(t))^2}{2\cdot 2!} + \frac{(\ln(t))^3}{3\cdot 3!} + \frac{(\ln(t))^4}{4\cdot 4!} + \cdots.$$

Use this series to compute numerically the value of $\int_2^x dt/\ln(t)$ for $x = 10$, 100, 1000, 10^4, 10^6, and 10^9. Compare the values you get with the values of $\pi(x)$ and $x/\ln(x)$ given in the table in Chapter 13. Which is closer to $\pi(x)$, the integral $\int_2^x dt/\ln(t)$ or the function $x/\ln(x)$? (This problem can be done with a simple calculator, but you'll probably prefer to use a computer or programmable calculator.)

(c) Differentiate the series in (b) and show that the derivative is actually equal to $1/\ln(t)$. (*Hint*. Use the series for e^x.)

Chapter 14

Mersenne Primes

In this chapter we will study primes that can be written in the form $a^n - 1$ with $n \geq 2$. For example, 31 is such a prime, since $31 = 2^5 - 1$. The first step is to look at some data.

$2^2 - 1 = 3$	$2^3 - 1 = 7$	$2^4 - 1 = 3 \cdot 5$	$2^5 - 1 = 31$
$3^2 - 1 = 2^3$	$3^3 - 1 = 2 \cdot 13$	$3^4 - 1 = 2^4 \cdot 5$	$3^5 - 1 = 2 \cdot 11^2$
$4^2 - 1 = 3 \cdot 5$	$4^3 - 1 = 3^2 \cdot 7$	$4^4 - 1 = 3 \cdot 5 \cdot 17$	$4^5 - 1 = 3 \cdot 11 \cdot 31$
$5^2 - 1 = 2^3 \cdot 3$	$5^3 - 1 = 2^2 \cdot 31$	$5^4 - 1 = 2^4 \cdot 3 \cdot 13$	$5^5 - 1 = 2^2 \cdot 11 \cdot 71$
$6^2 - 1 = 5 \cdot 7$	$6^3 - 1 = 5 \cdot 43$	$6^4 - 1 = 5 \cdot 7 \cdot 37$	$6^5 - 1 = 5^2 \cdot 311$
$7^2 - 1 = 2^4 \cdot 3$	$7^3 - 1 = 2 \cdot 3^2 \cdot 19$	$7^4 - 1 = 2^5 \cdot 3 \cdot 5^2$	$7^5 - 1 = 2 \cdot 3 \cdot 2801$
$8^2 - 1 = 3^2 \cdot 7$	$8^3 - 1 = 7 \cdot 73$	$8^4 - 1 = 3^2 \cdot 5 \cdot 7 \cdot 13$	$8^5 - 1 = 7 \cdot 31 \cdot 151$

An easy observation is that if a is odd then $a^n - 1$ is even, so it cannot be prime. Looking at the table, we also see that it appears that $a^n - 1$ is always divisible by $a - 1$. This observation is indeed true. We can prove that it is true by using the famous formula for the sum of a geometric series:

$$x^n - 1 = (x - 1)(x^{n-1} + x^{n-2} + \cdots + x^2 + x + 1). \qquad \textbf{Geometric Series}$$

To check this Geometric Series formula, we multiply out the product on the right. Thus,

$$\begin{aligned}(x - 1)&(x^{n-1} + x^{n-2} + \cdots + x^2 + x + 1) \\ &= x \cdot (x^{n-1} + x^{n-2} + \cdots + x^2 + x + 1) \\ &\qquad - 1 \cdot (x^{n-1} + x^{n-2} + \cdots + x^2 + x + 1)\end{aligned}$$

$$
\begin{aligned}
&= (x^n + x^{n-1} + \cdots + x^3 + x^2 + x) \\
&\qquad - (x^{n-1} + x^{n-2} + \cdots + x^2 + x + 1) \\
&= x^n - 1,
\end{aligned}
$$

since all the other terms cancel.

Using the Geometric Series formula with $x = a$, we see immediately that $a^n - 1$ is always divisible by $a - 1$. So $a^n - 1$ will be composite unless $a - 1 = 1$, that is, unless $a = 2$.

However, even if $a = 2$, the number $2^n - 1$ is frequently composite. Again we look at some data:

n	2	3	4	5	6	7	8	9	10
$2^n - 1$	3	7	$3 \cdot 5$	31	$3^2 \cdot 7$	127	$3 \cdot 5 \cdot 17$	$7 \cdot 73$	$3 \cdot 11 \cdot 31$

Even this short table suggests the following:

When n is even, $2^n - 1$ is divisible by $3 = 2^2 - 1$.
When n is divisible by 3, $2^n - 1$ is divisible by $7 = 2^3 - 1$.
When n is divisible by 5, $2^n - 1$ is divisible by $31 = 2^5 - 1$.

So we suspect that if n is divisible by m, then $2^n - 1$ will be divisible by $2^m - 1$.

Having made this observation, it is easy to verify that it is true. So suppose that n factors as $n = mk$. Then $2^n = 2^{mk} = (2^m)^k$. We use the Geometric Series formula with $x = 2^m$ to obtain

$$2^n - 1 = (2^m)^k - 1 = (2^m - 1)\big((2^m)^{k-1} + (2^m)^{k-2} + \cdots + (2^m)^2 + (2^m) + 1\big).$$

This shows that if n is composite then $2^n - 1$ is composite. We have verified the following fact.

Proposition 14.1. *If $a^n - 1$ is prime for some numbers $a \geq 2$ and $n \geq 2$, then a must equal 2 and n must be a prime.*

This means that if we are interested in primes of the form $a^n - 1$ we only need to consider the case that $a = 2$ and n is prime. Primes of the form

$$2^p - 1$$

are called *Mersenne primes*. The first few Mersenne primes are

$$2^2 - 1 = 3, \quad 2^3 - 1 = 7, \quad 2^5 - 1 = 31, \quad 2^7 - 1 = 127, \quad 2^{13} - 1 = 8191.$$

Of course, not every number $2^p - 1$ is prime. For example,

$$2^{11} - 1 = 2047 = 23 \cdot 89 \quad \text{and} \quad 2^{29} - 1 = 536870911 = 233 \cdot 1103 \cdot 2089.$$

The Mersenne primes are named after Father Marin Mersenne (1588–1648), who asserted in 1644 that $2^p - 1$ is prime for

$$p = 2, 3, 5, 7, 13, 17, 19, 31, 67, 127, 257$$

and that these are the only primes less than 258 for which $2^p - 1$ is prime. It is not known how Mersenne discovered these "facts," especially since it turns out that his list is not correct. The complete list of primes p less than 10000 for which $2^p - 1$ is prime is[1]

$$p = 2, 3, 5, 7, 13, 17, 19, 31, 61, 89, 107, 127, 521, 607, 1279,$$
$$2203, 2281, 3217, 4253, 4423, 9689, 9941.$$

It was only with the advent of computing machines that it became possible to check numbers with hundreds of digits for primality. Indeed, it wasn't until 1876 that E. Lucas proved conclusively that $2^{127} - 1$ is prime, and Lucas's number remained the largest known prime until the 1950s! Table 14.1 lists Mersenne primes that have been discovered in recent years using computers, together with the names of the people who made the discoveries.

The most recent Mersenne primes in Table 14.1 were unearthed using specialized software as part of Woltman's Great Internet Mersenne Prime Search. You, too, can take part in the search for world record primes[2] by downloading software from the GIMPS web site

www.mersenne.org/prime.htm

Further historical and topical information about Mersenne primes is available at

www.utm.edu/research/primes/mersenne.shtml

Of course, although it is interesting to see a list like this of the world's largest known primes, there is no huge mathematical significance in finding a few more Mersenne primes. Far more interesting from a mathematical perspective is the following question. The answer is not known.

Question 14.2. Are there infinitely many Mersenne primes, or does the list of Mersenne primes eventually stop?

---

[1]Notice that Father Mersenne made five mistakes, three of omission (61, 89, 107) and two of commission (67, 257).

[2]Andy Warhol opined that in the future everyone will be famous for 15 minutes. One route to such fame is to find the largest known (Mersenne) prime. And the quest for bigger and better primes continues.

p	Discovered by	Date
521 607 1279 2203 2281	Robinson	1952
3217	Riesel	1957
4253 4423	Hurwitz	1961
9689 9941 11213	Gillies	1963
19937	Tuckerman	1971
21707	Noll Nickel	1978
23209	Noll	1979
44497	Noll Slowinski	1979
86243	Slowinski	1982

p	Discovered by	Date
110503	Colquitt Welsch	1988
132049	Slowinski	1983
216091	Slowinski	1985
756839	Slowinski Gage	1992
859433	Slowinski Gage	1994
1257787	Slowinski Gage	1996
1398269	Armengaud	1996
2976221	Spence	1997
3021377	Clarkson	1998
6972593	Hajratwala	1999
13466917	Cameron	2001
20996011	Shafer	2003
24036583	Findley	2004

Table 14.1: Primes $p \geq 500$ for which $2^p - 1$ is known to be prime

Exercises

14.1. If $a^n + 1$ is prime for some numbers $a \geq 2$ and $n \geq 1$, show that n must be a power of 2.

14.2. Let $F_k = 2^{2^k} + 1$. For example, $F_1 = 5$, $F_2 = 17$, $F_3 = 257$, and $F_4 = 65537$. Fermat thought that all the F_k's might be prime, but Euler showed in 1732 that F_5 factors as $641 \cdot 6700417$, and in 1880 Landry showed that F_6 is composite. Primes of the form F_k are called *Fermat primes*. Show that if $k \neq m$, then the numbers F_k and F_m have no common factors; that is, show that $\gcd(F_k, F_m) = 1$. (*Hint.* If $k > m$, show that F_m divides $F_k - 2$.)

14.3. The numbers $3^n - 1$ are never prime (if $n \geq 2$), since they are always even. However, it sometimes happens that $(3^n - 1)/2$ is prime. For example, $(3^3 - 1)/2 = 13$ is prime.

(a) Find another prime of the form $(3^n - 1)/2$.

(b) If n is even, show that $(3^n - 1)/2$ is always divisible by 4, so it can never be prime.

(c) Use a similar argument to show that if n is a multiple of 5 then $(3^n - 1)/2$ is never a prime.

(d) Do you think that there are infinitely many primes of the form $(3^n - 1)/2$?

Chapter 15

Mersenne Primes and Perfect Numbers

The ancient Greeks observed that the number 6 has a surprising property. If you take the proper divisors of 6, that is, the divisors other than 6 itself, and add them up, you get back the number 6. Thus, the proper divisors of 6 are 1, 2, and 3, and when you add up these divisors, you get

$$1 + 2 + 3 = 6.$$

This property is rather rare, as can be seen by looking at a few examples:

n	Sum of Proper Divisors of n	
6	$1 + 2 + 3 = 6$	Sum is just right (perfect!).
10	$1 + 2 + 5 = 8$	Sum is too small.
12	$1 + 2 + 3 + 4 + 6 = 16$	Sum is too large.
15	$1 + 3 + 5 = 9$	Sum is too small.
20	$1 + 2 + 4 + 5 + 10 = 22$	Sum is too large.
28	$1 + 2 + 4 + 7 + 14 = 28$	Sum is just right (perfect!).
45	$1 + 3 + 5 + 9 + 15 = 33$	Sum is too small.

The Greeks called these special numbers perfect. That is, a *perfect number* is a number that is equal to the sum of its proper divisors. So far, we have discovered two perfect numbers, 6 and 28. Are there others?

The Greeks knew a method for finding some perfect numbers and, interestingly enough, their method is closely related to the Mersenne primes that we studied in the previous chapter. The following assertion occurs as Proposition 36 of Book IX of Euclid's *Elements*.

Theorem 15.1 (Euclid's Perfect Number Formula). *If $2^p - 1$ is a prime number, then $2^{p-1}(2^p - 1)$ is a perfect number.*

The first two Mersenne primes are $3 = 2^2 - 1$ and $7 = 2^3 - 1$. Euclid's Perfect Number Formula applied to these two Mersenne primes gives the two perfect numbers we already know,

$$2^{2-1}(2^2 - 1) = 6 \qquad \text{and} \qquad 2^{3-1}(2^3 - 1) = 28.$$

The next Mersenne prime is $2^5 - 1 = 31$, and Euclid's formula gives us a new perfect number,

$$2^{5-1}(2^5 - 1) = 496.$$

To check that 496 is perfect, we need to add up its proper divisors. Factoring $496 = 2^4 \cdot 31$, we see that the proper divisors of 496 are

$$1, 2, 2^2, 2^3, 2^4 \quad \text{and} \quad 31, 2 \cdot 31, 2^2 \cdot 31, 2^3 \cdot 31.$$

We could just add up these numbers, but to illustrate the general method we will sum them in two stages. First

$$1 + 2 + 2^2 + 2^3 + 2^4 = 31,$$

and second

$$31 + 2 \cdot 31 + 2^2 \cdot 31 + 2^3 \cdot 31 = 31(1 + 2 + 2^2 + 2^3) = 31 \cdot 15.$$

Now adding the two pieces gives $31 + 31 \cdot 15 = 31 \cdot 16 = 496$, so 496 is indeed perfect.

Using the same sort of idea, we can easily verify that Euclid's Perfect Number Formula is true in general. We let $q = 2^p - 1$, and we need to check that $2^{p-1}q$ is a perfect number. The proper divisors of $2^{p-1}q$ are

$$1, 2, 4, \ldots, 2^{p-1} \quad \text{and} \quad q, 2q, 4q, \ldots, 2^{p-2}q.$$

We add up these numbers using the formula for the Geometric Series from the previous chapter. The Geometric Series formula (slightly rearranged) says that

$$1 + x + x^2 + \cdots + x^{n-1} = \frac{x^n - 1}{x - 1}.$$

Putting $x = 2$ and $n = p$, we get

$$1 + 2 + 4 + \cdots + 2^{p-1} = \frac{2^p - 1}{2 - 1} = 2^p - 1 = q.$$

And we can use the formula with $x = 2$ and $n = p - 1$ to compute

$$\begin{aligned} q + 2q + 4q + \cdots + 2^{p-2}q &= q(1 + 2 + 4 + \cdots + 2^{p-2}) \\ &= q\left(\frac{2^{p-1} - 1}{2 - 1}\right) \\ &= q(2^{p-1} - 1). \end{aligned}$$

So if we add up all the proper divisors of $2^{p-1}q$, we get

$$1 + 2 + 4 + \cdots + 2^{p-1} + q + 2q + 4q + \cdots + 2^{p-2}q = q + q(2^{p-1} - 1) = 2^{p-1}q.$$

This shows that $2^{p-1}q$ is a perfect number.

We can use Euclid's Perfect Number Formula to write down many more perfect numbers. In fact, we get one perfect number for each Mersenne prime that we can find. The first few perfect numbers obtained in this fashion are listed in the following table. As you will observe, the numbers get large rather quickly.

p	2	3	5	7	13	17
$2^{p-1}(2^p - 1)$	6	28	496	8128	33550336	8589869056

We can also list perfect numbers that are incredibly huge. For example,

$$2^{756838}(2^{756839} - 1) \qquad \text{and} \qquad 2^{859432}(2^{859433} - 1)$$

are perfect numbers. The latter has more than half a million digits!

A natural question to ask at this point is whether Euclid's Perfect Number Formula actually describes all perfect numbers. In other words, does every perfect number look like $2^{p-1}(2^p - 1)$ with $2^p - 1$ prime, or are there other perfect numbers? Approximately 2000 years after Euclid's death, Leonhard Euler showed that Euclid's formula at least gives all *even* perfect numbers.

Theorem 15.2 (Euler's Perfect Number Theorem). *If n is an even perfect number, then n looks like*

$$n = 2^{p-1}(2^p - 1),$$

where $2^p - 1$ is a Mersenne prime.

We will prove Euler's theorem at the end of this chapter, but first we need to discuss a function that will be needed for the proof. This function, which is denoted with the Greek letter σ (sigma), is equal to

$$\sigma(n) = \text{sum of all divisors of } n \text{ (including 1 and } n).$$

Here are a few examples:

$$\begin{aligned}\sigma(6) &= 1+2+3+6 &&= 12\\ \sigma(8) &= 1+2+4+8 &&= 15\\ \sigma(18) &= 1+2+3+6+9+18 &&= 39.\end{aligned}$$

We can also give some general formulas. For example, if p is a prime number, then its only divisors are 1 and p, so $\sigma(p) = p+1$. More generally, the divisors of a prime power p^k are the numbers $1, p, p^2, \ldots, p^k$, so

$$\sigma(p^k) = 1+p+p^2+\cdots+p^k = \frac{p^{k+1}-1}{p-1}.$$

To study the sigma function further, we make a short table of its values.

$\sigma(1) = 1$	$\sigma(2) = 3$	$\sigma(3) = 4$	$\sigma(4) = 7$	$\sigma(5) = 6$
$\sigma(6) = 12$	$\sigma(7) = 8$	$\sigma(8) = 15$	$\sigma(9) = 13$	$\sigma(10) = 18$
$\sigma(11) = 12$	$\sigma(12) = 28$	$\sigma(13) = 14$	$\sigma(14) = 24$	$\sigma(15) = 24$
$\sigma(16) = 31$	$\sigma(17) = 18$	$\sigma(18) = 39$	$\sigma(19) = 20$	$\sigma(20) = 42$
$\sigma(21) = 32$	$\sigma(22) = 36$	$\sigma(23) = 24$	$\sigma(24) = 60$	$\sigma(25) = 31$
$\sigma(26) = 42$	$\sigma(27) = 40$	$\sigma(28) = 56$	$\sigma(29) = 30$	$\sigma(30) = 72$
$\sigma(31) = 32$	$\sigma(32) = 63$	$\sigma(33) = 48$	$\sigma(34) = 54$	$\sigma(35) = 48$
$\sigma(36) = 91$	$\sigma(37) = 38$	$\sigma(38) = 60$	$\sigma(39) = 56$	$\sigma(40) = 90$
$\sigma(41) = 42$	$\sigma(42) = 96$	$\sigma(43) = 44$	$\sigma(44) = 84$	$\sigma(45) = 78$
$\sigma(46) = 72$	$\sigma(47) = 48$	$\sigma(48) = 124$	$\sigma(49) = 57$	$\sigma(50) = 93$
$\sigma(51) = 72$	$\sigma(52) = 98$	$\sigma(53) = 54$	$\sigma(54) = 120$	$\sigma(55) = 72$
$\sigma(56) = 120$	$\sigma(57) = 80$	$\sigma(58) = 90$	$\sigma(59) = 60$	$\sigma(60) = 168$
$\sigma(61) = 62$	$\sigma(62) = 96$	$\sigma(63) = 104$	$\sigma(64) = 127$	$\sigma(65) = 84$

An examination of this table reveals that $\sigma(mn)$ is frequently equal to the product $\sigma(m)\sigma(n)$ and, after a little further analysis, we notice that this seems to be true when m and n are relatively prime. Thus, the sigma function appears to obey the same sort of multiplication formula as the phi function that we studied in Chapter 11. We record this rule, together with the formula for $\sigma(p^k)$.

Theorem 15.3 (Sigma Function Formulas). (a) *If p is a prime and $k \geq 1$, then*

$$\sigma(p^k) = 1+p+p^2+\cdots+p^k = \frac{p^{k+1}-1}{p-1}.$$

(b) *If* $\gcd(m,n) = 1$, *then*

$$\sigma(mn) = \sigma(m)\sigma(n).$$

Just as with the phi function, we can use the sigma function formulas to easily compute $\sigma(n)$ for large values of n. For example,

$$\begin{aligned}\sigma(16072) &= \sigma(2^3 \cdot 7^2 \cdot 41)\\ &= \sigma(2^3) \cdot \sigma(7^2) \cdot \sigma(41)\\ &= (1 + 2 + 2^2 + 2^3)(1 + 7 + 7^2)(1 + 41)\\ &= 15 \cdot 57 \cdot 42 = 35910,\end{aligned}$$

and

$$\begin{aligned}\sigma(800000) &= \sigma(2^8 \cdot 5^5)\\ &= \left(\frac{2^9 - 1}{2 - 1}\right)\left(\frac{5^6 - 1}{5 - 1}\right)\\ &= 511 \cdot \frac{15624}{4} = 1995966.\end{aligned}$$

At this point you probably expect that I will show you how to prove the multiplication formula for the sigma function. But I won't! You have now made enough progress in number theory that it is time for you to start acting as a mathematician yourself.[1] So I am going to ask you to prove the formula $\sigma(mn) = \sigma(m)\sigma(n)$ for relatively prime integers m and n. Don't be discouraged and give up if you don't succeed at first. One suggestion I can give you is to try to discover *why* the formula is true before you attempt to give a general proof. So, for example, first look at numbers like $21 = 3 \cdot 7$ and $65 = 5 \cdot 13$ that are products of two primes and list their divisors. This should enable you to prove that $\sigma(pq) = \sigma(p)\sigma(q)$ when p and q are distinct prime numbers. Then try some m's and n's that have two or three divisors each and try to see how the divisors of m and n fit together to give divisors of mn. If you can describe this precisely enough, you should be able to prove that $\sigma(mn) = \sigma(m)\sigma(n)$. Remember, though, that you'll need to use the fact that m and n are relatively prime.

How is the sigma function related to perfect numbers? A number n is perfect if the sum of its divisors, other than n itself, is equal to n. The sigma function $\sigma(n)$ is the sum of the divisors of n, including n, so it has an "extra" n. Therefore,

$$n \text{ is perfect exactly when } \sigma(n) = 2n.$$

We are now ready to prove Euler's formula for even perfect numbers, which we restate here for your convenience.

[1] Your mission, should you decide to accept it, is to prove the multiplication formula for the sigma function. Should you be captured or killed in this endeavor, we will be forced to deny all knowledge of your activities. Good luck!

Theorem 15.4 (Euler's Perfect Number Theorem). *If n is an even perfect number, then n looks like*

$$n = 2^{p-1}(2^p - 1),$$

where $2^p - 1$ is a Mersenne prime.

Proof. Suppose that n is an even perfect number. The fact that n is even means that we can factor it as

$$n = 2^k m \qquad \text{with } k \geq 1 \text{ and } m \text{ odd.}$$

Next we use the sigma function formulas to compute $\sigma(n)$,

$$\begin{aligned} \sigma(n) &= \sigma(2^k m) && \text{since } n = 2^k m, \\ &= \sigma(2^k)\sigma(m) && \text{using the multiplication formula for } \sigma \\ & && \text{and the fact that } \gcd(2^k, m) = 1, \\ &= (2^{k+1} - 1)\sigma(m) && \text{using the formula for } \sigma(p^k). \end{aligned}$$

But n is supposed to be perfect, which means that $\sigma(n) = 2n = 2^{k+1}m$. So we have two different expressions for $\sigma(n)$, and they must be equal,

$$2^{k+1}m = (2^{k+1} - 1)\sigma(m).$$

The number $2^{k+1} - 1$ is clearly odd, and $(2^{k+1} - 1)\sigma(m)$ is a multiple of 2^{k+1}, so 2^{k+1} must divide $\sigma(m)$. In other words, there is some number c so that $\sigma(m) = 2^{k+1}c$. We can substitute this into the above equation to get

$$2^{k+1}m = (2^{k+1} - 1)\sigma(m) = (2^{k+1} - 1)2^{k+1}c,$$

and then canceling 2^{k+1} from both sides gives $m = (2^{k+1} - 1)c$. To recapitulate, we have shown that there is an integer c so that

$$m = (2^{k+1} - 1)c \qquad \text{and} \qquad \sigma(m) = 2^{k+1}c.$$

We are going to show that $c = 1$ by assuming that $c > 1$ and deriving a false statement. So suppose that $c > 1$. Then $m = (2^{k+1} - 1)c$ would be divisible by the distinct numbers

$$1, \quad c, \quad \text{and} \quad m.$$

(N.B. The fact that our original number n was even means that $k \geq 1$, so c and m are different.) Of course, m is probably divisible by many other numbers, but in any case we find that

$$\sigma(m) \geq 1 + c + m = 1 + c + (2^{k+1} - 1)c = 1 + 2^{k+1}c.$$

However, we also know that $\sigma(m) = 2^{k+1}c$, so

$$2^{k+1}c \geq 1 + 2^{k+1}c.$$

Therefore, $0 \geq 1$, which is an absurdity. This contradiction shows that c must actually be equal to 1, which means that

$$m = (2^{k+1} - 1) \qquad \text{and} \qquad \sigma(m) = 2^{k+1} = m + 1.$$

Which numbers m have the property that $\sigma(m) = m + 1$? These are clearly the numbers whose only divisors are 1 and m, since otherwise the sum of their divisors would be larger. In other words, $\sigma(m) = m + 1$ exactly when m is prime. We have now prove that if n is an even perfect number then

$$n = 2^k(2^{k+1} - 1) \qquad \text{with } 2^{k+1} - 1 \text{ a prime number.}$$

We know from Chapter 14 that if $2^{k+1} - 1$ is prime then $k + 1$ must itself be prime, say $k + 1 = p$. So every even perfect number looks like $n = 2^{p-1}(2^p - 1)$ with $2^p - 1$ a Mersenne prime. This completes our proof of Euler's Perfect Number Theorem. □

Euler's Perfect Number Theorem gives an excellent description of all even perfect numbers, but it says nothing about odd perfect numbers.

Question 15.5 (Odd Perfect Number Quandary). Are there any odd perfect numbers?

To this day, no one has been able to discover any odd perfect numbers, although this is not through lack of trying. Many mathematicians have written many research papers (more than 50 papers in the last 50 years) studying these elusive creatures, and it is currently known that there are no odd perfect numbers less than 10^{300}. However, no one has yet been able to prove conclusively that none exist, so for now odd perfect numbers are like the little man in the poem:

Last night I met upon the stair,
A little man who wasn't there.
He wasn't there again today.
I wish to heck he'd go away.

Anonymous

If you do some experimentation with small numbers, you might suspect that $\sigma(n) < 2n$ for all odd numbers. If this were true, it would certainly prove that there are no odd perfect numbers, but unfortunately it is not true. The first odd number for which it is false is $n = 945 = 3^3 \cdot 5 \cdot 7$, which has $\sigma(945) = 1920$. This example should serve as a warning against believing a fact to be true simply because it has been checked for lots of small numbers. It is perfectly all right to make conjectures based on numerical data, but mathematicians insist on rigorous proofs precisely because such data can be misleading.

Exercises

15.1. If m and n are integers with $\gcd(m, n) = 1$, prove that $\sigma(mn) = \sigma(m)\sigma(n)$.

15.2. Compute the following values of the sigma function.
(a) $\sigma(10)$ (b) $\sigma(20)$ (c) $\sigma(1728)$

15.3. (a) Show that a power of 3 can never be a perfect number.
(b) More generally, if p is an odd prime, show that a power p^k can never be a perfect number.
(c) Show that a number of the form $3^i \cdot 5^j$ can never be a perfect number.
(d) More generally, if p is an odd prime number greater than 3, show that the product $3^i p^j$ can never be a perfect number.
(e) Even more generally, show that if p and q are distinct odd primes, then a number of the form $q^i p^j$ can never be a perfect number.

15.4. Show that a number of the form $3^m \cdot 5^n \cdot 7^k$ can never be a perfect number.

15.5. A perfect number is equal to the sum of its divisors (other than itself). If we look at the product instead of the sum, we could say that a number is *product perfect* if the product of all its divisors (other than itself) is equal to the original number. For example,

m	Product of factors	
6	$1 \cdot 2 \cdot 3 = 6$	product perfect
9	$1 \cdot 3 = 3$	product is too small
12	$1 \cdot 2 \cdot 3 \cdot 4 \cdot 6 = 144$	product is too large
15	$1 \cdot 3 \cdot 5 = 15$	product perfect.

So 6 and 15 are product perfect, while 9 and 12 are not product perfect.
(a) List all product perfect numbers between 2 and 50.
(b) Describe all product perfect numbers. Your description should be precise enough to enable you easily to solve problems such as "Is 35710 product perfect?" and "Find a product perfect number larger than 10000."
(c) Prove that your description in (b) is correct.

15.6. (a) Write a program to compute $\sigma(n)$, the sum of all the divisors of n (including 1 and n itself). You should compute $\sigma(n)$ by using a factorization of n into primes, not by actually finding all the divisors of n and adding them up.

(b) As you know, the Greeks called n *perfect* if $\sigma(n) = 2n$. They also called n *abundant* if $\sigma(n) > 2n$, and they called n *deficient* if $\sigma(n) < 2n$. Count how many n's between 2 and 100 are perfect, abundant, and deficient. Clearly, perfect numbers are very rare. Which do you think are more common, abundant numbers or deficient numbers? Extend your list for $100 < n \le 200$ and see if your guess still holds.

15.7. The Greeks called two numbers m and n an *amicable pair* if the sum of the proper divisors of m equals n and simultaneously the sum of the proper divisors of n equals m. (The proper divisors of a number n are all divisors of n excluding n itself.) The first amicable pair, and the only one (as far as we know) that was known in ancient Greece, is the pair $(220, 284)$. This pair is amicable since

$$284 = 1 + 2 + 4 + 5 + 10 + 11 + 20 + 22 + 44 + 55 + 110 \qquad \text{(divisors of 220)}$$
$$220 = 1 + 2 + 4 + 71 + 142 \qquad \text{(divisors of 284).}$$

(a) Show that m and n are an amicable pair if and only if $\sigma(n)$ and $\sigma(m)$ both equal $n + m$.

(b) Verify that each of the following pairs forms an amicable pair of numbers.

$$(220, 284), (1184, 1210), (2620, 2924), (5020, 5564), (6232, 6368),$$
$$(10744, 10856), (12285, 14595).$$

(c) There is a rule for generating amicable numbers, although it does not generate all of them. This rule was first discovered by Abu-l-Hasan Thabit ben Korrah around the ninth century and later rediscovered by many others, including Fermat and Descartes. The rule says to look at the three numbers

$$p = 3 \cdot 2^{e-1} - 1,$$
$$q = 2p + 1 = 3 \cdot 2^{e} - 1,$$
$$r = (p + 1)(q + 1) - 1 = 9 \cdot 2^{2e-1} - 1.$$

If all of p, q, and r happen to be odd primes, then $m = 2^e pq$ and $n = 2^e r$ are amicable. Prove that the method of Thabit ben Korrah gives amicable pairs.

(d) Taking $e = 2$ in Thabit ben Korrah's method gives the pair $(220, 284)$. Use his method to find a second pair. If you have access to a computer that will do factorizations for you, try to use Thabit ben Korrah's method to find additional amicable pairs.

15.8. Let

$$s(n) = \sigma(n) - n = \text{sum of proper divisors of } n,$$

that is, $s(n)$ is equal to the sum of all divisors of n other than n itself. So n is perfect if $s(n) = n$, and (m, n) are an amicable pair if $s(m) = n$ and $s(n) = m$. More generally, a collection of numbers $n_1, n_2, \ldots, n_t$ is called *sociable* (of order t) if

$$s(n_1) = n_2, \quad s(n_2) = n_3, \quad \ldots, \quad s(n_{t-1}) = n_t, \quad s(n_t) = n_1.$$

(An older name for a list of this sort is an *Aliquot cycle*.) For example, the numbers

14316, 19116, 31704, 47616, 83328, 177792, 295488,

629072, 589786, 294896, 358336, 418904, 366556, 274924,

275444, 243760, 376736, 381028, 285778, 152990, 122410,

97946, 48976, 45946, 22976, 22744, 19916, 17716

are a sociable collection of numbers of order 28.

(a) There is one other collection of sociable numbers that contains a number smaller than 16000. It has order 5. Find these five numbers.

(b) Up until 1970, the only known collections of sociable numbers of order at least 3 were these two examples of order 5 and 28. The next such collection has order 4, and its smallest member is larger than 1,000,000. Find it.

(c) Find a sociable collection of order 9 whose smallest member is larger than

800,000,000.

This is the only known example of order 9.

(d) Find a sociable collection of order 6 whose smallest member is larger than

90,000,000,000.

There are two known examples of order 6; this is the smallest.

Chapter 16

Powers Modulo m and Successive Squaring

How would you compute

$$5^{100000000000000} \pmod{12830603}?$$

If 12830603 were prime, you might try using Fermat's Little Theorem (Chapter 9), and even if it is not prime, Euler's Formula (Chapter 10) is available. In fact, it turns out that $12830603 = 3571 \cdot 3593$ and

$$\phi(12830603) = \phi(3571)\phi(3593) = 3570 \cdot 3592 = 12823440.$$

Euler's Formula tells us that

$$a^{\phi(m)} \equiv 1 \pmod{m} \quad \text{for any } a \text{ and } m \text{ with} \quad \gcd(a, m) = 1,$$

so we can use the fact that

$$100000000000000 = 7798219 \cdot 12823440 + 6546640$$

to "simplify" our problem,

$$\begin{aligned} 5^{100000000000000} &= (5^{12823440})^{7798219} \cdot 5^{6546640} \\ &\equiv 5^{6546640} \pmod{12830603}. \end{aligned}$$

Now we "only" have to compute the 6546640th power of 5 and then reduce it modulo 12830603. Unfortunately, the number $5^{6546640}$ has more than 4 million digits, so it would be difficult to calculate even with a computer. And later we will want to compute $a^k \pmod{m}$ for numbers a, k, and m having hundreds of digits, in which case the number of digits in a^k is larger than the number of subatomic particles in the known universe! We need to find a better method.

You may well be asking why anyone would want to compute such large powers. Aside from the intrinsic interest (if any) of being able to perform computations with large numbers,[1] there is a very practical reason. As we will see later, it is possible to use the computation of $a^k \pmod{m}$ to encode and decode messages. Amazingly enough, the resulting codes are so good that they are unbreakable by even the most sophisticated code-breaking techniques currently known. Having thus piqued your curiosity, we will spend the remainder of this chapter and the next discussing how to compute large powers and large roots modulo m. Then in Chapter 18 we will explain how to use such computations to create "unbreakable" codes.

The clever idea used to compute $a^k \pmod{m}$ is called the *Method of Successive Squaring*. Before describing the method in general, we illustrate it by computing

$$7^{327} \pmod{853}.$$

The first step is to create a table giving the values of $7, 7^2, 7^4, 7^8, 7^{16}, \ldots$ modulo 853. Notice that to get each successive entry in the list, we merely need to square the previous number. Furthermore, since we always reduce modulo 853 before squaring, we never have to work with any numbers larger than 852^2. Here's the table of 2^k-powers of 7 modulo 853.

$$\begin{aligned}
7^1 & & & \equiv 7 & & \equiv 7 & & \pmod{853} \\
7^2 &\equiv \left(7^1\right)^2 & &\equiv 7^2 & &\equiv 49 & &\equiv 49 \pmod{853} \\
7^4 &\equiv \left(7^2\right)^2 & &\equiv 49^2 & &\equiv 2401 & &\equiv 695 \pmod{853} \\
7^8 &\equiv \left(7^4\right)^2 & &\equiv 695^2 & &\equiv 483025 & &\equiv 227 \pmod{853} \\
7^{16} &\equiv \left(7^8\right)^2 & &\equiv 227^2 & &\equiv 51529 & &\equiv 349 \pmod{853} \\
7^{32} &\equiv \left(7^{16}\right)^2 & &\equiv 349^2 & &\equiv 121801 & &\equiv 675 \pmod{853} \\
7^{64} &\equiv \left(7^{32}\right)^2 & &\equiv 675^2 & &\equiv 455625 & &\equiv 123 \pmod{853} \\
7^{128} &\equiv \left(7^{64}\right)^2 & &\equiv 123^2 & &\equiv 15129 & &\equiv 628 \pmod{853} \\
7^{256} &\equiv \left(7^{128}\right)^2 & &\equiv 628^2 & &\equiv 394384 & &\equiv 298 \pmod{853}
\end{aligned}$$

The next step is to write the exponent 327 as a sum of powers of 2. This is called the *binary expansion* of 327. The largest power of 2 less than 327 is $2^8 = 256$, so we write $327 = 256 + 71$. Then the largest power of 2 less than 71 is $2^6 = 64$, so

[1] Question from a fourth grader: "What do mathematicians do, anyway, multiply really big numbers?"

$327 = 256 + 64 + 7$. And so on:

$$\begin{aligned} 327 &= 256 + 71 \\ &= 256 + 64 + 7 \\ &= 256 + 64 + 4 + 3 \\ &= 256 + 64 + 4 + 2 + 1. \end{aligned}$$

Now we use the binary expansion of 327 to compute

$$\begin{aligned} 7^{327} &= 7^{256+64+4+2+1} \\ &= 7^{256} \cdot 7^{64} \cdot 7^{4} \cdot 7^{2} \cdot 7^{1} \\ &\equiv 298 \cdot 123 \cdot 695 \cdot 49 \cdot 7 \pmod{853}. \end{aligned}$$

The numbers in the last line are taken from the table of powers of 7 that we computed earlier.

To complete the computation of $7^{327} \pmod{853}$, we just need to multiply the five numbers $298 \cdot 123 \cdot 695 \cdot 49 \cdot 7$ and reduce them modulo 853. And if the product of all five numbers is too large for our taste, we can just multiply the first two, reduce modulo 853, multiply by the third, reduce modulo 853, and so on. In this way, we still never need to work with any number larger than 852^2. Thus,

$$\begin{aligned} 298 \cdot 123 \cdot 695 \cdot 49 \cdot 7 &\equiv 828 \cdot 695 \cdot 49 \cdot 7 \equiv 538 \cdot 49 \cdot 7 \\ &\equiv 772 \cdot 7 \equiv 286 \pmod{853}. \end{aligned}$$

We're done!

$$7^{327} \equiv 286 \pmod{853}.$$

This may seem like a lot of work, but suppose that instead we try to compute $7^{327} \pmod{853}$ directly by first computing 7^{327} and then dividing by 853 and taking the remainder. It is possible to do this with a small computer, since

$$\begin{aligned} 7^{327} &= 2223612386895518058 2\underbrace{\ldots\ldots\ldots\ldots}_{\text{237 digits omitted}}3258493799550987954 3 \\ &\equiv 286 \pmod{853}, \end{aligned}$$

but, as you can see, the numbers get quite large. And it is completely infeasible to compute a^k exactly when k has, say, 20 digits, much less when k has the hundreds of digits required for the construction of secure codes.

On the other hand, the method of successive squaring can be used to compute $a^k \pmod{m}$ even when k has hundreds or thousands of digits, because a careful analysis of the method shows that it takes approximately $\log_2(k)$ steps to compute

$a^k \pmod{m}$. We will not perform this analysis here but will observe that $\log_2(k)$ is more-or-less 3.322 times the number of digits in k. So if k has, say, 1000 digits, then it takes approximately 3322 steps to compute $a^k \pmod{m}$. Admittedly, this is a lot of steps to do by hand, but it is the work of an instant on even a small desktop computer. To give you an idea of the times involved, my laptop computer (with a 1500-MHz Pentium chip, for those who are technically inclined) used successive squaring to compute

$$7^{10^{200,000}} \equiv 787 \pmod{853} \quad \text{in 0.36 seconds and}$$
$$7^{10^{2,000,000}} \equiv 303 \pmod{853} \quad \text{in 4.48 seconds.}$$

We now describe the general method of computing powers by successive squaring.

Algorithm 16.1 (Successive Squaring to Compute $a^k \pmod{m}$). *The following steps compute the value of $a^k \pmod{m}$:*

1. *Write k as a sum of powers of 2,*

$$k = u_0 + u_1 \cdot 2 + u_2 \cdot 4 + u_3 \cdot 8 + \cdots + u_r \cdot 2^r,$$

where each u_i is either 0 or 1. (This is called the binary expansion *of k.)*

2. *Make a table of powers of a modulo m using successive squaring.*

$$\begin{array}{llll} a^1 & & \equiv A_0 \pmod{m} \\ a^2 \equiv (a^1)^2 & \equiv A_0^2 & \equiv A_1 \pmod{m} \\ a^4 \equiv (a^2)^2 & \equiv A_1^2 & \equiv A_2 \pmod{m} \\ a^8 \equiv (a^4)^2 & \equiv A_2^2 & \equiv A_3 \pmod{m} \\ \vdots & & \vdots \\ a^{2^r} \equiv \left(a^{2^{r-1}}\right)^2 & \equiv A_{r-1}^2 & \equiv A_r \pmod{m} \end{array}$$

Note that to compute each line of the table you only need to take the number at the end of the previous line, square it, and then reduce it modulo m. Also note that the table has $r+1$ lines, where r is the highest exponent of 2 appearing in the binary expansion of k in Step 1.

3. *The product*

$$A_0^{u_0} \cdot A_1^{u_1} \cdot A_2^{u_2} \cdots A_r^{u_r} \pmod{m}$$

will be congruent to $a^k \pmod{m}$. Note that all the u_i's are either 0 or 1, so this number is really the product of those A_i's for which u_i equals 1.

Verification. Why does it work? We compute

$$\begin{aligned} a^k &= a^{u_0+u_1\cdot 2+u_2\cdot 4+u_3\cdot 8+\cdots+u_r\cdot 2^r} && \text{using Step 1,} \\ &= a^{u_0}\cdot(a^2)^{u_1}\cdot(a^4)^{u_2}\cdots(a^{2^r})^{u_r} \\ &\equiv A_0^{u_0}\cdot A_1^{u_1}\cdot A_2^{u_2}\cdots A_r^{u_r} \pmod{m} && \text{using the table from Step 2.} \end{aligned}$$ □

As mentioned earlier, computing large powers $a^k \pmod{m}$ has a real-world use in creating secure codes. To create these codes, it is necessary to find a few large primes, say primes with between 100 and 200 digits. This brings up the question of how to check whether or not a given number m is prime. A surefire but inefficient method is to try dividing by each number up to $\sqrt{m}$ and see if you find any factors. If not, then m is prime. Unfortunately, this method is not practical even for m's of moderate size.

Using successive squaring and Fermat's Little Theorem (Chapter 9), we can often show that a number m is composite without finding any factors at all! Here's how. Take any number a less than m. First compute $\gcd(a, m)$. If it is greater than 1, then you've found a factor of m, so m is composite and you're done. On the other hand, if $\gcd(a, m) = 1$, use successive squaring to compute

$$a^{m-1} \pmod{m}.$$

Fermat's Little Theorem says that if m is prime then the answer will be 1; so if the answer turns out to be anything other than 1, you know that m is composite without actually knowing any factors.

Here's an example. Using successive squaring we compute

$$2^{283976710803262} \equiv 280196559097287 \pmod{283976710803263},$$

so we know that 283976710803263 is definitely not a prime. In fact, its prime factorization is

$$283976710803263 = 104623 \cdot 90437 \cdot 30013.$$

Now consider $m = 630249099481$. Using successive squaring, we find that

$$2^{630249099480} \equiv 1 \pmod{630249099481}$$

and

$$3^{630249099480} \equiv 1 \pmod{630249099481}.$$

Does this mean that 630249099481 is prime? Not necessarily, but it certainly makes it likely. And if we check $a^{m-1} \pmod{m}$ for $a = 5$, 7, and 11 and again get 1 (which we do), then we would become even more convinced that 630249099481 is prime. Using Fermat's Little Theorem in this way, it is never possible to prove conclusively that a number is prime; but if $a^{m-1} \equiv 1 \pmod{m}$ for a lot of a's, then we would certainly suspect that m is indeed a prime. This is how Fermat's Little Theorem and successive squaring can be used to prove that certain numbers are composite and to strongly suggest that certain other numbers are prime. Unfortunately, there do exist composite numbers m such that $a^{m-1} \equiv 1 \pmod{m}$ for all a's with $\gcd(a, m) = 1$. Such m's are called *Carmichael numbers*. The smallest Carmichael number is 561, as you verified in Exercise 10.3. We investigate Carmichael numbers and primality testing further in Chapter 19.

Exercises

16.1. Use the method of successive squaring to compute each of the following powers.
(a) $5^{13} \pmod{23}$ (b) $28^{749} \pmod{1147}$

16.2. The method of successive squaring described in the text allows you to compute $a^k \pmod{m}$ quite efficiently, but it does involve creating a table of powers of a modulo m.

(a) Show that the following algorithm will also compute the value of $a^k \pmod{m}$. It is a more efficient way to do successive squaring, well-suited for implementation on a computer.

```
(1) Set b = 1
(2) Loop while k ≥ 1
(3)      If k is odd, set b = a · b (mod m)
(4)      Set a = a² (mod m).
(5)      Set k = k/2 (round down if k is odd)
(6) End of Loop
(7) Return the value of b (which equals a^k (mod m))
```

(b) Implement the above algorithm on a computer using the computer language of your choice.

(c) Use your program to compute the following quantities:
(i) $2^{1000} \pmod{2379}$ (ii) $567^{1234} \pmod{4321}$ (iii) $47^{258008} \pmod{1315171}$

16.3. (a) Compute $7^{7386} \pmod{7387}$ by the method of successive squaring. Is 7387 prime?
(b) Compute $7^{7392} \pmod{7393}$ by the method of successive squaring. Is 7393 prime?

16.4. Write a program to check if a number n is composite or probably prime as follows. Choose 10 random numbers $a_1, a_2, \ldots, a_{10}$ between 2 and $n-1$ and compute $a_i^{n-1} \bmod n$ for each a_i. If $a_i^{n-1} \not\equiv 1 \pmod{n}$ for any a_i, return the message "`n is composite`." If $a_i^{n-1} \equiv 1 \pmod{n}$ for all the a_i's, return the message "`n is probably prime`."

Incorporate this program into your factorization program (Exercise 7.6) as a way to check when a large number is prime.

16.5. Compute $2^{9990} \pmod{9991}$ by successive squaring and use your answer to say whether you believe that 9991 is prime.

Chapter 17

Computing k^{th} Roots Modulo m

In the last chapter we learned how to compute k^{th} powers modulo m when k and m are very large. Now we will travel in the opposite direction and try to compute k^{th} roots modulo m. In other words, suppose we are given a number b and told to find a solution to the congruence

$$x^k \equiv b \pmod{m}.$$

We could try substituting $x = 0, 1, 2, \ldots$ until we find a solution, but if m is large, this could take a long time. It turns out that if we know the value of $\phi(m)$ then we can compute the kth root of b modulo m fairly easily. As usual, we first illustrate the method with an example.

We are going to solve the congruence

$$x^{131} \equiv 758 \pmod{1073}.$$

The first step is to compute $\phi(1073)$. We can do this using the formulas for ϕ in Chapter 11 as soon as we factor 1073 into a product of primes. This is easily done; $1073 = 29 \cdot 37$, so $\phi(1073) = \phi(29)\phi(37) = 28 \cdot 36 = 1008$.

The next step is to find a solution in (positive) integers to the equation

$$ku - \phi(m)v = 1; \qquad \text{that is, to the equation} \qquad 131u - 1008v = 1.$$

We know a solution exists, since $\gcd(k, \phi(m)) = 1$, and the method described in Chapter 6 allows us to find the solution $u = 731$ and $v = 95$. (Actually, the method in Chapter 6 gives the solution

$$131 \cdot (-277) + 1008 \cdot 36 = 1.$$

To get positive values for u and v, we modify this solution,

$$u = -277 + 1008 = 731 \quad \text{and} \quad v = -36 + 131 = 95.)$$

The equation

$$131 \cdot 731 - 1008 \cdot 95 = 1$$

provides the key to solving the original problem.

We take x^{131} and raise it to the uth power, that is, to the 731st power. Notice that

$$\left(x^{131}\right)^{731} = x^{131 \cdot 731} = x^{1+1008 \cdot 95} = x \cdot \left(x^{1008}\right)^{95}.$$

But $1008 = \phi(1073)$, and Euler's formula (Chapter 10) tells us that

$$x^{1008} \equiv 1 \pmod{1073}.$$

This means that $\left(x^{131}\right)^{731} \equiv x \pmod{1073}$. So if we raise both sides of the congruence $x^{131} \equiv 758 \pmod{1073}$ to the 731st power, we get

$$x \equiv \left(x^{131}\right)^{731} \equiv 758^{731} \pmod{1073}.$$

Now we need merely use the method of successive squares (Chapter 16) to compute the number $758^{731} \pmod{1073}$. The answer we arrive at is $x \equiv 905 \pmod{1073}$. Finally, as a check, we can use successive squaring to verify that 905^{131} is indeed congruent to 758 modulo 1073.

Here, then, is the general method of computing roots modulo m.

Algorithm 17.1 (How to Compute k^{th} Roots Modulo m). *Let b, k, and m be given integers that satisfy*

$$\gcd(b, m) = 1 \qquad \textit{and} \qquad \gcd\bigl(k, \phi(m)\bigr) = 1.$$

The following steps give a solution to the congruence

$$x^k \equiv b \pmod{m}.$$

1. *Compute $\phi(m)$. (See Chapter 11.)*
2. *Find positive integers u and v that satisfy $ku - \phi(m)v = 1$. [See Chapter 6. Another way to say this is that u is a positive integer satisfying $ku \equiv 1 \pmod{\phi(m)}$, so u is actually the inverse of k modulo $\phi(m)$.]*
3. *Compute $b^u \pmod{m}$ by successive squaring. (See Chapter 16.) The value obtained gives the solution x.*

Why Does It Work? We need to check that $x = b^u$ is a solution to the congruence $x^k \equiv b \pmod{m}$.

$$\begin{aligned} x^k &= (b^u)^k && \text{substituting } x = b^u \text{ into } x^k, \\ &= b^{uk} \\ &= b^{1+\phi(m)v} && \text{since } ku - \phi(m)v = 1 \text{ from Step 2,} \\ &= b \cdot \left(b^{\phi(m)}\right)^v \\ &\equiv b \pmod{m} && \text{since } b^{\phi(m)} \equiv 1 \pmod{m} \text{ from Euler's} \\ & && \text{formula (Chapter 10).} \end{aligned}$$

This completes the verification that $x = b^u$ provides the desired solution to the congruence $x^k \equiv b \pmod{m}$. □

The successive squaring method described in Chapter 16 is a completely practical way to compute powers $a^k \pmod{m}$, even for very large numbers k and m. Is our method for finding k^{th} roots modulo m equally practical? In other words, how difficult is it, in practice, to solve $x^k \equiv b \pmod{m}$? We'll consider the three steps in reverse order. Step 3 says to compute $b^u \pmod{m}$ by successive squaring, so it causes no problem. Step 2 asks us to solve $ku - \phi(m)v = 1$. The method described in Chapter 6 for solving such equations is also quite practical, even for large values of k and $\phi(m)$, since it is based on the Euclidean algorithm.

Finally, we come to the innocuous-looking Step 1, which says to find the value of $\phi(m)$. If we know the factorization of m into primes, then it is easy to compute $\phi(m)$ using the formulas in Chapter 11. However, if m is very large, it may be extremely difficult, if not impossible, to factor m in any reasonable amount of time. For example, suppose that you are asked to solve the congruence

$$x^{3968039} \equiv 34781 \pmod{27040397}.$$

If you didn't have a computer, it might take you quite a while to discover that 27040397 factors as a product of two primes, $27040397 = 4409 \cdot 6133$, so

$$\phi(27040397) = 4408 \cdot 6132 = 27029856.$$

Having computed $\phi(m)$, we can do Step 2,

$$3968039 \cdot 17881559 - 27029856 \cdot 2625050 = 1,$$

and then Step 3,

$$x \equiv 34781^{17881559} \equiv 22929826 \pmod{27040397},$$

to find the solution.

Now imagine that rather than choosing an m with only 8 digits, I had instead taken two primes p and q, each of which has 100 digits, and set $m = pq$. Then it would be virtually impossible for you to solve $x^k \equiv b \pmod{m}$ unless I were to tell you the values of p and q, since if you don't know the values of p and q, then you won't be able to find the value of $\phi(m)$.

In summary, this chapter contains an efficient and practical method to solve

$$x^k \equiv b \pmod{m}$$

provided that we are able to calculate $\phi(m)$. It may seem unfortunate that the method does not work if we cannot calculate $\phi(m)$, but it is exactly this "weakness" that is exploited in the next chapter to construct extremely secure codes.

Exercises

17.1. Solve the congruence $x^{329} \equiv 452 \pmod{1147}$. (*Hint.* 1147 is not prime.)

17.2. (a) Solve the congruence $x^{113} \equiv 347 \pmod{463}$.
(b) Solve the congruence $x^{275} \equiv 139 \pmod{588}$.

17.3. In this chapter we described how to compute a kth root of b modulo m, but you may well have asked yourself if b can have more than one k^{th} root. Indeed it can! For example, if a is a square root of b modulo m, then clearly $-a$ is also a square root of b modulo m.

(a) Let b, k, and m be integers that satisfy

$$\gcd(b, m) = 1 \qquad \text{and} \qquad \gcd\big(k, \phi(m)\big) = 1.$$

Show that b has *exactly one* k^{th} root modulo m.

(b) Suppose instead that $\gcd\big(k, \phi(m)\big) > 1$. Show that either b has no k^{th} roots modulo m, or else it has at least two k^{th} roots modulo m.

(c) If $m = p$ is prime, look at some examples and try to find a formula for the number of kth roots of b modulo p (assuming that it has at least one).

17.4. Our method for solving $x^k \equiv b \pmod{m}$ is first to find integers u and v satisfying $ku - \phi(m)v = 1$, and then the solution is $x \equiv b^u \pmod{m}$. However, we only showed that this works provided that $\gcd(b, m) = 1$, since we used Euler's formula $b^{\phi(m)} \equiv 1 \pmod{m}$.

(a) If m is a product of distinct primes, show that $x \equiv b^u \pmod{m}$ is always a solution to $x^k \equiv b \pmod{m}$, even if $\gcd(b, m) > 1$.

(b) Show that our method does not work for the congruence $x^5 \equiv 6 \pmod{9}$.

17.5. (a) Try to use the methods in this chapter to compute the square root of 23 modulo 1279. (The number 1279 is prime.) What goes wrong?

(b) More generally, if p is an odd prime, explain why the methods in this chapter cannot be used to find square roots modulo p. We will investigate the problem of square roots modulo p in later chapters.

(c) Even more generally, explain why our method for computing k^{th} roots modulo m does not work if $\gcd(k, \phi(m))$ is greater than 1.

17.6. Write a program to solve $x^k \equiv b \pmod{m}$. Give the user the option of providing a factorization of m to be used for computing $\phi(m)$.

Chapter 18

Powers, Roots, and "Unbreakable" Codes

In the last two chapters we learned how to compute powers and roots of extremely large numbers modulo m. Briefly, we know how to compute $a^k \pmod{m}$ for any values of a, k, and m, and we know how to solve $x^k \equiv b \pmod{m}$ provided that we can calculate $\phi(m)$. Here's the basic idea that we use to encode and decode messages.[1]

The first step in encoding a message is to convert it into a string of numbers. We use the simplest possible method to do this. We set A $= 11$, B $= 12$, $\ldots$, Z $= 36$. Here's a convenient table to use:

A	B	C	D	E	F	G	H	I	J	K	L	M
11	12	13	14	15	16	17	18	19	20	21	22	23

N	O	P	Q	R	S	T	U	V	W	X	Y	Z
24	25	26	27	28	29	30	31	32	33	34	35	36

For example, the message "To be or not to be" becomes

T	O	B	E	O	R	N	O	T	T	O	B	E
30	25	12	15	25	28	24	25	30	30	25	12	15

[1]Technically, what we describe in this chapter is a cipher, not a code, so we are really enciphering and deciphering messages. Historically, the word code was reserved for methods in which entire words and phrases are replaced by a single symbol or number, while ciphers use individual letters as their basic units. More recently, the word code has acquired other mathematical meanings in different contexts. For ease of exposition, we use the terms code and cipher interchangeably.

So our message is the string of digits 30251215252824253030251215. Of course, in some sense the message is now encoded, since this string of digits serves to conceal the message. But even an amateur cryptographer would be able to break this simple code in just a few minutes.[2]

FOXTROT ©Bill Amend. Reprinted with permission of UNIVERSAL SYNDICATE.

Now we are ready to explain the crux of the encoding and decoding process. The first thing that we do is choose two (large) prime numbers p and q. Next we multiply them together to get a modulus $m = pq$. We can also compute $\phi(m) = \phi(p)\phi(q) = (p-1)(q-1)$, and we choose a number k that is relatively prime to $\phi(m)$. Now we publish the numbers m and k for the whole world to know, but we keep the values of p and q secret. Anyone who wants to send us a message uses the values of m and k to encode the material in the following manner.

First, they convert their message into a string of digits as described above. Next, they look at the number m and break their string of digits into numbers that are less than m. For example, if m is a number in the millions, they would write their message as a list of six-digit numbers. So now their message is a list of numbers $a_1, a_2, \ldots, a_r$. The next step is to use successive squaring to compute $a_1^k \pmod{m}$, $a_2^k \pmod{m}$, ..., $a_r^k \pmod{m}$. These values form a new list of numbers $b_1, b_2, \ldots, b_r$. This list is the encoded message. In other words, the message that is sent to us is the list of numbers $b_1, b_2, \ldots, b_r$.

How do we decode the message when we receive it? We have been sent the numbers $b_1, b_2, \ldots, b_r$, and we need to recover the numbers $a_1, a_2, \ldots, a_r$. Each b_i is congruent to $a_i^k \pmod{m}$, so to find a_i we need to solve the congruence $x^k \equiv b_i \pmod{m}$. This is exactly the problem we solved in the last chapter, assuming we were able to calculate $\phi(m)$. But we know the values of p and q

[2]We could have assigned a number, such as 99, to represent a space, and we could even have assigned numbers to represent various punctuation marks. But, to keep things simple, we ignore such niceties and just write our messages with all the letters squashed together.

with $m = pq$, so we easily compute

$$\phi(m) = \phi(p)\phi(q) = (p-1)(q-1) = pq - p - q + 1 = m - p - q + 1.$$

Now we just need to apply the method in Chapter 17 to solve each of the congruences $x^k \equiv b_i \pmod{m}$. The solutions are the numbers $a_1, a_2, \ldots, a_r$, and then it is easy to take this string of digits and recover the original message.

We illustrate the encoding and decoding procedure with the primes $p = 12553$ and $q = 13007$. We multiply them together to get the modulus $m = pq = 163276871$, and we also record for future use

$$\phi(m) = (p-1)(q-1) = 163251312.$$

We also need to choose a k that is relatively prime to $\phi(m)$, so we take $k = 79921$. In summary, we have chosen

$$p = 12553, \quad q = 13007, \quad m = pq = 163276871, \quad \text{and} \quad k = 79921.$$

Now suppose we want to send the message "To be or not to be." As described earlier, this message becomes the string of digits

$$30251215252824253030251215.$$

The number m is 9 digits long, so we break the message up into 8-digit numbers:

$$30251215, \quad 25282425, \quad 30302512, \quad 15.$$

Next we use the method of successive squares to raise each of these numbers to the kth power modulo m.

$$\begin{aligned} 30251215^{79921} &\equiv 149419241 \pmod{163276871} \\ 25282425^{79921} &\equiv 62721998 \pmod{163276871} \\ 30302512^{79921} &\equiv 118084566 \pmod{163276871} \\ 15^{79921} &\equiv 40481382 \pmod{163276871} \end{aligned}$$

The encoded message is the list of numbers

$$149419241, \quad 62721998, \quad 118084566, \quad 40481382.$$

Now let's try decoding a new message. It's after midnight, there's a knock at your door, and a mysterious messenger delivers the following cryptic missive:

$$145387828, \quad 47164891, \quad 152020614, \quad 27279275, \quad 35356191.$$

Without a moment of hesitation, you whip out your handy-dandy number theory decoding book and start to work. One number at a time, you use the methods from Chapter 17 to solve the congruences

$$\begin{aligned}
x^{79921} &\equiv 145387828 \pmod{163276871} &\Longrightarrow\quad x &\equiv 30182523\\
x^{79921} &\equiv 47164891 \pmod{163276871} &\Longrightarrow\quad x &\equiv 26292524\\
x^{79921} &\equiv 152020614 \pmod{163276871} &\Longrightarrow\quad x &\equiv 19291924\\
x^{79921} &\equiv 27279275 \pmod{163276871} &\Longrightarrow\quad x &\equiv 30282531\\
x^{79921} &\equiv 35356191 \pmod{163276871} &\Longrightarrow\quad x &\equiv 122215
\end{aligned}$$

This gives you the string of digits

301825232629252419291924302825311 22215,

and now you use the number-to-letter substitution table for the final decoding step.

30	18	25	23	26	29	25	24	19	29	19	24	30	28	25	31	12	22	15
T	H	O	M	P	S	O	N	I	S	I	N	T	R	O	U	B	L	E

Supplying the obvious word breaks and punctuation, you read

"Thompson is in trouble"

and off you go to the rescue.

Is this encoding scheme secure? Suppose that you intercept a message that you know has been encoded with the modulus m and the exponent k. How difficult would it be for you to break the code and read the message? At present, the only way to decode is to find the value of $\phi(m)$ and then use the decoding process just described. If m is the product of two primes p and q, then

$$\phi(m) = (p-1)(q-1) = pq - p - q + 1 = m - p - q + 1.$$

Since you already know the value of m, you just need to find the value of $p+q$. But if you can find $p+q$, then you can also determine p and q, since they are the roots of the quadratic equation

$$X^2 - (p+q)X + m = 0.$$

So in order to decode the intercepted message, you essentially need to find the factors p and q of m.

If m is not too large, say 5 or 10 digits, then a computer will find the factors almost immediately. Using more advanced methods from number theory, mathematicians have devised techniques that will factor much larger numbers, say those with 50 to 100 digits. So if you take primes p and q with less than 50 digits each, your code will not be secure. However, if you take primes with, say, 100 digits each, then no one at present will be able to decode your messages unless you reveal to them your values of p and q. Of course, it is possible that future mathematical advances will enable people to factor 200-digit numbers; but then you need merely take primes p and q with 200 digits each, and your 400-digit modulus m will again render your messages secure. The idea underlying the encoding scheme is thus a very simple one: It is easy to multiply large numbers together, but it is difficult to factor a large number.

The cryptographic method described in this chapter is called a *public key cryptosystem.* This name reflects the fact that the encoding key consisting of the modulus m and the exponent k can be distributed to the public while the decoding method remains secure. This idea, that it might be possible to have a code where knowledge of the encoding process does not enable one to decode messages, was propounded by Whitfield Diffie and Martin Hellman in 1976. Diffie and Hellman gave a theoretical description of how such a public key cryptosystem might work, and the following year Ron Rivest, Adi Shamir, and Leonard Adleman described a practical public key cryptosystem. Their idea, which we have described in this chapter, is called the *RSA public key cryptosystem* in honor of its three inventors.

Exercises

18.1. Decode the following message, which was sent using the modulus $m = 7081$ and the exponent $k = 1789$. (Note that you will first need to factor m.)

$$5192, \qquad 2604, \qquad 4222$$

18.2. It may appear that RSA decryption does not work if you are unlucky enough to choose a message a that is not relatively prime to m. Of course, if $m = pq$ and p and q are large, this is very unlikely to occur.

(a) Show that in fact RSA decryption does work for all messages a, regardless of whether or not they have a factor in common with m.

(b) More generally, show that RSA decryption works for all messages a as long as m is a product of distinct primes.

(c) Give an example with $m = 18$ and $a = 3$ where RSA decryption does not work. [Remember, k must be chosen relatively prime to $\phi(m) = 6$.]

18.3. Write a short report on one or more of the following topics.

(a) The history of public key cryptography

(b) The RSA public key cryptosystem

(c) Public key digital signatures

(d) The political and social consequences of the availability of inexpensive unbreakable codes and the government's response

18.4. Here are two longer messages to decode if you like to use computers.

(a) You have been sent the following message:

$$5272281348,\ 21089283929,\ 3117723025,\ 26844144908,\ 22890519533,$$
$$26945939925,\ 27395704341,\ 2253724391,\ 1481682985,\ 2163791130,$$
$$13583590307,\ 5838404872,\ 12165330281,\ 28372578777,\ 7536755222.$$

It has been encoded using $p = 187963$, $q = 163841$, $m = pq = 30796045883$, and $k = 48611$. Decode the message.

(b) You intercept the following message, which you know has been encoded using the modulus $m = 956331992007843552652604425031376690367$ and exponent $k = 12398737$. Break the code and decipher the message.

$$821566670681253393182493050080875560504,$$
$$87074173129046399720949786958511391052,$$
$$552100909946781566365272088688468880029,$$
$$491078995197839451033115784866534122828,$$
$$172219665767314444215921020847762293421.$$

(The material for this exercise is available on the *Friendly Introduction to Number Theory* home page listed in the Preface.)

18.5. Write a program to implement the RSA cryptosystem. Make your program as user friendly as possible. In particular, the person encoding a message should be able to type in their message as words, including spaces and punctuation; similarly, the decoder should see the message appear as words with spaces and punctuation.

18.6. The problem of factoring large numbers has been much studied in recent years because of its importance in cryptography. Find out about one of the following factorization methods and write a short description of how it works. (Information on these methods is available in number theory textbooks and on the web.)

(a) Pollard's ρ method (that is the Greek letter rho)
(b) Pollard's $p - 1$ method
(c) The quadratic sieve factorization method
(d) Lenstra's elliptic curve factorization method
(e) The number field sieve

(The last two methods require advanced ideas, so you will need to learn about elliptic curves or number fields before you can understand them.) The number field sieve is the most powerful factorization method currently known. It is capable of factoring numbers of more than 150 digits.

18.7. Write a computer program implementing one of the factorization methods that you studied in the previous exercise, such as Pollard's ρ method, Pollard's $p - 1$ method, or the quadratic sieve. Use your program to factor the following numbers.

(a) 47386483629775753
(b) 1834729514979351371768185745442640443774091

Chapter 19

Primality Testing and Carmichael Numbers

Prime numbers are the fundamental building blocks of the integers. Within the infinitude of prime numbers we see displayed some of the deepest and most beautiful patterns in all of number theory, and indeed in all of mathematics. And prime numbers, especially large prime numbers, have their practical side as well, as we saw when we constructed the RSA cryptosystem in Chapter 18. This leads us inexorably to the following question:

How can we tell if a (large) number is prime?

For small numbers n such as

$$8629, \qquad 8633, \qquad \text{and} \qquad 8641,$$

we can simply check all possible (prime) divisors up to $\sqrt{n}$, and either we find a divisor or we know when we're done that n is prime. Thus we find that 8629 and 8641 are prime numbers, but 8633 factors as $89 \cdot 97$, so it is not prime.

For larger numbers such as

$$m = 113736947625310405231177973028344375862964001$$

and

$$n = 113736947625310405231177973028344375862953603$$

it is too much work, even with a computer, to try all possible divisors up to the square root. However, we saw in Chapter 16 that it is not very difficult (on a computer) to raise numbers to very high powers modulo very large numbers. For example it takes very little time for a computer to calculate

$$\begin{aligned} 2^m &= 2^{1137369476253104052311779730283443758629640 01} \\ &\equiv 3924197081539349906012004369263061596179002 0 \pmod{m}. \end{aligned}$$

At first glance, this seems like a completely useless calculation to make, but in fact it has tremendous practical significance.

To explain why, we recall Fermat's Little Theorem (see Chapter 9), which says that if p is a prime number then

$$a^p \equiv a \pmod{p} \qquad \text{for every integer } a.$$

Thus the fact that 2^m is *not* congruent to 2 modulo m tells us that m is definitely *not* a prime number. We can state this unequivocally: the incongruence $2^m \not\equiv 2 \pmod{m}$ constitutes a proof that m is not prime. It is worth reflecting for a moment on the surprising strength of our conclusion. We have proved that m is not prime, even though we do not know how to factor m; and indeed our proof that m is composite provides no clues[1] to aid us in finding a factor! The lesson to be learned is that it is often possible to establish that a number is composite without being able to factor it.

Now consider the other number

$$n = 1137369476253104052311779730283443758629536 03.$$

If we perform a similar calculation, we find that

$$2^n = 2^{1137369476253104052311779730283443758629536 03} \equiv 2 \pmod{n}.$$

Can we use Fermat's Little Theorem to conclude that n is prime? The answer is absolutely not, Fermat's Little Theorem doesn't work in that direction. So we try a few more numbers, say up to $a = 100$, and we find that

$$\begin{gathered} 3^n \equiv 3 \pmod{n}, \quad 4^n \equiv 4 \pmod{n}, \quad 5^n \equiv 5 \pmod{n}, \\ \ldots \quad 100^n \equiv 100 \pmod{n}. \end{gathered}$$

We still cannot use Fermat's Little Theorem to conclude that n is prime, but the fact that $a^n \equiv a \pmod{n}$ for 99 different values of a certainly suggests that n is "probably" prime.

[1]The number m is the product of the following two rather large prime numbers:
40103836670582470495139653 and 28360615110109983 17.

This is a rather odd assertion; how can a number be "probably prime"? Either it is a prime or else it isn't a prime; it can't be prime on Tuesdays and Thursdays and composite the rest of the week.[2]

Suppose that we think of the number n as a natural phenomenon and we study n in the spirit of an experimental scientist. We perform experiments by choosing different values for the number a and computing the value of

$$a^n \pmod{n}.$$

If even a single experiment results in any number other than a, we conclude that n is definitely composite. So it is reasonable to believe that each time we perform an experiment and do obtain the value a we have gathered some "evidence" that n is prime.

We can put this reasoning on a firm footing by looking at those values of a whose n^{th} power is different from a. We say that the number a is a *witness for* n if

$$a^n \not\equiv a \pmod{n}.$$

This is an excellent name for a since, if the number n is trying to impersonate a prime, the prosecuting attorney can put a on the witness stand to prove that n is actually composite.

If n is prime, then it obviously has no witnesses. The table on page 126, in which we have listed the witnesses for all numbers n up to 20, suggests that composite numbers tend to have quite a few witnesses.

To further bolster this observation, we selected some random composite numbers between 100 and 1000 and counted how many witnesses they have. We also give the percentage of the numbers between 1 and n that serve as witnesses.

n	287	190	314	586	935	808	728	291
# of witnesses	278	150	310	582	908	804	720	282
% of witnesses	96.9%	78.9%	98.7%	99.3%	97.1%	99.5%	98.9%	96.9%

It seems that if n is composite, then most values of a serve as witnesses. For example, if $n = 287$, and if we choose a random value for a, then there is a 96.9%

[2]"When I use a number," Humpty Dumpty said in a rather scornful tone, "it means just what I choose it to mean—neither more nor less."

"The question is," said Alice, "whether you *can* make numbers mean different things."

"The question is," said Humpty Dumpty, "which is to be master—that's all."

Or, as Hamlet was wont to say, "I am but mad north-north-west: when the wind is southerly I know a prime from a composite."

n	Witnesses for n
3	prime
4	$2, 3$
5	prime
6	$2, 5$
7	prime
8	$2, 3, 4, 5, 6, 7$
9	$2, 3, 4, 5, 6, 7$
10	$2, 3, 4, 7, 8, 9$
11	prime
12	$2, 3, 5, 6, 7, 8, 10, 11$
13	prime
14	$2, 3, 4, 5, 6, 9, 10, 11, 12, 13$
15	$2, 3, 7, 8, 12, 13$
16	$2, 3, 4, 5, 6, 7, 8, 9, 10, 11, 12, 13, 14, 15$
17	prime
18	$2, 3, 4, 5, 6, 7, 8, 11, 12, 13, 14, 15, 16, 17$
19	prime
20	$2, 3, 4, 6, 7, 8, 9, 10, 11, 12, 13, 14, 15, 17, 18, 19$

chance that a is a witness for the compositeness of n. Thus it will not take very many experiments to prove that n is composite.

All our evidence and also common sense suggest that composite numbers have lots of witnesses. But is this really true? If we start to make a list of all numbers with their witnesses, we eventually run into the sad case of $n = 561$. This is a composite number, since $561 = 3 \cdot 11 \cdot 17$, but unfortunately 561 doesn't have even a single witness! One way to verify that 561 has no witnesses is to compute $a^n \pmod{n}$ for all 561 values of a. We take an easier approach. To prove that

$$a^{561} \equiv a \pmod{561},$$

it is enough to prove that

$$a^{561} \equiv a \pmod{3}, \quad a^{561} \equiv a \pmod{11}, \quad \text{and} \quad a^{561} \equiv a \pmod{17},$$

since if a number is divisible by 3, by 11, and by 17, then it is divisible by their product $3 \cdot 11 \cdot 17$. For the first congruence, we observe that if 3 divides a then both sides are 0, while if 3 does not divide a, we can use Fermat's Little Theorem $a^2 \equiv 1 \pmod 3$ to compute

$$a^{561} = a^{2 \cdot 280+1} = (a^2)^{280} \cdot a \equiv 1 \cdot a \equiv a \pmod 3.$$

The second and third congruences are checked in a similar fashion. Thus either 11 divides a and both sides are 0 modulo 11, or else we use the congruence $a^{10} \equiv 1 \pmod{11}$ to compute

$$a^{561} = a^{10 \cdot 56+1} = (a^{10})^{56} \cdot a \equiv 1 \cdot a \equiv a \pmod{11}.$$

Finally, either 17 divides a and both sides are equal to 0 modulo 17, or else we use $a^{16} \equiv 1 \pmod{17}$ to compute

$$a^{561} = a^{16 \cdot 35+1} = (a^{16})^{35} \cdot a \equiv 1 \cdot a \equiv a \pmod{17}.$$

Hence there are no witnesses for the composite number 561.

This example and 14 others were first noted by R.D. Carmichael in 1910, so they are named in his honor. A *Carmichael number* is a composite number n with the property that

$$a^n \equiv a \pmod n \qquad \text{for every integer } 1 \le a \le n.$$

In other words, a Carmichael number is a composite number that can masquerade as a prime, because there are no witnesses to its composite nature. We have seen that 561 is a Carmichael number, and in fact it is the smallest one.

Here is the complete list of all Carmichael numbers up to 10000.

$$561, \quad 1105, \quad 1729, \quad 2465, \quad 2821, \quad 6601, \quad 8911.$$

Factoring them,

$$\begin{aligned} 561 &= 3 \cdot 11 \cdot 17 & \qquad 2821 &= 7 \cdot 13 \cdot 31 \\ 1105 &= 5 \cdot 13 \cdot 17 & 6601 &= 7 \cdot 23 \cdot 41 \\ 1729 &= 7 \cdot 13 \cdot 19 & 8911 &= 7 \cdot 19 \cdot 67 \\ 2465 &= 5 \cdot 17 \cdot 29 \end{aligned}$$

we immediately observe that each number in our list is the product of three distinct odd primes. So we might make the conjecture that Carmichael numbers are always the product of three distinct odd primes.

Our conjecture doesn't fare too well, since

$$62745 = 3 \cdot 5 \cdot 47 \cdot 89$$

is a Carmichael number with four prime factors. This does not mean we should abandon our conjecture, merely that we must make some modifications. Notice that our conjecture was really three conjectures: that a Carmichael number has exactly three prime factors, that the prime factors are distinct, and that the prime factors are odd. So we drop the part that is false and state the other two parts separately:

(A) Every Carmichael number is odd.

(B) Every Carmichael number is a product of distinct primes.

Let's prove these two assertions. For (A), we use the Carmichael congruence

$$a^n \equiv a \pmod{n}$$

with $a = n - 1 \equiv -1 \pmod{n}$ to get

$$(-1)^n \equiv -1 \pmod{n}.$$

This implies that n is odd (or $n = 2$).

Next we prove (B). Suppose that n is a Carmichael number. Let p be a prime number dividing n.

p^{e+1} be the largest power of p dividing n.

We want to show that e is 0. The fact that n is a Carmichael number means that $a^n \equiv a \pmod{n}$ for every value of a. In particular, this is true for $a = p^e$, so

$$p^{en} \equiv p^e \pmod{n}.$$

Thus n divides the difference $p^{en} - p^e$, and by assumption p^{e+1} divides n, so we conclude that

$$p^{e+1} \text{ divides } p^{en} - p^e.$$

Therefore,

$$\frac{p^{en} - p^e}{p^{e+1}} = \frac{p^{en-e} - 1}{p} \quad \text{is an integer.}$$

The only way that this can be true is if $e = 0$, which completes the proof of (B).

The two properties (A) and (B) of Carmichael numbers are useful, but it would be even more useful if we could devise a simple method for checking whether or not a number is a Carmichael number. Our earlier verification that 561 is a Carmichael number provides a clue. Rather than verifying that $a^n \equiv a \pmod{n}$, we instead checked that $a^n \equiv a \pmod{p}$ for each prime p dividing a. Then the congruence modulo p was compared with Fermat's Little Theorem to give us a relationship between p and n. The upshot is a criterion for Carmichael numbers that we now formally state and prove.

Theorem 19.1 (Korselt's Criterion for Carmichael Numbers). *Let n be a composite number. Then n is a Carmichael number if and only if it is odd and every prime p dividing n satisfies the following two conditions:*
(1) *p^2 does not divide n.*
(2) *$p-1$ divides $n-1$.*

Verification. Suppose first that n is a composite number, and further suppose that every prime divisor p of n satisfies conditions (1) and (2). We want to prove that n is a Carmichael number. Our proof uses the same arguments that we used to prove that 561 is a Carmichael number.

We factor n as

$$n = p_1 p_2 p_3 \cdots p_r$$

into a product of primes. From condition (1) we know that $p_1, p_2, \ldots, p_r$ are all different. We also know from condition (2) that each $p_i - 1$ divides $n-1$, so for each i we can factor

$$n - 1 = (p_i - 1)k_i \qquad \text{for some integer } k_i.$$

Now take any integer a. We compute the value of a^n modulo p_i as follows. First, if p_i divides a, then clearly

$$a^n \equiv 0 \equiv a \pmod{p_i}.$$

Otherwise p_i does not divide a and we can use Fermat's Little Theorem to compute

$$\begin{aligned}
a^n &= a^{(p_i-1)k_i+1} && \text{since } n-1 = (p_i-1)k_i, \\
&= \left(a^{p_i-1}\right)^{k_i} \cdot a \\
&\equiv 1^{k_i} \cdot a \pmod{p_i} && \text{by Fermat's Little Theorem, which} \\
& && \text{tells us that } a^{p_i-1} \equiv 1 \pmod{p_i}, \\
&\equiv a \pmod{p_i}.
\end{aligned}$$

We have now proved that

$$a^n \equiv a \pmod{p_i} \quad \text{for each } i = 1, 2, \ldots, r.$$

In other words, $a^n - a$ is divisible by each of the primes $p_1, p_2, \ldots, p_r$, and hence it is divisible by their product $n = p_1 p_2 \cdots p_r$. (Notice that this is where we are using the fact that $p_1, p_2, \ldots, p_t$ are all different.) Therefore,

$$a^n \equiv a \pmod{n},$$

and since we have shown that this is true for every integer a, we have completed the proof that n is a Carmichael number.

This proves one half of Korselt's Criterion: an odd composite number satisfying conditions (1) and (2) is a Carmichael number. For the other direction, we proved earlier that every Carmichael number satisfies condition (1), and in Exercise 19.1 we ask you to show that Carmichael numbers also satisfy condition (2). □

To illustrate the power of Korselt's Criterion, we verify that two of the examples given previously are actually Carmichael numbers. First, Korselt's Criterion tells us that $1729 = 7 \cdot 13 \cdot 19$ is a Carmichael number, since

$$\frac{1729-1}{7-1} = 288, \qquad \frac{1729-1}{13-1} = 144, \qquad \text{and} \qquad \frac{1729-1}{19-1} = 96.$$

Second, $62745 = 3 \cdot 5 \cdot 47 \cdot 89$ is a Carmichael number, since

$$\frac{62745-1}{3-1} = 31372, \qquad \frac{62745-1}{5-1} = 15686,$$
$$\frac{62745-1}{47-1} = 1364, \qquad \frac{62745-1}{89-1} = 713.$$

In his 1910 paper, Carmichael conjectured that there are infinitely many Carmichael numbers. (Of course, he didn't call them Carmichael numbers!) This conjecture remained unproved for more than 70 years. It was finally verified in 1994 by W.R. Alford, A. Granville, and C. Pomerance.

The fact that Carmichael numbers exist means that we need a better method for checking if a number is composite. The Rabin–Miller test for composite numbers is based on the following fact.

Theorem 19.2 (A Property of Prime Numbers). *Let p be an odd prime and write*

$$p - 1 = 2^k q \qquad \text{with } q \text{ odd.}$$

Let a be any number not divisible by p. Then one of the following two conditions is true:

(i) *a^q is congruent to 1 modulo p.*

(ii) *One of the numbers $a^q,\ a^{2q},\ a^{4q}, \ldots, a^{2^{k-1}q}$ is congruent to -1 modulo p.*

Verification. Fermat's Little Theorem tells us that $a^{p-1} \equiv 1 \pmod{p}$. This means that, when we look at the list of numbers

$$a^q,\ a^{2q},\ a^{4q},\ldots,\ a^{2^{k-1}q},\ a^{2^k q},$$

we know that the last number in the list is congruent to 1 modulo p (since $2^k q$ equals $p-1$). Furthermore, each number in the list is the square of the previous number. Therefore, one of the following two possibilities must be true:

(i) The first number in the list is congruent to 1 modulo p.

(ii) Some number in the list is not congruent to 1 modulo p, but when squared, it becomes congruent to 1 modulo p. The only number fitting this description is -1 modulo p, so in this case the list contains -1 modulo p.

This completes the verification. □

Turning the preceding property of prime numbers on its head, we obtain a test for composite numbers called the Rabin–Miller test. Thus, if n is an odd number and if n does not have the aforementioned prime number property, then we know it must be a composite number. Furthermore, if n does have the prime number property for a lot of different values of a, then it is likely that n is prime.

Theorem 19.3 (Rabin–Miller Test for Composite Numbers). *Let n be an odd integer and write $n-1=2^k q$ with q odd. If both of the following conditions are true for some a not divisible by n, then n is a composite number.*
(a) $a^q \not\equiv 1 \pmod{n}$
(b) $a^{2^i q} \not\equiv -1 \pmod{n}$ *for all* $i = 0, 1, \ldots, k-1$

We have already verified that the Rabin–Miller test works, since if n satisfies (a) and (b), then it does not satisfy the prime number property, so it must be composite. Note that the Rabin–Miller test is very fast and easy to implement on a computer, since, after computing $a^q \pmod{n}$, we simply compute a few squares modulo n.

For any particular choice of a, the Rabin–Miller test either conclusively proves that n is composite, or it suggests that n might be prime. A *Rabin–Miller witness* for the compositeness of n is a number a for which the Rabin–Miller test successfully proves that n is composite. The reason that the Rabin–Miller test is so useful is due to the following fact, which is proved in more advanced texts.

If n is an odd composite number, then at least 75% of the numbers a between 1 and $n-1$ act as Rabin–Miller witnesses for n.

In other words, every composite number has lots of Rabin–Miller witnesses to its compositeness, so there aren't any "Carmichael-type numbers" for the Rabin–Miller test.

For example, if we randomly choose 100 different values for a, and if none of them are Rabin–Miller witnesses for n, then the probability of n being composite is less than 0.75^{100}, which is approximately $3 \cdot 10^{-13}$. And if you feel that this is taking too much of a risk, you can always try another few hundred values for a. In practice, if n is composite, then just a few Rabin–Miller tests virtually always reveal this fact.

To illustrate, we apply the Rabin–Miller test with $a = 2$ to the number $n = 561$, which you may recall is a Carmichael number. We have $n-1 = 560 = 2^4 \cdot 35$, so we compute

$$\begin{aligned} 2^{35} &\equiv 263 \pmod{561}, \\ 2^{2\cdot 35} &\equiv 263^2 \equiv 166 \pmod{561}, \\ 2^{4\cdot 35} &\equiv 166^2 \equiv 67 \pmod{561}, \\ 2^{8\cdot 35} &\equiv 67^2 \equiv 1 \pmod{561}. \end{aligned}$$

The first number $2^{35} \pmod{561}$ is neither 1 nor -1, and the other numbers are not -1, so 2 is a Rabin–Miller witness to the fact that 561 is a composite number.

As a second example, we take the larger number $n = 172947529$. We have

$$n - 1 = 172947528 = 2^3 \cdot 21618441.$$

We apply the Rabin–Miller test with $a = 17$, and at the first step we get

$$17^{21618441} \equiv 1 \pmod{172947529}.$$

So 17 is not a Rabin–Miller witness for n. Next we try $a = 3$, but unfortunately

$$3^{21618441} \equiv -1 \pmod{172947529},$$

so 3 also fails to be a Rabin–Miller witness. At this point we might suspect that n is prime, but if we try another value, such as $a = 23$, we find

$$\begin{aligned} 23^{21618441} &\equiv 40063806 \pmod{172947529}, \\ 23^{2\cdot 21618441} &\equiv 2257065 \pmod{172947529}, \\ 23^{4\cdot 21618441} &\equiv 1 \pmod{172947529}, \end{aligned}$$

so 23 is a Rabin–Miller witness that n is actually composite. In fact, n is a Carmichael number, but it's not so easy to factor (by hand).

Exercises

19.1. Let n be a Carmichael number and let p be a prime number that divides n.

(a) Finish the proof of Korselt's Criterion by proving that $p - 1$ divides $n - 1$. [*Hint.* Use the Carmichael congruence $a^n \equiv a \pmod{n}$ with a taken to be a primitive root modulo p.]

(b) Prove that $p - 1$ actually divides the smaller number $\frac{n}{p} - 1$.

19.2. Are there any Carmichael numbers that have only two prime factors? Either find an example or prove that none exists.

19.3. Use Kursolt's Criterion to determine which of the following numbers are Carmichael numbers.

(a) 1105 (b) 1235 (c) 2821 (d) 6601
(e) 8911 (f) 10659 (g) 19747 (h) 105545
(i) 126217 (j) 162401 (k) 172081 (l) 188461

19.4. Suppose that k is chosen so that the three numbers

$$6k + 1, \quad 12k + 1, \quad 18k + 1$$

are all prime numbers.

(a) Prove that their product $n = (6k + 1)(12k + 1)(18k + 1)$ is a Carmichael number.

(b) Find the first five values of k for which this method works and give the Carmichael numbers produced by the method.

19.5. Find a Carmichael number that is the product of five primes.

19.6. (a) Write a computer program that uses Korselt's Criterion to check if a number n is a Carmichael number.

(b) Earlier we listed all Carmichael numbers that are less than 10,000. Use your program to extend this list up to 100,000.

(c) Use your program to find the smallest Carmichael number larger than 1,000,000.

19.7. (a) Let $n = 1105$, so $n - 1 = 2^4 \cdot 69$. Compute the values of

$$2^{69} \pmod{1105}, \quad 2^{2\cdot 69} \pmod{1105}, \quad 2^{4\cdot 69} \pmod{1105}, \quad 2^{8\cdot 69} \pmod{1105},$$

and use the Rabin–Miller test to conclude that n is composite.

(b) Use the Rabin–Miller test with $a = 2$ to prove that $n = 294409$ is composite. Then find a factorization of n and show that it is a Carmichael number.

(c) Repeat (b) with $n = 118901521$.

19.8. Program the Rabin–Miller test with multiprecision integers and use it to investigate which of the following numbers are composite.

(a) 155196355420821961
(b) 155196355420821889
(c) 285707540662569884530199015485750433489
(d) 285707540662569884530199015485751094149

Chapter 20

Euler's Phi Function and Sums of Divisors

When we studied perfect numbers in Chapter 15, we used the sigma function $\sigma(n)$, where $\sigma(n)$ is defined to be the sum of all the divisors of n. We now propose to conduct what may seem like a strange experiment. We take all the divisors of n, apply Euler's phi function to each divisor, add the values of Euler's phi function and see what we get.

We start with an example, say $n = 15$. The divisors of 15 are 1, 3, 5, and 15. We first evaluate Euler's phi function at the numbers 1, 3, 5, and 15,

$$\phi(1) = 1, \qquad \phi(3) = 2, \qquad \phi(5) = 4, \qquad \phi(15) = 8.$$

Next we add the values to get

$$\phi(1) + \phi(3) + \phi(5) + \phi(15) = 1 + 2 + 4 + 8 = 15.$$

The result is 15, the number we started with; but surely that's just a coincidence.

Let's try a larger number that has lots of factors, say $n = 315$. The divisors of 315 are

$$1,\ 3,\ 5,\ 7,\ 9,\ 15,\ 21,\ 35,\ 45,\ 63,\ 105,\ 315,$$

and if we evaluate Euler's phi function and add the values, we get

$$\begin{aligned} &\phi(1) + \phi(3) + \phi(5) + \phi(7) + \phi(9) + \phi(15) + \phi(21) + \phi(35) \\ &\qquad\qquad + \phi(45) + \phi(63) + \phi(105) + \phi(315) \\ &= 1 + 2 + 4 + 6 + 6 + 8 + 12 + 24 + 24 + 36 + 48 + 144 \\ &= 315. \end{aligned}$$

Again we end up with the number we started with. This is beginning to look like more than a coincidence. We might even make the following guess.

Guess. Let $d_1, d_2, \ldots, d_r$ be the numbers that divide n, including both 1 and n. Then

$$\phi(d_1) + \phi(d_2) + \cdots + \phi(d_r) = n.$$

How might we go about proving that our guess is correct? The easiest case to check would be when n has very few divisors. For example, suppose that we take $n = p$, where p is a prime. The divisors of p are 1 and p, and we know that $\phi(1) = 1$ and $\phi(p) = p - 1$. Adding these gives

$$\phi(1) + \phi(p) = 1 + (p - 1) = p.$$

So we have verified our guess when n is a prime.

Next we try $n = p^2$. The divisors of p^2 are 1, p, and p^2, and we know from Chapter 11 that $\phi(p^2) = p^2 - p$, so we find that

$$\phi(1) + \phi(p) + \phi(p^2) = 1 + (p - 1) + (p^2 - p) = p^2.$$

Notice how the terms cancel until only p^2 is left.

Emboldened by these successes, let's try to verify our guess when $n = p^k$ is any power of a prime. The divisors of p^k are $1, p, p^2, \ldots, p^k$. In Chapter 11 we found a formula for Euler's phi function at a prime power: $\phi(p^i) = p^i - p^{i-1}$. Using this formula enables us to compute

$$\begin{aligned}
&\phi(1) + \phi(p) + \phi(p^2) + \cdots + \phi(p^{k-1}) + \phi(p^k) \\
&\quad = 1 + (p - 1) + (p^2 - p) + \cdots + (p^{k-1} - p^{k-2}) + (p^k - p^{k-1}) \\
&\quad = p^k.
\end{aligned}$$

Again the terms cancel, leaving exactly p^k. We have now verified that our guess is true whenever n is a prime power.

If n is not a power of a prime, the situation is somewhat more complicated. As always, we start with the simplest case. Suppose that $n = pq$ is the product of two different primes. Then the divisors of n are 1, p, q, and pq; so we need to sum

$$\phi(1) + \phi(p) + \phi(q) + \phi(pq).$$

We showed in Chapter 11 that Euler's phi function satisfies a multiplication formula $\phi(mn) = \phi(m)\phi(n)$ provided that m and n are relatively prime. In particular, p and q are relatively prime, so $\phi(pq) = \phi(p)\phi(q)$. This means that

$$\begin{aligned}
\phi(1) + \phi(p) + \phi(q) + \phi(pq) &= 1 + \phi(p) + \phi(q) + \phi(p)\phi(q) \\
&= \bigl(1 + \phi(p)\bigr)\bigl(1 + \phi(q)\bigr) \\
&= pq,
\end{aligned}$$

which is exactly what we want.

Using this example as a guide, we are now ready to tackle the general case. For any number n, we define a function $F(n)$ by the formula

$$F(n) = \phi(d_1) + \phi(d_2) + \cdots + \phi(d_r), \qquad \text{where } d_1, d_2, \ldots, d_r \text{ are the divisors of } n.$$

Our goal is to show that $F(n) = n$ for every number n. The first step is to check that the function F satisfies a multiplication formula.

Claim 20.1. *If* $\gcd(m,n) = 1$, *then* $F(mn) = F(m)F(n)$.

Verification. Let

$$d_1, d_2, \ldots, d_r \text{ be the divisors of } n,$$

and

$$e_1, e_2, \ldots, e_s \text{ be the divisors of } m.$$

The fact that m and n are relatively prime means that the divisors of mn are precisely the various products

$$d_1e_1, d_1e_2, \ldots, d_1e_s, d_2e_1, d_2e_2, \ldots, d_2e_s, \ldots, d_re_1, d_re_2, \ldots, d_re_s.$$

Furthermore, every d_i is relatively prime to every e_j, so $\phi(d_ie_j) = \phi(d_i)\phi(e_j)$. Using these facts, we can compute

$$\begin{aligned}
F(mn) &= \phi(d_1e_1) + \cdots + \phi(d_1e_s) + \phi(d_2e_1) + \cdots + \phi(d_2e_s) \\
&\qquad + \cdots + \phi(d_re_1) + \cdots + \phi(d_re_s) \\
&= \phi(d_1)\phi(e_1) + \cdots + \phi(d_1)\phi(e_s) + \phi(d_2)\phi(e_1) + \cdots + \phi(d_2)\phi(e_s) \\
&\qquad + \cdots + \phi(d_r)\phi(e_1) + \cdots + \phi(d_r)\phi(e_s) \\
&= \big(\phi(d_1) + \phi(d_2) + \cdots + \phi(d_r)\big) \cdot \big(\phi(e_1) + \phi(e_2) + \cdots + \phi(e_s)\big) \\
&= F(m)F(n).
\end{aligned}$$

This completes the verification of our claim. □

Using the claim, it is now a simple matter to prove the following summation formula for Euler's phi function.

Theorem 20.2 (Euler's Phi Function Summation Formula). *Let* $d_1, d_2, \ldots, d_r$ *be the divisors of* n. *Then*

$$\phi(d_1) + \phi(d_2) + \cdots + \phi(d_r) = n.$$

Proof. We let $F(n) = \phi(d_1) + \phi(d_2) + \cdots + \phi(d_r)$, and we need to verify that $F(n)$ always equals n. We already checked that $F(p^k) = p^k$ for prime powers. Now factor n into a product of prime powers, say $n = p_1^{k_1} p_2^{k_2} \cdots p_r^{k_r}$. The different prime powers are relatively prime to one another, so we can use the multiplication formula for F to compute

$$\begin{aligned} F(n) &= F\big(p_1^{k_1} p_2^{k_2} \cdots p_r^{k_r}\big) \\ &= F\big(p_1^{k_1}\big) F\big(p_2^{k_2}\big) \cdots F\big(p_r^{k_r}\big) && \text{from the multiplication formula,} \\ &= p_1^{k_1} p_2^{k_2} \cdots p_r^{k_r} && \text{since } F(p^k) = p^k \text{ for prime powers,} \\ &= n. \end{aligned}$$ □

Exercises

20.1. Liouville's lambda function $\lambda(n)$ is defined by factoring n into a product of primes, $n = p_1^{k_1} p_2^{k_2} \cdots p_r^{k_r}$, and then setting

$$\lambda(n) = (-1)^{k_1+k_2+\cdots+k_r}.$$

[Also, we let $\lambda(1) = 1$.] For example, to compute $\lambda(1728)$, we factor $1728 = 2^6 \cdot 3^3$, and then $\lambda(1728) = (-1)^{6+3} = (-1)^9 = -1$.

(a) Compute the following values of Liouville's function: $\lambda(30)$; $\lambda(504)$; $\lambda(60750)$.

(b) We use Liouville's lambda function to define a new function $G(n)$ by the formula

$$G(n) = \lambda(d_1) + \lambda(d_2) + \cdots + \lambda(d_r), \qquad \text{where } d_1, d_2, \ldots, d_r \text{ are the divisors of } n.$$

Compute the value of $G(n)$ for all $1 \le n \le 18$.

(c) Use your computations in (b), and additional computations if necessary, to make a guess as to the value of $G(n)$. Check your guess for a few more values of n. Use your guess to find the value of $G(62141689)$ and $G(60119483)$.

(d) Prove that your guess in (c) is correct.

20.2. A function $f(n)$ that satisfies the multiplication formula

$$f(mn) = f(m)f(n) \qquad \text{for all numbers } m \text{ and } n \text{ with } \gcd(m,n) = 1$$

is called a *multiplicative function*. For example, we have seen that Euler's phi function $\phi(n)$ is multiplicative (Chapter 11) and that the sum of divisors function $\sigma(n)$ is multiplicative (Chapter 15).

(a) Show that Liouville's lambda function $\lambda(n)$ described in Exercise 20.1 is a multiplicative function.

(b) Suppose now that $f(n)$ is any multiplicative function, and define a new function

$$g(n) = f(d_1) + f(d_2) + \cdots + f(d_r), \quad \text{where } d_1, d_2, \ldots, d_r \text{ are the divisors of } n.$$

Prove that $g(n)$ is a multiplicative function.

20.3. Let $d_1, d_2, \ldots, d_r$ be the numbers that divide n, including 1 and n. The tth *power sigma function* $\sigma_t(n)$ is equal to the sum of the tth powers of the divisors of n,

$$\sigma_t(n) = d_1^t + d_2^t + \cdots + d_r^t.$$

For example, $\sigma_2(10) = 1^2 + 2^2 + 5^2 + 10^2 = 130$. Of course, $\sigma_1(n)$ is just our old friend, the sigma function $\sigma(n)$.

(a) Compute the values of $\sigma_2(12)$, $\sigma_3(10)$, and $\sigma_0(18)$.

(b) Show that if $\gcd(m, n) = 1$, then $\sigma_t(mn) = \sigma_t(m)\sigma_t(n)$. In other words, show that σ_t is a multiplicative function. Is this formula still true if m and n are not relatively prime?

(c) We showed in Chapter 15 that $\sigma(p^k) = (p^{k+1} - 1)/(p - 1)$. Find a similar formula for $\sigma_t(p^k)$, and use it to compute $\sigma_4(2^6)$.

(d) The function $\sigma_0(n)$ counts the number of different divisors of n. Does your formula in (c) work for σ_0? If not, give a correct formula for $\sigma_0(p^k)$. Use your formula and (b) to find the value of $\sigma_0(42336000)$.

Chapter 21

Powers Modulo p and Primitive Roots

If a and p are relatively prime, Fermat's Little Theorem (Chapter 9) tells us that

$$a^{p-1} \equiv 1 \pmod{p}.$$

Of course, it's quite possible that some smaller power of a is congruent to 1 modulo p. For example, $2^3 \equiv 1 \pmod{7}$. On the other hand, there may be some values of a that require the full $(p-1)^{\text{st}}$ power. For example, the powers of 3 modulo 7 are

$$\begin{array}{lll} 3^1 \equiv 3 \pmod{7}, & 3^2 \equiv 2 \pmod{7}, & 3^3 \equiv 6 \pmod{7}, \\ 3^4 \equiv 4 \pmod{7}, & 3^5 \equiv 5 \pmod{7}, & 3^6 \equiv 1 \pmod{7}. \end{array}$$

Thus, the full 6th power of 3 is required before we get to 1 modulo 7.

Let's look at some more examples to see if we can spot a pattern. Table 21.1 lists the smallest power of a that is congruent to 1 modulo p for the primes $p = 5, 7$, and 11 and for each a between 1 and $p-1$. We might make two observations.

1. The smallest exponent e so that $a^e \equiv 1 \pmod{p}$ seems to divide $p-1$.

2. There are always some a's that require the exponent $p-1$.

Since we are studying this smallest exponent in this chapter, we give it a name. The *order of a modulo p* is the quantity

$$e_p(a) = \left(\begin{array}{l} \text{the smallest exponent } e \geq 1 \\ \text{so that } a^e \equiv 1 \pmod{p} \end{array} \right)$$

$p = 5$
$1^1 \equiv 1 \pmod 5$
$2^4 \equiv 1 \pmod 5$
$3^4 \equiv 1 \pmod 5$
$4^2 \equiv 1 \pmod 5$

$p = 7$
$1^1 \equiv 1 \pmod 7$
$2^3 \equiv 1 \pmod 7$
$3^6 \equiv 1 \pmod 7$
$4^3 \equiv 1 \pmod 7$
$5^6 \equiv 1 \pmod 7$
$6^2 \equiv 1 \pmod 7$

$p = 11$
$1^1 \equiv 1 \pmod{11}$
$2^{10} \equiv 1 \pmod{11}$
$3^5 \equiv 1 \pmod{11}$
$4^5 \equiv 1 \pmod{11}$
$5^5 \equiv 1 \pmod{11}$
$6^{10} \equiv 1 \pmod{11}$
$7^{10} \equiv 1 \pmod{11}$
$8^{10} \equiv 1 \pmod{11}$
$9^5 \equiv 1 \pmod{11}$
$10^2 \equiv 1 \pmod{11}$

Table 21.1: Smallest power of a that equals 1 modulo p

(Note that we only allow values of a that are relatively prime to p.)

Referring to Table 21.1, we see, for example, that $e_5(2) = 4$, $e_7(4) = 3$, and $e_{11}(7) = 10$. Fermat's Little Theorem says that $a^{p-1} \equiv 1 \pmod p$, so we know that $e_p(a) \le p - 1$. Our first observation was that $e_p(a)$ seems to divide $p - 1$. Our second observation was that there always seem to be some a's with $e_p(a) = p - 1$. We are going to check that both of these observations are true. We begin with the first, which is the easier of the two.

Theorem 21.1 (Order Divisibility Property). *Let a be an integer not divisible by the prime p, and suppose that $a^n \equiv 1 \pmod p$. Then the order $e_p(a)$ divides n.*

In particular, the order $e_p(a)$ always divides $p - 1$.

Verification. The definition of the order $e_p(a)$ tells us that

$$a^{e_p(a)} \equiv 1 \pmod p,$$

and we are assuming that $a^n \equiv 1 \pmod p$. Let $G = \gcd\big(e_p(a), n\big)$, and let (u, v) be a solution in positive integers to the equation

$$e_p(a)u - nv = G.$$

(The Linear Equation Theorem from Chapter 6 says that there is a solution.) Now we compute the quantity $a^{e_p(a)u} \pmod{p}$ in two different ways:

$$a^{e_p(a)u} = \left(a^{e_p(a)}\right)^u \equiv 1^u \equiv 1 \pmod{p},$$
$$a^{e_p(a)u} = a^{nv+G} = (a^n)^v \cdot a^G \equiv 1^v \cdot a^G \equiv a^G \pmod{p}.$$

This shows that $a^G \equiv 1 \pmod{p}$. But $e_p(a)$ is the smallest power of a that is congruent to 1 modulo p, so we must have $G \geq e_p(a)$. On the other hand, $G = \gcd(e_p(a), n)$, so G divides both $e_p(a)$ and n. In particular, $G \leq e_p(a)$. The only possibility is that $G = e_p(a)$, which shows that $e_p(a)$ divides n.

Finally, Fermat's Little Theorem (Chapter 9) tells us that $a^{p-1} \equiv 1 \pmod{p}$, so taking $n = p - 1$, we conclude that $e_p(a)$ divides $p - 1$. □

Our next task is to look at the numbers that have the largest possible order: $e_p(a) = p - 1$. If a is such a number, then the powers

$$a,\ a^2,\ a^3, \ldots, a^{p-3},\ a^{p-2},\ a^{p-1} \pmod{p}$$

must all be different modulo p. [If the powers are not all different, then we would have $a^i \equiv a^j \pmod{p}$ for some exponents $1 \leq i < j \leq p-1$, which would mean that $a^{j-i} \equiv 1 \pmod{p}$, where the exponent $j - i$ is less than $p - 1$.] The numbers that require the largest exponent are of sufficient importance for us to give them a name.

> A number g with maximum order
> $e_p(g) = p - 1$
> is called a *primitive root modulo p*.

Looking back at the tables for $p = 5, 7, 11$, we see that 2 and 3 are primitive roots modulo 5, that 3 and 5 are primitive roots modulo 7, and that 2, 6, 7, and 8 are primitive roots modulo 11.

We now come to the most important result in this chapter.

Theorem 21.2 (Primitive Root Theorem). *Every prime p has a primitive root. More precisely, there are exactly $\phi(p-1)$ primitive roots modulo p.*

For example, the Primitive Root Theorem says that there are $\phi(10) = 4$ primitive roots modulo 11 and, sure enough, we saw that the primitive roots modulo 11 are the numbers 2, 6, 7, and 8. Similarly, the theorem says that there are $\phi(36) = 12$ primitive roots modulo 37 and that there are $\phi(9906) = 3024$ primitive roots modulo 9907. In fact, the primitive roots modulo 37 are the 12 numbers $2, 5, 13, 15, 17, 18, 19, 20, 22, 24, 32, 35$. We won't waste the space to list the 3024 primitive roots modulo 9907. One drawback of the Primitive Root Theorem is that it doesn't give a method for actually finding a primitive root modulo p. All we can do is start checking $a = 2$, $a = 3$, $a = 5$, $a = 6$, ... until we find a value of a with $e_p(a) = p - 1$. (Do you see why 4 can never be a primitive root?) However, once we find one primitive root modulo p, it is not hard to find all the others (see Exercise 21.6).

Proof of the Primitive Root Theorem. We prove the Primitive Root Theorem using one of the most powerful techniques available in number theory: COUNTING. You may wonder how counting can be so powerful. After all, it's one of the first things taught in kindergarten.[1] What we do is to take a certain set of numbers. We will count how many numbers are in the set in *two different ways*. It is this idea of counting things in two different ways and then comparing the results that has wide applicability in number theory, and indeed in all mathematics.

For each number a between 1 and $p - 1$, we know that the order $e_p(a)$ divides $p - 1$. So for each number d dividing $p - 1$, we might ask how many a's have their order $e_p(a)$ equal to d. We call this number $\psi(d)$. In other words,

$$\psi(d) = (\text{the number of } a\text{'s with } 1 \le a < p \text{ and } e_p(a) = d).$$

In particular, $\psi(p - 1)$ is the number of primitive roots modulo p.

Let n be any number dividing $p - 1$, say $p - 1 = nk$. Then we can factor the polynomial $X^{p-1} - 1$ as

$$\begin{aligned} X^{p-1} - 1 &= X^{nk} - 1 \\ &= (X^n)^k - 1 \\ &= (X^n - 1)\big((X^n)^{k-1} + (X^n)^{k-2} + \cdots + (X^n)^2 + X^n + 1\big). \end{aligned}$$

We count how many roots these polynomials have modulo p.

First we observe that

$$X^{p-1} - 1 \equiv 0 \pmod{p} \qquad \text{has exactly } p - 1 \text{ solutions,}$$

since Fermat's Little Theorem tells us that $X = 1, 2, 3, \ldots, p - 1$ are all solutions. On the other hand,

$$X^n - 1 \equiv 0 \pmod{p}$$

has at most n solutions, and

[1] Yet another illustration of the principle that *Everything I Ever Needed To Know I Learned in Kindergarten*, although proving theorems in number theory probably isn't one of the basic skills that Robert Fulghum had in mind when he wrote his book.

$$(X^n)^{k-1} + (X^n)^{k-2} + \cdots + X^n + 1 \equiv 0 \pmod{p}$$

has at most $nk - n$ solutions. More generally, if $F(X)$ is a polynomial of degree D with integer coefficients, then the congruence $F(X) \equiv 0 \pmod{p}$ has at most D solutions modulo p. We leave this fact as an exercise for you to prove.

So we now know that

$$\underbrace{X^{p-1} - 1}_{\substack{\text{exactly } p-1=nk \\ \text{roots mod } p}} = \underbrace{(X^n - 1)}_{\substack{\text{at most } n \\ \text{roots mod } p}} \times \underbrace{\left((X^n)^{k-1} + (X^n)^{k-2} + \cdots + X^n + 1\right)}_{\text{at most } nk-n \text{ roots mod } p}$$

The only way for this to be true is if $X^n - 1$ has exactly n roots modulo p, since otherwise the right-hand side won't have enough roots. This proves the following important fact:

> If n divides $p - 1$, then the congruence
> $$X^n - 1 \equiv 0 \pmod{p}$$
> has exactly n solutions with $0 \le X < p$.

Now let's count the number of solutions to $X^n - 1 \equiv 0 \pmod{p}$ in a different way. If $X = a$ is a solution, then $a^n \equiv 1 \pmod{p}$, so by the Order Divisibility Property, we know that $e_p(a)$ divides n. So if we look at the divisors of n and if for each divisor d of n we take those a's with $e_p(a) = d$, then we end up with all the solutions of the congruence $X^n - 1 \equiv 0 \pmod{p}$. In other words, if $d_1, d_2, \ldots, d_r$ are the divisors of n, then the number of solutions to $X^n - 1 \equiv 0 \pmod{p}$ is equal to

$$\psi(d_1) + \psi(d_2) + \cdots + \psi(d_r).$$

We have now counted the number of solutions to $X^n - 1 \equiv 0 \pmod{p}$ in two different ways. First, we showed that there are n solutions, and second we showed that there are $\psi(d_1) + \cdots + \psi(d_r)$ solutions. These numbers must be the same, so merely by counting the number of solutions, we have proved the following beautiful formula:

> Let n divide $p-1$ and let $d_1, d_2, \ldots, d_r$ be the divisors of n, including both 1 and n. Then
> $$\psi(d_1) + \psi(d_2) + \cdots + \psi(d_r) = n.$$

This formula should look familiar; it's exactly the same as the formula we proved for Euler's phi function in Chapter 20. We now use the fact that ϕ and ψ both satisfy this formula to show that ϕ and ψ are actually equal.

Our first observation is that $\phi(1) = 1$ and $\psi(1) = 1$, so we're okay for $n = 1$. Next we check that $\phi(q) = \psi(q)$ when $n = q$ is a prime. The divisors of q are 1 and q, so

$$\phi(q) + \phi(1) = q = \psi(q) + \psi(1).$$

But we know that $\phi(1) = \psi(1) = 1$, so subtracting 1 from both sides gives $\phi(q) = \psi(q)$.

How about $n = q^2$? The divisors of q^2 are 1, q and q^2, so

$$\phi(q^2) + \phi(q) + \phi(1) = q^2 = \psi(q^2) + \psi(q) + \psi(1).$$

But we already know that $\phi(q) = \psi(q)$ and $\phi(1) = \psi(1)$, so canceling them from both sides gives $\phi(q^2) = \psi(q^2)$.

Similarly, if $n = q_1 q_2$ for two different primes q_1 and q_2, then the divisors of n are 1, q_1, q_2, and $q_1 q_2$. This gives

$$\begin{aligned}\phi(q_1 q_2) + \phi(q_1) + \phi(q_2) + \phi(1) &= q_1 q_2 \\ &= \psi(q_1 q_2) + \psi(q_1) + \psi(q_2) + \psi(1),\end{aligned}$$

and canceling the terms that we already know are equal leaves $\phi(q_1 q_2) = \psi(q_1 q_2)$.

These examples illustrate how to prove that $\phi(n) = \psi(n)$ for every n by working up from small values of n to larger values of n. More formally, we can give a proof by induction. (You might want to look back at the proof of the Fundamental Theorem of Arithmetic in Chapter 7 for another induction proof.) So we assume that we have already proved that $\phi(d) = \psi(d)$ for all numbers $d < n$, and we attempt to prove that $\phi(n) = \psi(n)$. Let $d_1, d_2, \ldots, d_r$ be the divisors of n as usual. One of these divisors is n itself, so relabeling them, we may as well assume that $d_1 = n$. Using the summation formulas for ϕ and ψ, we find that

$$\begin{aligned}\phi(n) + \phi(d_2) + \phi(d_3) + \cdots + \phi(d_r) &= n \\ &= \psi(n) + \psi(d_2) + \psi(d_3) + \cdots + \psi(d_r).\end{aligned}$$

But all of the numbers $d_2, d_3, \ldots, d_r$ are strictly less than n, so our assumption tells us that $\phi(d_i) = \psi(d_i)$ for each $i = 2, 3, \ldots, r$. This means we can cancel these values from both sides of the equation, which leaves the desired equality $\phi(n) = \psi(n)$.

To recapitulate, we have proved that for each number n dividing $p - 1$ there are exactly $\phi(n)$ numbers a with $e_p(a) = n$. Taking $n = p - 1$, we see that there are exactly $\phi(p-1)$ numbers a with $e_p(a) = p - 1$. But a's with $e_p(a) = p - 1$ are precisely primitive roots modulo p, so we have proved that there are $\phi(p-1)$ primitive roots modulo p. Since the number $\phi(p-1)$ is always at least 1, we see that every prime has at least one primitive root. This completes our verification of the Primitive Root Theorem. $\square$

The Primitive Root Theorem tells us that there are lots of primitive roots modulo p, in fact, precisely $\phi(p-1)$ of them. Unfortunately, it doesn't give us any information at all about which specific numbers are primitive roots. Suppose we turn the question around, fix a number a, and ask for which primes p is a a primitive root. For example, for which primes p is 2 a primitive root? The Primitive Root Theorem gives us no information at all!

Here is a list of the order $e_p(2)$ for all primes up to 100, where we write e_p instead of $e_p(2)$ to save space.

$e_3 = 2$	$e_5 = 4$	$e_7 = 3$	$e_{11} = 10$	$e_{13} = 12$	$e_{17} = 8$
$e_{19} = 18$	$e_{23} = 11$	$e_{29} = 28$	$e_{31} = 5$	$e_{37} = 36$	$e_{41} = 20$
$e_{43} = 14$	$e_{47} = 23$	$e_{53} = 52$	$e_{59} = 58$	$e_{61} = 60$	$e_{67} = 66$
$e_{71} = 35$	$e_{73} = 9$	$e_{79} = 39$	$e_{83} = 82$	$e_{89} = 11$	$e_{97} = 48$

Looking at this list, we see that 2 is a primitive root for the primes

$$p = 3, 5, 11, 13, 19, 29, 37, 53, 59, 61, 67, 83.$$

Do you see any pattern? Don't be discouraged if you don't; no one else has yet found a simple pattern, either. However, in the 1920s Emil Artin made the following conjecture.

Conjecture 21.3 (Artin's Conjecture). *There are infinitely many primes p such that 2 is a primitive root modulo p.*

Of course, there's really nothing too special about the number 2, so Artin also made the following conjecture.

Conjecture 21.4 (The Generalized Artin Conjecture). *Let a be any integer that is not a perfect square and is not equal to -1. Then there are infinitely many primes p such that a is a primitive root modulo p.*

Artin's Conjecture is still unsolved, although much progress has been made on it in recent years. For example, in 1967 Christopher Hooley proved that, if a certain other conjecture called the Generalized Riemann Hypothesis is true, then the Generalized Artin Conjecture is also true. Equally striking, Rajiv Gupta, M. Ram Murty, and Roger Heath-Brown proved in 1985 that there are at most three pairwise relatively prime values of a for which the Generalized Artin Conjecture is false. Of course, these three putative "bad values" of a probably don't exist, but no one yet knows how to prove that they don't exist. And no one has been able to prove that $a = 2$ is not a bad value, so even Artin's original conjecture remains unproved!

Exercises

21.1. Let p be a prime number.

(a) What is the value of $1 + 2 + 3 + \cdots + (p-1) \pmod p$?

(b) What is the value of $1^2 + 2^2 + 3^2 + \cdots + (p-1)^2 \pmod p$?

(c) For any positive integer k, find the value of

$$1^k + 2^k + 3^k + \cdots + (p-1)^k \pmod p$$

and prove that your answer is correct.

21.2. For any integers a and m with $\gcd(a, m) = 1$, we let $e_m(a)$ be the smallest exponent $e \geq 1$ so that $a^e \equiv 1 \pmod m$. We call $e_m(a)$ the *order of a modulo m*.

(a) Compute the following values of $e_m(a)$:

(i) $e_9(2)$ (ii) $e_{15}(2)$ (iii) $e_{16}(3)$ (iv) $e_{10}(3)$

(b) Show that $e_m(a)$ always divides $\phi(m)$.

21.3. In this exercise you will investigate the value of $e_m(2)$ for odd integers m. To save space, we write e_m instead of $e_m(2)$, so for this exercise e_m is the smallest power of 2 that is congruent to 1 modulo m.

(a) Compute the value of e_m for each odd number $11 \leq m \leq 19$.

(b) Here is a table giving the values of e_m for all odd numbers between 3 and 149 [except for $11 \leq m \leq 19$ which you did in part (a)].

$e_3 = 2$	$e_5 = 4$	$e_7 = 3$	$e_9 = 6$	$e_{11} = **$	$e_{13} = **$	$e_{15} = **$
$e_{17} = **$	$e_{19} = **$	$e_{21} = 6$	$e_{23} = 11$	$e_{25} = 20$	$e_{27} = 18$	$e_{29} = 28$
$e_{31} = 5$	$e_{33} = 10$	$e_{35} = 12$	$e_{37} = 36$	$e_{39} = 12$	$e_{41} = 20$	$e_{43} = 14$
$e_{45} = 12$	$e_{47} = 23$	$e_{49} = 21$	$e_{51} = 8$	$e_{53} = 52$	$e_{55} = 20$	$e_{57} = 18$
$e_{59} = 58$	$e_{61} = 60$	$e_{63} = 6$	$e_{65} = 12$	$e_{67} = 66$	$e_{69} = 22$	$e_{71} = 35$
$e_{73} = 9$	$e_{75} = 20$	$e_{77} = 30$	$e_{79} = 39$	$e_{81} = 54$	$e_{83} = 82$	$e_{85} = 8$
$e_{87} = 28$	$e_{89} = 11$	$e_{91} = 12$	$e_{93} = 10$	$e_{95} = 36$	$e_{97} = 48$	$e_{99} = 30$
$e_{101} = 100$	$e_{103} = 51$	$e_{105} = 12$	$e_{107} = 106$	$e_{109} = 36$	$e_{111} = 36$	$e_{113} = 28$
$e_{115} = 44$	$e_{117} = 12$	$e_{119} = 24$	$e_{121} = 110$	$e_{123} = 20$	$e_{125} = 100$	$e_{127} = 7$
$e_{129} = 14$	$e_{131} = 130$	$e_{133} = 18$	$e_{135} = 36$	$e_{137} = 68$	$e_{139} = 138$	$e_{141} = 46$
$e_{143} = 60$	$e_{145} = 28$	$e_{147} = 42$	$e_{149} = 148$			

Using this table, find (i.e., guess) a formula for e_{mn} in terms of e_m and e_n whenever $\gcd(m, n) = 1$.

(c) Use your conjectural formula from (b) to find the value of e_{11227}. (Note that $11227 = 103 \cdot 109$.)

(d) Prove that your conjectural formula in (b) is true.

(e) Use the table to guess a formula for e_{p^k} in terms of e_p, p, and k, where p is an odd prime. Use your formula to find the value of e_{68921}. (Note that $68921 = 41^3$.)

(f) Can you prove that your conjectural formula for e_{p^k} in (e) is correct?

21.4. (a) Find all primitive roots modulo 13.
(b) For each number d dividing 12, list the a's with $1 \le a < 13$ and $e_{13}(a) = d$.

21.5. (a) Let $F(X) = X^D + A_1X^{D-1} + A_2X^{D-2} + \cdots + A_{D-1}X + A_D$ be a polynomial whose coefficients $A_1, A_2, \ldots, A_D$ are integers. Let p be a prime number. Show that the congruence $F(X) \equiv 0 \pmod{p}$ has at most D solutions with $0 \le X < p$.
(b) How many solutions are there to the congruence $X^4 + 5X^3 + 4X^2 - 6X - 4 \equiv 0 \pmod{11}$ with $0 \le X < 11$? Are there four solutions, or are there fewer than four solutions?
(c) How many solutions are there to the congruence $X^2 - 1 \equiv 0 \pmod{8}$ with $0 \le X < 8$? Since there are more than two solutions, why doesn't this show that (a) is incorrect?

21.6. (a) If g is a primitive root modulo 37, which of the numbers $g^2, g^3, \ldots, g^8$ are primitive roots modulo 37?
(b) If g is a primitive root modulo p, develop an easy to use rule for determining if g^k is a primitive root modulo p, and prove that your rule is correct.
(c) Suppose that g is a primitive root modulo the prime $p = 21169$. Use your rule from (b) to determine which of the numbers $g^2, g^3, \ldots, g^{20}$ are primitive roots modulo 21169.

21.7. (a) Find all primes less than 20 for which 3 is a primitive root.
(b) If you know how to program a computer, find all primes less than 100 for which 3 is a primitive root.

21.8. If $a = b^2$ is a perfect square and p is an odd prime, explain why it is impossible for a to be a primitive root modulo p.

21.9. Let p be a prime, let k be a number not divisible by p, and let b be a number that has a kth root modulo p. Find a formula for the number of kth roots of b modulo p and prove that your formula is correct. (*Hint*. Your formula should only depend on p and k, not on b.)

21.10. Write a program to compute $e_p(a)$, which is the smallest positive exponent e such that $a^e \equiv 1 \pmod{p}$. [Be sure to use the fact that if $a^e \not\equiv 1 \pmod{p}$ for all $1 \le e < p/2$, then $e_p(a)$ is automatically equal to $p - 1$.]

21.11. Write a program that finds the smallest primitive root for a given prime p. Make a list of all primes between 100 and 200 for which 2 is a primitive root.

21.12. Write a program that takes as input a polynomial $f(X)$ (having integer coefficients) and a number m and produces as output all the solutions to the congruence

$$f(X) \equiv 0 \pmod{m}.$$

(Don't be fancy; just substitute in $X = 0, 1, 2, \ldots m - 1$ and see which values are solutions.)

21.13. If a is relatively prime to both m and n and if $\gcd(m, n) = 1$, find a formula for $e_{mn}(a)$ in terms of $e_m(a)$ and $e_n(a)$.

21.14. For any number $m \geq 2$, not necessarily prime, we say that g is a *primitive root modulo* m if the smallest power of g that is congruent to 1 modulo m is the $\phi(m)^{\text{th}}$ power. In other words, g is a primitive root modulo m if $\gcd(g, m) = 1$ and $g^k \not\equiv 1 \pmod{m}$ for all powers $1 \leq k < \phi(m)$.

(a) For each number $2 \leq m \leq 25$, determine if there are any primitive roots modulo m. (If you have a computer, do the same for all $m \leq 50$.)

(b) Use your data from (a) to make a conjecture as to which m's have primitive roots and which ones do not.

(c) Prove that your conjecture in (a) is correct.

Chapter 22

Primitive Roots and Indices

The beauty of a primitive root g modulo a prime p is the appearance of every nonzero number modulo p as a power of g. So for any number $1 \le a < p$, we can pick out exactly one of the powers

$$g, g^2, g^3, g^4, \ldots, g^{p-3}, g^{p-2}, g^{p-1}$$

as being congruent to a modulo p. The exponent is called the *index of a modulo p for the base g*. Assuming that p and g have been specified, we write $I(a)$ for the index.

For example, if we use the primitive root 2 as base for the prime 13, then $I(3) = 4$, since $2^4 = 16 \equiv 3 \pmod{13}$. Similarly, $I(5) = 9$, since $2^9 = 512 \equiv 5 \pmod{13}$. To find the index of any particular number, such as 7, we just compute the powers $2, 2^2, 2^3, \ldots$ modulo 13 until we get to a number that is congruent to 7.

Another approach is to make a table of all powers of 2 modulo 13. Then we can read any information we want from the table.

I	1	2	3	4	5	6	7	8	9	10	11	12
$2^I \pmod{13}$	2	4	8	3	6	12	11	9	5	10	7	1

Powers of 2 Modulo 13

For example, to find $I(11)$, we scan the second row of the table until we find the number 11, and then the index $I(11) = 7$ can be read from the first row.

This suggests another way to arrange the data that might be more useful. What we do is rearrange the numbers so that the second row is in numerical order from 1 to 12, and then we switch the first and second rows. The resulting table has the numbers from 1 to 12 in order in the first row, and below each number is its index.

a	1	2	3	4	5	6	7	8	9	10	11	12
$I(a)$	12	1	4	2	9	5	11	3	8	10	7	6

Table of Indices Modulo 13 for the Base 2

Now it's even easier to read off the index of any given number, such as $I(8) = 3$ and $I(10) = 10$.

In the past, number theorists compiled tables of indices to be used for numerical calculations.[1] The reason that indices are useful for calculations is highlighted by the following theorem.

Theorem 22.1 (Rules for Indices). *Indices satisfy the following rules*:

(a) $I(ab) \equiv I(a) + I(b) \pmod{p-1}$ [Product Rule]

(b) $I(a^k) \equiv kI(a) \pmod{p-1}$ [Power Rule]

Verification. These rules are nothing more than the usual laws of exponents, combined with the fact that g is a primitive root. Thus, to check (a), we compute

$$g^{I(ab)} \equiv ab \equiv g^{I(a)}g^{I(b)} \equiv g^{I(a)+I(b)} \pmod{p}.$$

This means that $g^{I(ab)-I(a)-I(b)} \equiv 1 \pmod{p}$. But g is a primitive root, so $I(ab) - I(a) - I(b)$ must be a multiple of $p-1$. This completes the verification of (a). To check (b), we perform a similar computation,

$$g^{I(a^k)} \equiv a^k \equiv (g^{I(a)})^k \equiv g^{kI(a)} \pmod{p}.$$

This implies that $I(a^k) - kI(a)$ is a multiple of $p-1$, which is (b). $\square$

One of the most common mistakes made when working with indices is to reduce them modulo p instead of modulo $p-1$. It is important to keep in mind that indices appear as exponents, and the exponent in Fermat's Little Theorem is $p-1$, not p. We reiterate:

Always Reduce Indices Modulo $p-1$.

[1] In 1839, Karl Jacobi published a *Canon Arithmeticus* containing a table of indices for all primes less than 1000. More recently, an extensive table containing all primes up to 50021 was compiled by Western and Miller and published by the Royal Society at Cambridge University in 1968.

I want to explain briefly how the index rules and a table of indices can be used to simplify calculations and solve congruences. For that purpose, here is a table of indices for the prime $p = 37$ and the base $g = 2$.

a	1	2	3	4	5	6	7	8	9	10	11	12	13	14	15	16	17	18
$I(a)$	36	1	26	2	23	27	32	3	16	24	30	28	11	33	13	4	7	17

a	19	20	21	22	23	24	25	26	27	28	29	30	31	32	33	34	35	36
$I(a)$	35	25	22	31	15	29	10	12	6	34	21	14	9	5	20	8	19	18

Table of Indices Modulo 37 for the Base 2

If we want to compute $23 \cdot 19 \pmod{37}$, rather than multiplying 23 and 19, we can instead add their indices. Thus,

$$I(23 \cdot 19) \equiv I(23) + I(19) \equiv 15 + 35 \equiv 50 \equiv 14 \pmod{36}.$$

Note that the computation is done modulo $p - 1$, in this case, modulo 36. Looking at the table, we find that $I(30) = 14$ and conclude that $23 \cdot 19 \equiv 30 \pmod{37}$.

"Wait a minute," you are probably protesting, "using indices to compute the product $23 \cdot 19 \pmod{37}$ is lot of work." It would be easier to just multiply 23 by 19, divide the product by 37, and take the remainder. There is a somewhat stronger case to be made for using indices to compute powers. For example,

$$I(29^{14}) \equiv 14 \cdot I(29) \equiv 14 \cdot 21 \equiv 294 \equiv 6 \pmod{36}.$$

From the table we see that $I(27) = 6$, so $29^{14} \equiv 27 \pmod{37}$. Here the number 29^{14} has 21 digits, so we wouldn't want to compute the exact value of 29^{14} by hand and then reduce modulo 37. On the other hand, we know how to compute $29^{14} \pmod{37}$ quite rapidly using the method of successive squares (Chapter 16). So are indices actually useful for anything? The answer is that the real power of a table of indices lies not in its use for direct computations, but rather as a tool for solving congruences. We give two illustrations.

For our first example, consider the congruence

$$19x \equiv 23 \pmod{37}.$$

If x is a solution, then the index of $19x$ is equal to the index of 23. Using the product rule and taking values from the table of indices, we can compute

$$\begin{aligned} I(19x) &= I(23) \\ I(19) + I(x) &\equiv I(23) \pmod{36} \\ 35 + I(x) &\equiv 15 \pmod{36} \\ I(x) &\equiv -20 \equiv 16 \pmod{36}. \end{aligned}$$

Thus, the index of the solution is $I(x) = 16$, and looking again at the table, we find that $x \equiv 9 \pmod{37}$. You should compare this solution of $19x \equiv 23 \pmod{37}$ with the more cumbersome method described in Chapter 8. Of course, the index method won't work unless you have a table of indices already compiled, so the Linear Congruence Theorem in Chapter 8 is certainly not obsolete.

For our second example we'll solve a problem that until now would have required a great deal of tedious computation. We ask for all solutions to the congruence

$$3x^{30} \equiv 4 \pmod{37}.$$

We start by taking the index of both sides and using the product and power rules.

$$\begin{aligned} I(3x^{30}) &= I(4) \\ I(3) + 30I(x) &\equiv I(4) \pmod{36} \\ 26 + 30I(x) &\equiv 2 \pmod{36} \\ 30I(x) &\equiv -24 \equiv 12 \pmod{36}. \end{aligned}$$

So we need to solve the congruence $30I(x) \equiv 12 \pmod{36}$ for $I(x)$. [*Warning*: Do not divide both sides by 6 to get $5I(x) \equiv 2 \pmod{36}$, you'll lose some of the answers.] We saw in Chapter 8 how to solve a congruence of this sort. In general, the congruence $ax \equiv c \pmod{m}$ has $\gcd(a, m)$ solutions if $\gcd(a, m)$ divides c, otherwise it has no solutions. In our case $\gcd(30, 36) = 6$ does divide 12, so there should be 6 solutions. Using the methods of Chapter 8, or just by trial and error, we find that

$$30I(x) \equiv 12 \pmod{36}$$

for

$$I(x) \equiv 4, 10, 16, 22, 28, \text{ and } 34 \pmod{36}.$$

Finally, we look back at the table of indices to get the corresponding values of x,

$$\begin{aligned} I(16) &= 4, & I(25) &= 10, & I(9) &= 16, \\ I(21) &= 22, & I(12) &= 28, & I(28) &= 34. \end{aligned}$$

Thus, the congruence $3x^{30} \equiv 4 \pmod{37}$ has six solutions,

$$x \equiv 16, 25, 9, 21, 12, 28 \pmod{37}.$$

The computational advantages of using indices are easily stated. The Index Rules convert multiplication into addition and exponentiation into multiplication.

This undoubtedly sounds familiar, since it is exactly the same as the rules satisfied by logarithms:

$$\log(ab) = \log(a) + \log(b) \qquad \text{and} \qquad \log(a^k) = k\log(a).$$

For this reason, the index is also known as the *discrete logarithm*. And just as logarithm tables were used to do computations in days of yore before the proliferation of inexpensive calculators, just so were tables of indices used for computations in number theory. Nowadays, with the availability of desktop computers, index tables are used less frequently for numerical computations, but indices retain their usefulness as a theoretical tool.

However, there is another aspect to discrete lograrithms that makes them of great interest in modern cryptography. Suppose that you are given a large prime number p and two numbers a and g modulo p. The *Discrete Logarithm Problem* (DLP) is the problem of finding the exponent k so that

$$g^k \equiv a \pmod{p}.$$

In other words, the discrete logarithm problem asks you to find the index of a modulo p for the base g. As we saw in Chapter 16, it is relatively easy to compute $g^k \pmod{p}$ if you know g and k. However, if p is large, it is quite difficult to find the value of k if you're given the value of $g^k \pmod{p}$. This dichotomy can be used to construct public key cryptosystems, similar to the way that the difficulty of factoring numbers was used to construct the RSA cryptosystem in Chapter 18. We describe one such construction in Exercise 22.6.

Exercises

22.1. Use the table of indices modulo 37 to find all solutions to the following congruences.

(a) $12x \equiv 23 \pmod{37}$ (c) $x^{12} \equiv 11 \pmod{37}$
(b) $5x^{23} \equiv 18 \pmod{37}$ (d) $7x^{20} \equiv 34 \pmod{37}$

22.2. (a) Create a table of indices modulo 17 using the primitive root 3.
(b) Use your table to solve the congruence $4x \equiv 11 \pmod{17}$.
(c) Use your table to find all solutions to the congruence $5x^6 \equiv 7 \pmod{17}$.

22.3. (a) If a and b satisfy the relation $ab \equiv 1 \pmod{p}$, how are the indices $I(a)$ and $I(b)$ related to one another?
(b) If a and b satisfy the relation $a + b \equiv 0 \pmod{p}$, how are the indices $I(a)$ and $I(b)$ related to one another?
(c) If a and b satisfy the relation $a + b \equiv 1 \pmod{p}$, how are the indices $I(a)$ and $I(b)$ related to one another?

22.4. (a) If k divides $p - 1$, show that the congruence $x^k \equiv 1 \pmod{p}$ has exactly k distinct solutions modulo p.

(b) More generally, consider the congruence

$$x^k \equiv a \pmod{p}.$$

Find a simple way to use the values of k, p, and the index $I(a)$ to determine how many solutions this congruence has.

(c) The number 3 is a primitive root modulo the prime 1987. How many solutions are there to the congruence $x^{111} \equiv 729 \pmod{1987}$? (*Hint*. $729 = 3^6$.)

22.5. Write a program that takes as input a prime p, a primitive root g for p, and a number a, and produces as output the index $I(a)$. Use your program to make a table of indices for the prime $p = 47$ and the primitive root $g = 5$.

22.6. In this exercise we describe a public key cryptosystem called the ElGamal Cryptosystem that is based on the difficulty of solving the discrete logarithm problem. Let p be a large prime number and let g be a primitive root modulo p. Here's how Alice creates a key and Bob sends Alice a message.

The first step is for Alice to choose a number k to be her secret key. She computes the number $a \equiv g^k \pmod{p}$. She publishes this number a, which is the public key that Bob (or anyone else) will use to send her messages.

Now suppose that Bob wants to send Alice the message m, where m is a number between 2 and $p - 1$. He randomly chooses a number r and computes the two numbers

$$e_1 \equiv g^r \pmod{p} \qquad \text{and} \qquad e_2 \equiv ma^r \pmod{p}.$$

Bob sends Alice the pair of numbers (e_1, e_2).

Finally, Alice needs to decrypt the message. She first uses her secret key k to compute $c \equiv e_1^k \pmod{p}$. Next she computes $u \equiv c^{-1} \pmod{p}$. [That is, she solves $cu \equiv 1 \pmod{p}$ for u, using the method in Chapter 8.] Finally, she computes $v \equiv ue_2 \pmod{p}$. We can summarize Alice's computation by the formula

$$v \equiv e_2 \cdot (e_1^k)^{-1} \pmod{p}.$$

(a) Show that when Alice finishes her computation the number v that she computes equals Bob's message m.

(b) Show that if someone knows how to solve the discrete logarithm problem for the prime p and base g then he or she can read Bob's message.

22.7. For this exercise, use the ElGamal cryptosystem described in Exercise 22.6.

(a) Bob wants to use Alice's public key $a = 22695$ for the prime $p = 163841$ and base $g = 3$ to send her the message $m = 39828$. He chooses to use the random number $r = 129381$. Compute the encrypted message (e_1, e_2) he should send to Alice.

(b) Suppose that Bob sends the same message to Alice, but he chooses a different value for r. Will the encrypted message be the same?

(c) Alice has chosen the secret key $k = 278374$ for the prime $p = 380803$ and the base $g = 2$. She receives a message (consisting of three message blocks)

$$(61745, 206881), \qquad (255836, 314674), \qquad (108147, 350768)$$

from Bob. Decrypt the message and convert it to letters using the number-to-letter conversion table in Chapter 18.

Chapter 23

Squares Modulo p

We learned long ago how to solve linear congruences

$$ax \equiv c \pmod{m}$$

(see Chapter 8), and now it's time to take the plunge and move on to quadratic equations. We devote the next three chapters to answering the following types of questions:

- Is 3 congruent to the square of some number modulo 7?
- Does the congruence $x^2 \equiv -1 \pmod{13}$ have a solution?
- For which primes p does the congruence $x^2 \equiv 2 \pmod{p}$ have a solution?

We can answer the first two questions right now. To see if 3 is congruent to the square of some number modulo 7, we just square each of the numbers from 0 to 6, reduce modulo 7, and see if any of them are equal to 3. Thus,

$$\begin{aligned}
0^2 &\equiv 0 \pmod{7}\\
1^2 &\equiv 1 \pmod{7}\\
2^2 &\equiv 4 \pmod{7}\\
3^2 &\equiv 2 \pmod{7}\\
4^2 &\equiv 2 \pmod{7}\\
5^2 &\equiv 4 \pmod{7}\\
6^2 &\equiv 1 \pmod{7}.
\end{aligned}$$

So we see that 3 is not congruent to a square modulo 7. In a similar fashion, if we square each number from 0 to 12 and reduce modulo 13, we find that the congruence $x^2 \equiv -1 \pmod{13}$ has two solutions, $x \equiv 5 \pmod{13}$ and $x \equiv 8 \pmod{13}$.[1]

As always, we need to look at some data before we can even begin to look for patterns and make conjectures. Here are some tables giving all the squares modulo p for $p = 5$, 7, 11, and 13.

b	b^2
0	0
1	1
2	4
3	4
4	1

Modulo 5

b	b^2
0	0
1	1
2	4
3	2
4	2
5	4
6	1

Modulo 7

b	b^2
0	0
1	1
2	4
3	9
4	5
5	3
6	3
7	5
8	9
9	4
10	1

Modulo 11

b	b^2
0	0
1	1
2	4
3	9
4	3
5	12
6	10
7	10
8	12
9	3
10	9
11	4
12	1

Modulo 13

Many interesting patterns are already apparent from these lists. For example, each number (other than 0) that appears as a square seems to appear exactly twice. Thus, 5 is both 4^2 and 7^2 modulo 11, and 3 is both 4^2 and 9^2 modulo 13. In fact, if we fold each list over in the middle, the same numbers appear as squares on the top and on the bottom.

How can we describe this pattern with a formula? We are saying that the square of the number b and the square of the number $p - b$ are the same modulo p. But now that we've expressed our pattern by a formula, it's easy to prove. Thus,

$$(p-b)^2 = p^2 - 2pb + b^2 \equiv b^2 \pmod{p}.$$

[1]For many years during the nineteenth century, mathematicians were uneasy with the idea of the number $\sqrt{-1}$. Its current appellation "imaginary number" still reflects that disquiet. But if you work modulo 13, for example, then there's nothing mysterious about $\sqrt{-1}$. In fact, 5 and 8 are both square roots of -1 modulo 13.

So if we want to list all the (nonzero) numbers that are squares modulo p, we only need to compute half of them:

$$1^2 \pmod{p}, \quad 2^2 \pmod{p}, \quad 3^2 \pmod{p}, \ldots, \quad \left(\frac{p-1}{2}\right)^2 \pmod{p}.$$

Our goal is to find patterns that can be used to distinguish squares from nonsquares modulo p. Ultimately, we will be led to one of the most beautiful theorems in all number theory, the Law of Quadratic Reciprocity, but first we must perform the mundane task of assigning some names to the numbers we want to study.

> A nonzero number that is congruent to a square modulo p is called a *quadratic residue modulo p*. A number that is not congruent to a square modulo p is called a (*quadratic*) *nonresidue modulo p*. We abbreviate these long expressions by saying that a quadratic residue is a QR and a quadratic nonresidue is an NR. A number that is congruent to 0 modulo p is neither a residue nor a nonresidue.

To illustrate this terminology using the data from our tables, 3 and 12 are QRs modulo 13, while 2 and 5 are NRs modulo 13. Note that 2 and 5 are NRs because they do not appear in the list of squares modulo 13. The full set of QRs modulo 13 is $\{1, 3, 4, 9, 10, 12\}$, and the full set of NRs modulo 13 is $\{2, 5, 6, 7, 8, 11\}$. Similarly, the set of QRs modulo 7 is $\{1, 2, 4\}$ and the set of NRs modulo 7 is $\{3, 5, 6\}$.

Notice that there are 6 quadratic residues and 6 nonresidues modulo 13, and there are 3 quadratic residues and 3 nonresidues modulo 7. Using our earlier observation that $(p-b)^2 \equiv b^2 \pmod{p}$, we can easily verify that there are an equal number of quadratic residues and nonresidues modulo any (odd) prime.

Theorem 23.1. *Let p be an odd prime. Then there are exactly* $(p-1)/2$ *quadratic residues modulo p and exactly* $(p-1)/2$ *nonresidues modulo p.*

Verification. The quadratic residues are the nonzero numbers that are squares modulo p, so they are the numbers

$$1^2, 2^2, \ldots, (p-1)^2 \pmod{p}.$$

But, as we noted above, we only need to go halfway,

$$1^2, 2^2, \ldots, \left(\frac{p-1}{2}\right)^2 \pmod{p},$$

since the same numbers are repeated in reverse order if we square the remaining numbers

$$\left(\frac{p+1}{2}\right)^2, \ldots, (p-2)^2, (p-1)^2 \pmod{p}.$$

So in order to show that there are exactly $(p-1)/2$ quadratic residues, we need to check that the numbers $1^2, 2^2, \ldots, \left(\frac{p-1}{2}\right)^2$ are all different modulo p.

Suppose that b_1 and b_2 are numbers between 1 and $(p-1)/2$, and suppose that $b_1^2 \equiv b_2^2 \pmod{p}$. We want to show that $b_1 = b_2$. The fact that $b_1^2 \equiv b_2^2 \pmod{p}$ means that

$$p \quad \text{divides} \quad b_1^2 - b_2^2 = (b_1 - b_2)(b_1 + b_2).$$

However, $b_1 + b_2$ is between 2 and $p-1$, so it can't be divisible by p. Thus p must divide $b_1 - b_2$. But $|b_1 - b_2| < (p-1)/2$, so the only way for $b_1 - b_2$ to be divisible by p is to have $b_1 = b_2$. This shows that the numbers $1^2, 2^2, \ldots, \left(\frac{p-1}{2}\right)^2$ are all different modulo p, so there are exactly $(p-1)/2$ quadratic residues modulo p. Now we need only observe that there are $p-1$ numbers between 1 and $p-1$, so if half of them are quadratic residues, the other half must be nonresidues. □

Suppose that we take two quadratic residues and multiply them together. Do we get a QRor an NR, or do we sometimes get one and sometimes the other? For example, 3 and 10 are QRs modulo 13, and their product $3 \cdot 10 = 30 \equiv 4$ is again a QR modulo 13. Actually, this should have been clear without any computation, since if we multiply two squares, we should get a square. We can formally verify this in the following way. Suppose that a_1 and a_2 are both QRs modulo p. This means that there are numbers b_1 and b_2 so that $a_1 \equiv b_1^2 \pmod{p}$ and $a_2 \equiv b_2^2 \pmod{p}$. Multiplying these two congruences together, we find that $a_1 a_2 \equiv (b_1 b_2)^2 \pmod{p}$, which shows that $a_1 a_2$ is a QR.

The situation is much less clear if we multiply a QR by an NRor if we multiply two NRs together. Here are some examples using the data in our tables:

QR × NR ≡ ?? (mod p)	
$2 \times 5 \equiv 3 \pmod{7}$	NR
$5 \times 6 \equiv 8 \pmod{11}$	NR
$4 \times 5 \equiv 7 \pmod{13}$	NR
$10 \times 7 \equiv 5 \pmod{13}$	NR

NR × NR ≡ ?? (mod p)	
$3 \times 5 \equiv 1 \pmod{7}$	QR
$6 \times 7 \equiv 9 \pmod{11}$	QR
$5 \times 11 \equiv 3 \pmod{13}$	QR
$7 \times 11 \equiv 12 \pmod{13}$	QR

Thus, multiplying a quadratic residue and a nonresidue seems to yield a nonresidue, while the product of two nonresidues always seems to be a residue. Symbolically, we might write that

$$\text{QR} \times \text{QR} = \text{QR}, \qquad \text{QR} \times \text{NR} = \text{NR}, \qquad \text{NR} \times \text{NR} = \text{QR}.$$

We've already seen that the first relation is true. Before checking the other two, we make a brief digression to discuss the relationship between primitive roots and quadratic residues.

Let g be a primitive root for p. The Primitive Root Theorem (Chapter 21) assures us that there is at least one such primitive root. Then the powers of g,

$$g, g^2, g^3, \dots, g^{p-3}, g^{p-2}, g^{p-1},$$

give all the nonzero numbers modulo p. We know that half of them are residues and half are nonresidues. How can we tell which are which?

Certainly, g^2 is a QR, since it's obviously a square. Similarly, g^4 is a QR, since it equals $(g^2)^2$, and g^6 is a QR, since it equals $(g^3)^2$. Continuing in this fashion, we see that any even power of g, say g^{2k}, is a QR, since it equals $(g^k)^2$. So the even powers of g are certainly QRs.

In the list $g, g^2, \dots, g^{p-1}$, exactly half of the exponents are even and half of the exponents are odd. We observed that there are exactly $(p-1)/2$ quadratic residues modulo p, so the even powers of g must give all the quadratic residues. The remaining numbers, namely the odd powers of g, must then be the nonresidues.[2]

Here's another way to say the same thing. Recall that the index of a modulo p (for a primitive root g) is the power $I(a)$ with the property that $a \equiv g^{I(a)} \pmod{p}$. So the index of a is even if a is congruent to an even power of g, and the index of a is odd if a is congruent to an odd power of g.

The *Quadratic Residues* are those numbers a whose index $I(a)$ is even.	The *Nonresidues* are those numbers a whose index $I(a)$ is odd.

Using this description of residues and nonresidues, it is now a simple matter to verify the multiplication rules for quadratic residues.

Theorem 23.2 (Quadratic Residue Multiplication Rule). (Version 1) *Let p be an odd prime. Then:*

(i) *The product of two quadratic residues modulo p is a quadratic residue.*

(ii) *The product of a quadratic residue and a nonresidue is a nonresidue.*

(iii) *The product of two nonresidues is a quadratic residue.*

These three rules can by summarized symbolically by the formulas

$$QR \times QR = QR, \qquad QR \times NR = NR, \qquad NR \times NR = QR.$$

Verification. For any numbers a and b relatively prime to p! the Product Rule for Indices (Chapter 22) says that

$$I(ab) \equiv I(a) + I(b) \pmod{p-1}.$$

[2] "When you have eliminated all of the quadratic residues, the remaining numbers, no matter how improbable, must be the nonresidues!" (with apologies to Sherlock Holmes and Arthur Conan Doyle).

Notice that $p - 1$ is even, since we have assumed that p is an odd prime. So in particular it is true that

$$I(ab) \equiv I(a) + I(b) \pmod{2}.$$

[In other words, we know that $p-1$ divides $I(ab) - I(a) - I(b)$, and $p-1$ is even, so it is certainly true that 2 divides $I(ab) - I(a) - I(b)$.] Now we can consider the three cases for a and b.

(i) If a and b are both quadratic residues, then $I(a)$ and $I(b)$ are even, so

$$I(ab) \equiv I(a) + I(b) \equiv 0 + 0 \equiv 0 \pmod{2}.$$

Therefore, $I(ab)$ is even, so ab is a quadratic residue.

(ii) If a is a quadratic residue and b is a nonresidue, then $I(a)$ is even and $I(b)$ is odd, so

$$I(ab) \equiv I(a) + I(b) \equiv 0 + 1 \equiv 1 \pmod{2}.$$

Therefore, $I(ab)$ is odd, so ab is a nonresidue.

(iii) Finally, if a and b are both nonresidues, then $I(a)$ and $I(b)$ are odd, so

$$I(ab) \equiv I(a) + I(b) \equiv 1 + 1 \equiv 0 \pmod{2}.$$

Therefore, $I(ab)$ is even, so ab is a quadratic residue. □

This completes the verification of the quadratic residue multiplication rules. Now take a minute to stare at

$$\text{QR} \times \text{QR} = \text{QR}, \qquad \text{QR} \times \text{NR} = \text{NR}, \qquad \text{NR} \times \text{NR} = \text{QR}.$$

Do these rules remind you of anything? If not, here's a hint. Suppose that we try to replace the symbols QR and NR with numbers. What numbers would work? That's right, the symbol QR behaves like $+1$ and the symbol NR behaves like -1. Notice that the somewhat mysterious third rule, the one that says that the product of two nonresidues is a quadratic residue, reflects the equally mysterious rule $(-1) \times (-1) = +1$.[3]

Having observed that QRs behave like $+1$ and NRs behave like -1, Adrien-Marie Legendre introduced the following extremely useful notation.

The *Legendre symbol* of a modulo p is

$$\left(\frac{a}{p}\right) = \begin{cases} 1 & \text{if } a \text{ is a quadratic residue modulo } p, \\ -1 & \text{if } a \text{ is a nonresidue modulo } p. \end{cases}$$

[3]You may no longer consider the formula $(-1) \times (-1) = +1$ mysterious, since it's so familiar to you. But you should have found it mysterious the first time you saw it. And if you stop to think about it, there is no obvious reason why the product of two negative numbers should equal a positive number. Can you come up with a convincing argument that $(-1) \times (-1)$ must equal $+1$?

For example, using data from our earlier tables, we have

$$\left(\frac{3}{13}\right) = 1, \qquad \left(\frac{11}{13}\right) = -1, \qquad \left(\frac{2}{7}\right) = 1, \qquad \left(\frac{3}{7}\right) = -1.$$

Using the Legendre symbol, our quadratic residue multiplication rules can be given by a single formula.

Theorem 23.3 (Quadratic Residue Multiplication Rule). (Version 2) *Let p be an odd prime. Then*

$$\left(\frac{a}{p}\right)\left(\frac{b}{p}\right) = \left(\frac{ab}{p}\right).$$

The Legendre symbol is useful for making calculations. For example, suppose that we want to know if 75 is a square modulo 97. We can compute

$$\left(\frac{75}{97}\right) = \left(\frac{3 \cdot 5 \cdot 5}{97}\right) = \left(\frac{3}{97}\right)\left(\frac{5}{97}\right)\left(\frac{5}{97}\right) = \left(\frac{3}{97}\right).$$

Notice that it doesn't matter whether $\left(\frac{5}{97}\right)$ is $+1$ or -1, since it appears twice, and $(+1)^2 = (-1)^2 = 1$. Now we observe that $10^2 \equiv 3 \pmod{97}$, so 3 is a QR. Hence,

$$\left(\frac{75}{97}\right) = \left(\frac{3}{97}\right) = 1.$$

Of course, we were lucky in being able to recognize 3 as a QR modulo 97. Is there some way to evaluate a Legendre symbol like $\left(\frac{3}{97}\right)$ without relying on luck or trial and error? The answer is yes, but that's a topic for another chapter.

Exercises

23.1. Make a list of all the quadratic residues and all the nonresidues modulo 19.

23.2. 💻 Write a program that takes as input a prime p and produces as output the two numbers

$$A = \text{sum of all } 1 \le a < p \text{ such that } a \text{ is a quadratic residue modulo } p,$$
$$B = \text{sum of all } 1 \le a < p \text{ such that } a \text{ is a nonresidue modulo } p.$$

For example, if $p = 11$, then the quadratic residues are

$$1^2 \equiv 1 \pmod{11}, \qquad 2^2 \equiv 4 \pmod{11}, \qquad 3^2 \equiv 9 \pmod{11},$$
$$4^2 \equiv 5 \pmod{11}, \qquad 5^2 \equiv 3 \pmod{11},$$

so

$$A = 1 + 4 + 9 + 5 + 3 = 22 \qquad \text{and} \qquad B = 2 + 6 + 7 + 8 + 10 = 33.$$

(a) Make a list of A and B for all primes $p < 100$.
(b) What is the value of $A + B$? Prove that your guess is correct.
(c) Compute $A \bmod p$ and $B \bmod p$. Find a pattern and prove that it is correct.
(d) For which primes is it true that $A = B$? After reading Chapter 24, prove that your guess is correct.
(e) If $A \neq B$, which one tends to be larger, A or B? Try to prove that your guess is correct, but be forewarned that this is a very difficult problem.

23.3. A number a is called a *cubic residue modulo* p if it is congruent to a cube modulo p [that is, if there is a number b so that $a \equiv b^3 \pmod{p}$].
(a) Make a list of all the cubic residues modulo 5, modulo 7, modulo 11, and modulo 13.
(b) Find two numbers a_1 and b_1 so that neither a_1 nor b_1 is a cubic residue modulo 19, but $a_1 b_1$ is a cubic residue modulo 19. Similarly, find two numbers a_2 and b_2 so that none of the three numbers a_2, b_2, or $a_2 b_2$ is a cubic residue modulo 19.
(c) If $p \equiv 2 \pmod{3}$, make a conjecture as to which a's are cubic residues. Prove that your conjecture is correct.
(d) If $p \equiv 1 \pmod{3}$, show that a is a cubic residue modulo p exactly when its index $I(a)$ is divisible by 3.

Chapter 24

Is -1 a Square Modulo p? Is 2?

In the previous chapter we took various primes p and looked at the a's that were quadratic residues and the a's that were nonresidues. For example, we made a table of squares modulo 13 and used the table to see that 3 and 12 are QRs modulo 13, while 2 and 5 are NRs modulo 13.

In keeping with all the best traditions of mathematics, we now turn this problem on its head. Rather than taking a particular prime p and listing the a's that are QRs and NRs, we instead fix an a and ask for which primes p is a a QR. To make it clear exactly what we're asking, we start with the particular value $a = -1$. The question that we want to answer is as follows:

For which primes p is -1 a QR?

We can rephrase this question in other ways, such as "For which primes p does the congruence $x^2 \equiv -1 \pmod{p}$ have a solution?" and "For which primes p is $\left(\frac{-1}{p}\right) = 1$?"

As always, we need some data before we can make any hypotheses. We can answer our question for small primes in the usual mindless way by making a table of $1^2, 2^2, 3^2, \ldots \pmod{p}$ and checking if any of the numbers are congruent to -1 modulo p. So, for example, -1 is not a square modulo 3, since $1^2 \not\equiv -1 \pmod{3}$ and $2^2 \not\equiv -1 \pmod{3}$, while -1 is a square modulo 5, since $2^2 \equiv -1 \pmod{5}$. Here's a more extensive list.

p	3	5	7	11	13	17	19	23	29	31
Solution(s) to $x^2 \equiv -1 \pmod{p}$	NR	$2, 3$	NR	NR	$5, 8$	$4, 13$	NR	NR	$12, 17$	NR

Reading from this table, we compile the following data:

$$-1 \text{ is a quadratic residue for } p = 5, 13, 17, 29.$$
$$-1 \text{ is a nonresidue for } p = 3, 7, 11, 19, 23, 31.$$

It's not hard to discern the pattern. If p is congruent to 1 modulo 4, then -1 seems to be a quadratic residue modulo p, and if p is congruent to 3 modulo 4, then -1 seems to be a nonresidue. We can express this guess using Legendre symbols,

$$\left(\frac{-1}{p}\right) \stackrel{?}{=} \begin{cases} 1 & \text{if } p \equiv 1 \pmod 4, \\ -1 & \text{if } p \equiv 3 \pmod 4. \end{cases}$$

Let's check our conjecture on the next few cases. The next two primes, 37 and 41, are both congruent to 1 modulo 4 and, sure enough,

$$x^2 \equiv -1 \pmod{37} \text{ has the solutions } x \equiv 6 \text{ and } 31 \pmod{37}, \text{ and}$$
$$x^2 \equiv -1 \pmod{41} \text{ has the solutions } x \equiv 9 \text{ and } 32 \pmod{41}.$$

Similarly, the next two primes 43 and 47 are congruent to 3 modulo 4, and we check that -1 is a nonresidue for 43 and 47. Our guess is looking good!

The tool that we use to verify our conjecture might be called the "Square Root of Fermat's Little Theorem." How, you may well ask, does one take the square root of a theorem? Recall that Fermat's Little Theorem (Chapter 9) says

$$a^{p-1} \equiv 1 \pmod p.$$

We won't really be taking the square root of this theorem, of course. Instead, we take the square root of the quantity a^{p-1} and ask for its value. So we want to answer the following question:

Let $A = a^{(p-1)/2}$. What is the value of A modulo p?

One thing is obvious. If we square A, then Fermat's Little Theorem tells us that

$$A^2 = a^{p-1} \equiv 1 \pmod p.$$

Hence, p divides $A^2 - 1 = (A-1)(A+1)$, so either p divides $A - 1$ or p divides $A + 1$. (Notice how we are using the property of prime numbers proved on page 44.) Thus A must be congruent to either $+1$ or -1.

Here are a few random values of p, a, and A. For comparison purposes, we have also included the value of the Legendre symbol $\left(\frac{a}{p}\right)$. Do you see a pattern?

p	11	31	47	97	173	409	499	601	941	1223
a	3	7	10	15	33	78	33	57	222	129
$A \pmod{p}$	1	1	-1	-1	1	-1	1	-1	1	1
$\left(\frac{a}{p}\right)$	1	1	-1	-1	1	-1	1	-1	1	1

It certainly appears that $A \equiv 1 \pmod{p}$ when a is a quadratic residue and that $A \equiv -1 \pmod{p}$ when a is a nonresidue. In other words, it looks like $A \pmod{p}$ has the same value as the Legendre symbol $\left(\frac{a}{p}\right)$. We can use primitive roots to verify this assertion, which goes by the name of Euler's Criterion.

Theorem 24.1 (Euler's Criterion). *Let p be an odd prime. Then*

$$a^{(p-1)/2} \equiv \left(\frac{a}{p}\right) \pmod{p}.$$

Verification. Let g be a primitive root modulo p. Every number a is congruent to some power of g, and we also know that a is a quadratic residue precisely when it is congruent to an even power of g. Let's see what happens first when a is a quadratic residue and then when a is a nonresidue.

So suppose that a is a QR, which means that $\left(\frac{a}{p}\right) = 1$. The fact that a is a QR means that a is an even power of g, say $a \equiv g^{2k} \pmod{p}$. Now we can use Fermat's Little Theorem (Chapter 9) to compute

$$a^{(p-1)/2} \equiv (g^{2k})^{(p-1)/2} \equiv (g^{p-1})^k \equiv 1^k \equiv 1 \pmod{p}.$$

Thus, if a is a QR, then $a^{(p-1)/2} \equiv \left(\frac{a}{p}\right) \pmod{p}$.

Next we suppose that a is a NR, so $\left(\frac{a}{p}\right) = -1$. The fact that a is a NR means that a is an odd power of g, say $a \equiv g^{2k+1} \pmod{p}$. Again we use Fermat's Little Theorem to compute

$$a^{(p-1)/2} \equiv (g^{2k+1})^{(p-1)/2} \equiv (g^{p-1})^k \cdot g^{(p-1)/2} \equiv g^{(p-1)/2} \pmod{p}.$$

Of course, $g^{(p-1)/2}$ has to be congruent to either $+1$ or -1 modulo p. But g is a primitive root, so the smallest power of g that is congruent to $+1$ is the $(p-1)^{\text{st}}$ power. This means that $g^{(p-1)/2}$ must be congruent to -1 modulo p, which shows that

$$a^{(p-1)/2} \equiv -1 \pmod{p}.$$

This completes the verification that $a^{(p-1)/2} \equiv \left(\frac{a}{p}\right) \pmod{p}$ when a is a NR. □

Using Euler's Criterion, it is very easy to determine if -1 is a quadratic residue modulo p. For example, if we want to know whether -1 is a square modulo the prime $p = 6911$, we just need to compute

$$(-1)^{(6911-1)/2} = (-1)^{3455} = -1.$$

Euler's Criterion then tells us that

$$\left(\frac{-1}{6911}\right) \equiv -1 \pmod{6911}.$$

But $\left(\frac{a}{p}\right)$ is always either $+1$ or -1, so in this case we must have $\left(\frac{-1}{6911}\right) = -1$. Hence, -1 is a nonresidue modulo 6911.

Similarly, for the prime $p = 7817$ we find that

$$(-1)^{(7817-1)/2} = (-1)^{3908} = 1.$$

Hence, $\left(\frac{-1}{7817}\right) = 1$, so -1 is a quadratic residue modulo 7817. Observe that, although we now know that the congruence

$$x^2 \equiv -1 \pmod{7817}$$

is solvable, we still don't have any efficient way to find a solution. The solutions turn out to be $x \equiv 2564 \pmod{7817}$ and $x \equiv 5253 \pmod{7817}$.

As these two examples make clear, Euler's Criterion can be used to determine exactly which primes have -1 as a quadratic residue. The elegant result, which answers the initial question in the title of this chapter, is the first part of the Law of Quadratic Reciprocity.

Theorem 24.2 (Quadratic Reciprocity). (Part I) *Let p be an odd prime. Then*

$$\begin{aligned} -1 \text{ is a quadratic residue modulo } p \quad & \text{if } p \equiv 1 \pmod{4}, \text{ and} \\ -1 \text{ is a nonresidue modulo } p \quad & \text{if } p \equiv 3 \pmod{4}. \end{aligned}$$

In other words, using the Legendre symbol,

$$\left(\frac{-1}{p}\right) = \begin{cases} 1 & \text{if } p \equiv 1 \pmod{4}, \\ -1 & \text{if } p \equiv 3 \pmod{4}. \end{cases}$$

Verification. Euler's Criterion says that

$$(-1)^{(p-1)/2} \equiv \left(\frac{-1}{p}\right) \pmod{p}.$$

Suppose first that $p \equiv 1 \pmod{4}$, say $p = 4k + 1$. Then

$$(-1)^{(p-1)/2} = (-1)^{2k} = 1, \quad \text{so } 1 \equiv \left(\frac{-1}{p}\right) \pmod{p}.$$

But $\left(\frac{-1}{p}\right)$ is either $+1$ or -1, so it must equal 1. This proves that if $p \equiv 1 \pmod{4}$ then $\left(\frac{-1}{p}\right) = 1$.

Next we suppose that $p \equiv 3 \pmod{4}$, say $p = 4k + 3$. Then

$$(-1)^{(p-1)/2} = (-1)^{2k+1} = -1, \quad \text{so} \quad -1 \equiv \left(\frac{-1}{p}\right) \pmod{p}.$$

This shows that $\left(\frac{-1}{p}\right)$ must equal -1, which completes the verification of Quadratic Reciprocity (Part I). □

We can use the first part of quadratic reciprocity to answer a question left over from Chapter 12. As you may recall, we showed that there are infinitely many primes that are congruent to 3 modulo 4, but we left unanswered the analogous question for primes congruent to 1 modulo 4.

Theorem 24.3 (Primes 1 (Mod 4) Theorem). *There are infinitely many primes that are congruent to 1 modulo 4.*

Proof. Suppose we are given a list of primes $p_1, p_2, \ldots, p_r$, all of which are congruent to 1 modulo 4. We are going to find a new prime, not in our list, that is congruent to 1 modulo 4. Repeating this process gives a list of any desired length.

Consider the number

$$A = (2p_1p_2 \cdots p_r)^2 + 1.$$

We know that A can be factored into a product of primes, say

$$A = q_1q_2 \cdots q_s.$$

It is clear that $q_1, q_2, \ldots, q_s$ are not in our original list, since none of the p_i's divide A. So all we need to do is show that at least one of the q_i's is congruent to 1 modulo 4. In fact, we'll see that all of them are.

First we note that A is odd, so all the q_i's are odd. Next, each q_i divides A, so

$$(2p_1p_2 \cdots p_r)^2 + 1 = A \equiv 0 \pmod{q_i}.$$

This means that $x = 2p_1p_2 \cdots p_r$ is a solution to the congruence

$$x^2 \equiv -1 \pmod{q_i},$$

so -1 is a quadratic residue modulo q_i. Now Quadratic Reciprocity tells us that $q_i \equiv 1 \pmod{4}$. □

We can use the procedure described in this proof to produce a list of primes that are congruent to 1 modulo 4. Thus, if we start with $p_1 = 5$, then we form $A = (2p_1)^2 + 1 = 101$, so our second prime is $p_2 = 101$. Then

$$A = (2p_1p_2)^2 + 1 = 1020101,$$

which is again prime, so our third prime is $p_3 = 1020101$. We'll go one more step,

$$\begin{aligned} A &= (2p_1p_2p_3)^2 + 1 \\ &= 1061522231810040101 \\ &= 53 \cdot 1613 \cdot 12417062216309. \end{aligned}$$

Notice that all the primes 53, 1613, and 12417062216309 are congruent to 1 modulo 4, just as predicted by the theory.

Having successfully answered the first question in the title of this chapter, we move on to the second question and consider $a = 2$, that "oddest" of all primes. Just as we did with $a = -1$, we are looking for some simple characterization for the primes p such that 2 is a quadratic residue modulo p. Can you find the pattern in the following data, where the line labeled $x^2 \equiv 2$ gives the solutions to $x^2 \equiv 2 \pmod{p}$ if 2 is a quadratic residue modulo p and is marked NR if 2 is a nonresidue?

p	3	5	7	11	13	17	19	23	29	31
$x^2 \equiv 2$	NR	NR	$3, 4$	NR	NR	$6, 11$	NR	$5, 18$	NR	$8, 23$

p	37	41	43	47	53	59	61	67	71	73
$x^2 \equiv 2$	NR	$17, 24$	NR	$7, 40$	NR	NR	NR	NR	$12, 59$	$32, 41$

p	79	83	89	97	101	103	107	109	113	127
$x^2 \equiv 2$	$9, 70$	NR	$25, 64$	$14, 83$	NR	$38, 65$	NR	NR	$51, 62$	$16, 111$

Here's the list of primes separated according to whether 2 is a residue or a nonresidue.

$$\begin{aligned} &2 \text{ is a quadratic residue for } p = 7, 17, 23, 31, 41, 47, 71, 73, \\ &\qquad\qquad 79, 89, 97, 103, 113, 127 \\ &2 \text{ is a nonresidue for } p = 3, 5, 11, 13, 19, 29, 37, 43, 53, 59, \\ &\qquad\qquad 61, 67, 83, 101, 107, 109 \end{aligned}$$

For $a = -1$, it turned out that the congruence class of p modulo 4 was crucial. Is there a similar pattern if we reduce these two lists of primes modulo 4? Here's what happens if we do.

$$\begin{aligned} &7,17,23,31,41,47,71,73,79,89,97,103,113,127 \\ &\qquad\qquad \equiv 3,1,3,3,1,3,3,1,3,1,1,3,1,3 \pmod 4, \\ &3,5,11,13,19,29,37,43,53,59,61,67,83,101,107,109 \\ &\qquad\qquad \equiv 3,1,3,1,3,1,1,3,1,3,1,3,3,1,3,1 \pmod 4. \end{aligned}$$

This doesn't look too promising. Maybe we should try reducing modulo 3.

$$\begin{aligned} &7,17,23,31,41,47,71,73,79,89,97,103,113,127 \\ &\qquad\qquad \equiv 1,2,2,1,2,2,2,1,1,2,1,1,2,1 \pmod 3 \\ &3,5,11,13,19,29,37,43,53,59,61,67,83,101,107,109 \\ &\qquad\qquad \equiv 0,2,2,1,1,2,1,1,2,2,1,1,2,2,2,1 \pmod 3. \end{aligned}$$

This doesn't look any better. Let's make one more attempt before we give up. What happens if we reduce modulo 8?

$$\begin{aligned} &7,17,23,31,41,47,71,73,79,89,97,103,113,127 \\ &\qquad\qquad \equiv 7,1,7,7,1,7,7,1,7,1,1,7,1,7 \pmod 8 \\ &3,5,11,13,19,29,37,43,53,59,61,67,83,101,107,109 \\ &\qquad\qquad \equiv 3,5,3,5,3,5,5,3,5,3,5,3,3,5,3,5 \pmod 8. \end{aligned}$$

Eureka! It surely can't be a coincidence that the first line is all 1's and 7's and the second line is all 3's and 5's. This suggests the general rule that 2 is a quadratic residue modulo p if p is congruent to 1 or 7 modulo 8 and that 2 is a nonresidue if p is congruent to 3 or 5 modulo 8. In terms of Legendre symbols, we would write

$$\left(\frac{2}{p}\right) \stackrel{?}{=} \begin{cases} 1 & \text{if } p \equiv 1 \text{ or } 7 \pmod 8, \\ -1 & \text{if } p \equiv 3 \text{ or } 5 \pmod 8. \end{cases}$$

Can we use Euler's Criterion to verify our guess? Unfortunately, the answer is no, or at least not in any obvious way, since there doesn't seem to be an easy method to calculate $2^{(p-1)/2} \pmod p$. However, if you go back and examine our proof of Fermat's Little Theorem in Chapter 9, you'll see that we took the numbers $1, 2, \ldots, p-1$, multiplied each one by a, and then multiplied them all together. This gave us a factor of a^{p-1} to pull out. In order to use Euler's Criterion, we only want $\frac{1}{2}(p-1)$ factors of a to pull out, so rather than starting with all the numbers from 1 to p, we just take the numbers from 1 to $\frac{1}{2}(p-1)$. We illustrate this idea, which is due to Gauss, to determine if 2 is a quadratic residue modulo 13.

We begin with half the numbers from 1 to 12: $1, 2, 3, 4, 5, 6$. If we multiply each by 2 and then multiply them together, we get

$$\begin{aligned} 2 \cdot 4 \cdot 6 \cdot 8 \cdot 10 \cdot 12 &= (2 \cdot 1)(2 \cdot 2)(2 \cdot 3)(2 \cdot 4)(2 \cdot 5)(2 \cdot 6) \\ &= 2^6 \cdot 1 \cdot 2 \cdot 3 \cdot 4 \cdot 5 \cdot 6 \\ &= 2^6 \cdot 6!. \end{aligned}$$

Notice the factor of $2^6 = 2^{(13-1)/2}$, which is the number we're really interested in.

Gauss's idea is to take the numbers $2, 4, 6, 8, 10, 12$ and reduce each of them modulo 13 to get a number lying between -6 and 6. The first three stay the same, but we need to subtract 13 from the last three to get them into this range. Thus,

$$\begin{array}{lll} 2 \equiv 2 \pmod{13} & 4 \equiv 4 \pmod{13} & 6 \equiv 6 \pmod{13} \\ 8 \equiv -5 \pmod{13} & 10 \equiv -3 \pmod{13} & 12 \equiv -1 \pmod{13}. \end{array}$$

Multiplying these numbers together, we find that

$$\begin{aligned} 2 \cdot 4 \cdot 6 \cdot 8 \cdot 10 \cdot 12 &\equiv 2 \cdot 4 \cdot 6 \cdot (-5) \cdot (-3) \cdot (-1) \\ &\equiv (-1)^3 \cdot 2 \cdot 4 \cdot 6 \cdot 5 \cdot 3 \cdot 1 \\ &\equiv -6! \pmod{13}. \end{aligned}$$

Equating these two values of $2 \cdot 4 \cdot 6 \cdot 8 \cdot 10 \cdot 12 \pmod{13}$, we see that

$$2^6 \cdot 6! \equiv -6! \pmod{13}.$$

This implies that $2^6 \equiv -1 \pmod{13}$, so Euler's Criterion tells us that 2 is a non-residue modulo 13.

Let's briefly use the same ideas to check if 2 is a quadratic residue modulo 17. We take the numbers from 1 to 8, multiply each by 2, multiply them together, and calculate the product in two different ways. The first way gives

$$2 \cdot 4 \cdot 6 \cdot 8 \cdot 10 \cdot 12 \cdot 14 \cdot 16 = 2^8 \cdot 8!.$$

For the second way, we reduce modulo 17 to bring the numbers into the range from -8 to 8. Thus,

$$\begin{array}{lll} 2 \equiv 2 \pmod{17} & 4 \equiv 4 \pmod{17} & 6 \equiv 6 \pmod{17} \\ 8 \equiv 8 \pmod{17} & 10 \equiv -7 \pmod{17} & 12 \equiv -5 \pmod{17} \\ 14 \equiv -3 \pmod{17} & 16 \equiv -1 \pmod{17}. & \end{array}$$

Multiplying these together gives

$$\begin{aligned} 2 \cdot 4 \cdot 6 \cdot 8 \cdot 10 \cdot 12 \cdot 14 \cdot 16 &\equiv 2 \cdot 4 \cdot 6 \cdot 8 \cdot (-7) \cdot (-5) \cdot (-3) \cdot (-1) \\ &\equiv (-1)^4 \cdot 8! \pmod{17}. \end{aligned}$$

Therefore, $2^8 \cdot 8! \equiv (-1)^4 \cdot 8! \pmod{17}$, so $2^8 \equiv 1 \pmod{17}$, and hence 2 is a quadratic residue modulo 17.

Now let's think about Gauss's method a little more generally. Let p be any odd prime. To make our formulas simpler, we let

$$P = \frac{p-1}{2}.$$

We start with the even numbers $2, 4, 6, \ldots, p-1$. Multiplying them together and factoring out a 2 from each number gives

$$2 \cdot 4 \cdot 6 \cdots (p-1) = 2^{(p-1)/2} \cdot 1 \cdot 2 \cdot 3 \cdots \frac{p-1}{2} = 2^P \cdot P!.$$

The next step is to take the list $2, 4, 6, \ldots, p-1$ and reduce each number modulo p so that it lies in the range from $-P$ to P, that is, between $-(p-1)/2$ and $(p-1)/2$. The first few numbers won't change, but at some point in the list we'll start hitting numbers that are larger than $(p-1)/2$, and each of these large numbers needs to have p subtracted from it. Notice that the number of minus signs introduced is exactly the number of times we need to subtract p. In other words,

$$\text{Number of minus signs} = \begin{pmatrix} \text{Number of integers in the list} \\ 2, 4, 6, \ldots, (p-1) \\ \text{that are larger than } \frac{1}{2}(p-1) \end{pmatrix}.$$

The following illustration may help to explain this procedure.

$$\underbrace{2 \cdot 4 \cdot 6 \cdot 8 \cdot 10 \cdot 12 \cdots}_{\substack{\text{Numbers } \le (p-1)/2 \\ \text{are left unchanged.}}} \;\Big|\; \underbrace{\cdots (p-5) \cdot (p-3) \cdot (p-1)}_{\substack{\text{Numbers } > (p-1)/2. \\ \text{Need to substract } p \text{ from each.}}}$$

Comparing the two products, we get

$$2^P \cdot P! = 2 \cdot 4 \cdot 6 \cdots (p-1) \equiv (-1)^{(\text{Number of minus signs})} \cdot P! \pmod{p},$$

so canceling $P!$ from each side gives the fundamental formula

$$2^{(p-1)/2} \equiv (-1)^{(\text{Number of minus signs})} \pmod{p}.$$

Using this formula, it is easy to verify our earlier guess, thereby answering the second question in the chapter title.

Theorem 24.4 (Quadratic Reciprocity). (Part II) *Let p be an odd prime. Then 2 is a quadratic residue modulo p if p is congruent to 1 or 7 modulo 8, and 2 is a* *nonresidue modulo p if p is congruent to 3 or 5 modulo 8. In terms of the Legendre symbol,*

$$\left(\frac{2}{p}\right) = \begin{cases} 1 & \textit{if } p \equiv 1 \textit{ or } 7 \pmod 8, \\ -1 & \textit{if } p \equiv 3 \textit{ or } 5 \pmod 8. \end{cases}$$

Verification. There are actually four cases to consider, depending on the value of $p \pmod 8$. We do two of them and leave the other two for you.

We start with the case that $p \equiv 3 \pmod 8$, say $p = 8k + 3$. We need to list the numbers $2, 4, \ldots, p-1$ and determine how many of them are larger than $\frac{1}{2}(p-1)$. In this case, $p - 1 = 8k + 2$ and $\frac{1}{2}(p-1) = 4k + 1$, so the cutoff is as indicated in the following diagram:

$$2 \cdot 4 \cdot 6 \cdots 4k \quad \Big| \quad (4k+2) \cdot (4k+4) \cdots (8k+2).$$

We need to count how many numbers there are to the right of the vertical bar. In other words, how many even numbers are there between $4k + 2$ and $8k + 2$? The answer is $2k + 1$. (If this isn't clear to you, try a few values for k and you'll see why it's correct.) This shows that there are $2k + 1$ minus signs, so the fundamental formula given above tells us that

$$2^{(p-1)/2} \equiv (-1)^{2k+1} \equiv -1 \pmod p.$$

Now Euler's Criterion says that 2 is a nonresidue, so we have proved that 2 is a nonresidue for any prime p that is congruent to 3 modulo 8.

Next let's look at the primes that are congruent to 7 modulo 8, say $p = 8k + 7$. Now the even numbers $2, 4, \ldots, p - 1$ are the numbers from 2 to $8k + 6$, and the midpoint is $\frac{1}{2}(p-1) = 4k + 3$. The cutoff in this case is

$$2 \cdot 4 \cdot 6 \cdots (4k+2) \quad \Big| \quad (4k+4) \cdot (4k+6) \cdots (8k+6).$$

There are exactly $2k + 2$ numbers to the right of the vertical bar, so we get $2k + 2$ minus signs. This yields

$$2^{(p-1)/2} \equiv (-1)^{2k+2} \equiv 1 \pmod p,$$

so Euler's criterion tells us that 2 is a quadratic residue. This proves that 2 is a quadratic residue for any prime p that is congruent to 7 modulo 8. □

Exercises

24.1. Determine whether each of the following congruences has a solution. (All of the moduli are primes.)

(a) $x^2 \equiv -1 \pmod{5987}$ (c) $x^2 + 14x - 35 \equiv 0 \pmod{337}$
(b) $x^2 \equiv 6780 \pmod{6781}$ (d) $x^2 - 64x + 943 \equiv 0 \pmod{3011}$

[*Hint*. For (c), use the quadratic formula to find out what number you need to take the square root of modulo 337, and similarly for (d).]

24.2. Use the procedure described in the Primes 1 (Mod 4) Theorem to generate a list of primes congruent to 1 modulo 4, starting with the seed $p_1 = 17$.

24.3. Here is a list of the first few primes for which 3 is a quadratic residue and a nonresidue.

Quadratic Residue: $p = 11, 13, 23, 37, 47, 59, 61, 71, 73, 83, 97, 107, 109$
Nonresidue: $p = 5, 7, 17, 19, 29, 31, 41, 43, 53, 67, 79, 89, 101, 103, 113, 127$

Try reducing this list modulo m for various m's until you find a pattern, and make a conjecture explaining which primes have 3 as a quadratic residue.

24.4. Finish the verification of Quadratic Reciprocity Part II for the other two cases: primes congruent to 1 modulo 8 and primes congruent to 5 modulo 8.

24.5. Use the same ideas we used to verify Quadratic Reciprocity (Part II) to verify the following two assertions.

(a) If p is congruent to 1 modulo 5, then 5 is a quadratic residue modulo p.
(b) If p is congruent to 2 modulo 5, then 5 is a nonresidue modulo p.

[*Hint*. Reduce the numbers $5, 10, 15, \ldots, \frac{5}{2}(p-1)$ so that they lie in the range from $-\frac{1}{2}(p-1)$ to $\frac{1}{2}(p-1)$ and check how many of them are negative.]

24.6. Suppose that q is a prime number that is congruent to 1 modulo 4, and suppose that the number $p = 2q + 1$ is also a prime number. (For example, q could equal 5 and p equal 11.) Show that 2 is a primitive root modulo p.

Chapter 25

Quadratic Reciprocity

Our current quest is to determine, for a given number a, exactly which primes p have a as a quadratic residue. In the previous chapter we solved this problem for $a = -1$ and $a = 2$. In both cases we found that we could determine whether a is a quadratic residue modulo p by looking at p modulo m for some small m, more specifically for $m = 4$ or $m = 8$.

Now we want to tackle the question of the Legendre symbol $\left(\frac{a}{p}\right)$ for other values of a. For example, suppose we want to compute $\left(\frac{70}{p}\right)$. We can use the Quadratic Residue Multiplication Rules (Chapter 23) to compute

$$\left(\frac{70}{p}\right) = \left(\frac{2 \cdot 5 \cdot 7}{p}\right) = \left(\frac{2}{p}\right)\left(\frac{5}{p}\right)\left(\frac{7}{p}\right).$$

We already know how to find $\left(\frac{2}{p}\right)$, so we're left with the problem of determining $\left(\frac{5}{p}\right)$ and $\left(\frac{7}{p}\right)$.

In general, if we want to compute $\left(\frac{a}{p}\right)$ for any number a, we can start by factoring a into a product of primes, say

$$a = q_1 q_2 \cdots q_r.$$

(It's okay if some of the q_i's are the same.) Then the Quadratic Residue Multiplication Rules give

$$\left(\frac{a}{p}\right) = \left(\frac{q_1}{p}\right)\left(\frac{q_2}{p}\right)\cdots\left(\frac{q_r}{p}\right).$$

The moral of this story: If we know how to compute $\left(\frac{q}{p}\right)$ for primes q, then we know how to compute $\left(\frac{a}{p}\right)$ for every a.[1] Since nothing we have done so far tells us

[1] Yet another instance of the principle that primes are the basic building blocks of number theory, so if you can solve a problem for primes, you're usually well on your way to solving it for all numbers.

anything about $\left(\frac{q}{p}\right)$ (for fixed q and varying p), the time has come[2] to compile some data and use it to make some conjectures. The following table gives the value of the Legendre symbol $\left(\frac{q}{p}\right)$ for all odd primes $p, q \le 37$.

$p \backslash q$	3	5	7	11	13	17	19	23	29	31	37
3		−1	1	−1	1	−1	1	−1	−1	1	1
5	−1		−1	1	−1	−1	1	−1	1	1	−1
7	−1	−1		1	−1	−1	−1	1	1	−1	1
11	1	1	−1		−1	−1	−1	1	−1	1	1
13	1	−1	−1	−1		1	−1	1	1	−1	−1
17	−1	−1	−1	−1	1		1	−1	−1	−1	−1
19	−1	1	1	1	−1	1		1	−1	−1	−1
23	1	−1	−1	−1	1	−1	−1		1	1	−1
29	−1	1	1	−1	1	−1	−1	1		−1	−1
31	−1	1	1	−1	−1	−1	1	−1	−1		−1
37	1	−1	1	1	−1	−1	−1	−1	−1	−1	

The Value of the Legendre Symbol $\left(\frac{q}{p}\right)$

Before reading further, you should take some time to study this table and try to find some patterns. Don't worry if you don't immediately discover the answer; the most important pattern concealed in this table is somewhat subtle. But you will find that it is well worth the effort to uncover the design on your own, since you then share the thrill of discovery with Legendre and Gauss.

Now that you've formulated your own conjectures, we'll examine the table together. We are going to compare the rows with the columns or, what amounts to the same thing, we are going to compare the entries when we reflect across the diagonal of the table. For example, the row with $p = 5$ reads

q	3	5	7	11	13	17	19	23	29	31	37
$\left(\frac{q}{5}\right)$	−1		−1	1	−1	−1	1	−1	1	1	−1

[2]"The time has come," the Walrus said, "to talk of many things, of shoes, and primes, and residues, and cabbages and kings."

Similarly, the column with $q = 5$ (turned sideways to save space) is

p	3	5	7	11	13	17	19	23	29	31	37
$\left(\frac{5}{p}\right)$	-1		-1	1	-1	-1	1	-1	1	1	-1

They match! So we might guess that

$$\left(\frac{5}{p}\right) = \left(\frac{p}{5}\right)$$

for all primes p. Do you see how useful a rule like this would be? We are looking for a method to calculate the Legendre symbol $\left(\frac{5}{p}\right)$, a difficult problem, but the Legendre symbol $\left(\frac{p}{5}\right)$ is easy to compute, because it only depends on p modulo 5. In other words, we know that

$$\left(\frac{p}{5}\right) = \begin{cases} 1 & \text{if } p \equiv 1 \text{ or } 4 \pmod{5}, \\ -1 & \text{if } p \equiv 2 \text{ or } 3 \pmod{5}. \end{cases}$$

So if our guess that $\left(\frac{5}{p}\right) = \left(\frac{p}{5}\right)$ is correct, then we would know, for example, that 5 is a nonresidue modulo 3593, since

$$\left(\frac{5}{3593}\right) = \left(\frac{3593}{5}\right) = \left(\frac{3}{5}\right) = -1.$$

Similarly,

$$\left(\frac{5}{3889}\right) = \left(\frac{3889}{5}\right) = \left(\frac{4}{5}\right) = 1,$$

so 5 should be a quadratic residue modulo 3889, and sure enough we find that $5 \equiv 2901^2 \pmod{3889}$.

Emboldened by this success, we might guess that

$$\left(\frac{q}{p}\right) = \left(\frac{p}{q}\right)$$

for all primes p and q. Unfortunately, this isn't even true for the first row and column of the table. For example,

$$\left(\frac{3}{7}\right) = -1 \qquad \text{and} \qquad \left(\frac{7}{3}\right) = 1.$$

So sometimes $\left(\frac{q}{p}\right)$ is equal to $\left(\frac{p}{q}\right)$, and sometimes it is equal to $-\left(\frac{p}{q}\right)$. The following table will help us find a rule explaining when they are the same and when they are opposites.

$p \backslash q$	3	5	7	11	13	17	19	23	29	31	37
3		♡	★	★	♡	♡	★	★	♡	★	♡
5	♡		♡	♡	♡	♡	♡	♡	♡	♡	♡
7	★	♡		★	♡	♡	★	★	♡	★	♡
11	★	♡	★		♡	♡	★	★	♡	★	♡
13	♡	♡	♡	♡		♡	♡	♡	♡	♡	♡
17	♡	♡	♡	♡	♡		♡	♡	♡	♡	♡
19	★	♡	★	★	♡	♡		★	♡	★	♡
23	★	♡	★	★	♡	♡	★		♡	★	♡
29	♡	♡	♡	♡	♡	♡	♡	♡		♡	♡
31	★	♡	★	★	♡	♡	★	★	♡		♡
37	♡	♡	♡	♡	♡	♡	♡	♡	♡	♡	

Table with ♡ if $\left(\frac{q}{p}\right) = \left(\frac{p}{q}\right)$ and ★ if $\left(\frac{q}{p}\right) = -\left(\frac{p}{q}\right)$

Looking at this table, we can pick out the primes that have ♡-filled rows and columns:

$$p = 5, 13, 17, 29, 37.$$

The primes whose rows and columns are not exactly the same (i.e., the rows and columns containing ★'s) are

$$p = 3, 7, 11, 19, 23, 31.$$

With our previous experience, there is no mystery about these lists; the former consists of the primes that are congruent to 1 modulo 4, and the latter contains the primes that are congruent to 3 modulo 4.

So our first conjecture might be that if $p \equiv 1 \pmod 4$ or if $q \equiv 1 \pmod 4$ then the rows and columns are the same. We can write this in terms of Legendre symbols.

Conjecture: If $p \equiv 1 \pmod 4$ or $q \equiv 1 \pmod 4$, then $\left(\dfrac{q}{p}\right) = \left(\dfrac{p}{q}\right)$.

What happens if both p and q are congruent to 3 modulo 4? Looking at the table, we find in every instance that $\left(\frac{q}{p}\right)$ and $\left(\frac{p}{q}\right)$ are opposites. So we are led to make a further guess.

Conjecture: If $p \equiv 3 \pmod 4$ and $q \equiv 3 \pmod 4$, then $\left(\frac{q}{p}\right) = -\left(\frac{p}{q}\right)$.

These two conjectural relations form the heart of the Law of Quadratic Reciprocity.

Theorem 25.1 (Law of Quadratic Reciprocity). *Let p and q be distinct odd primes.*

$$\left(\frac{-1}{p}\right) = \begin{cases} 1 & \textit{if } p \equiv 1 \pmod 4 \\ -1 & \textit{if } p \equiv 3 \pmod 4 \end{cases}$$

$$\left(\frac{2}{p}\right) = \begin{cases} 1 & \textit{if } p \equiv 1 \textit{ or } 7 \pmod 8 \\ -1 & \textit{if } p \equiv 3 \textit{ or } 5 \pmod 8 \end{cases}$$

$$\left(\frac{q}{p}\right) = \begin{cases} \left(\frac{p}{q}\right) & \textit{if } p \equiv 1 \pmod 4 \textit{ or } q \equiv 1 \pmod 4 \\ -\left(\frac{p}{q}\right) & \textit{if } p \equiv 3 \pmod 4 \textit{ and } q \equiv 3 \pmod 4 \end{cases}$$

We are content with having proved the Law of Quadratic Reciprocity for $\left(\frac{-1}{p}\right)$ and $\left(\frac{2}{p}\right)$ and do not give the general proof for $\left(\frac{q}{p}\right)$. There are literally dozens of different proofs of Quadratic Reciprocity, including one that is very similar to our proof for $\left(\frac{2}{p}\right)$ in the previous chapter.[3] Euler and Lagrange were the first to formulate the Law of Quadratic Reciprocity, but it remained for Gauss to give the first proof in his famous monograph *Disquisitiones arithmeticae* in 1801. Gauss discovered the law for himself when he was 19, and during his lifetime he found seven different proofs! Mathematicians during the nineteenth century subsequently formulated and proved Cubic and Quartic Reciprocity Laws, and these in turn were subsumed into the Class Field Theory developed by David Hilbert, Emil Artin, and others from the 1890s through the 1920s and 1930s. During the 1960s and 1970s a number of mathematicians formulated a series of conjectures that vastly generalize Class Field Theory and that today go by the name of the Langlands Program. The fundamental theorem proved by Andrew Wiles in 1995 is a small piece of the Langlands Program, yet it sufficed to solve Fermat's 350-year-old "Last Theorem."

Karl Friedrich Gauss (1777–1855) Karl Friedrich Gauss was one of the greatest mathematicians of all time, and arguably the finest number theorist to have ever lived. As a child, he was a mathematical prodigy whose feats impressed his family, friends, and teachers, and his mathematical talents only

[3]You can find this proof in Chapter III of Davenport's wonderful book *The Higher Arithmetic*, Cambridge University Press, 1952 (7th edition, 1999).

grew as he matured. His most influential work in number theory was published in 1801 under the title of *Disquisitiones arithmeticae*. It contains, among other things, the theory of quadratic reciprocity and the representation of numbers by binary forms. Much of the material in Gauss's *Disquisitiones* was far ahead of its time and, as such, furnished paths for number theorists to follow during the subsequent century and a half. In addition to his work in number theory, Gauss made fundamental contributions to many other areas of mathematics, including geometry and differential equations. He also made many discoveries in physics and astronomy, including a method for computing orbits that he used to compute the position of the newly discovered asteroid Ceres in 1801. He published major papers in areas as diverse as crystallography, optics, and the physics of fluids, and he invented an electromagnetic telegraph with Wilhelm Weber in 1833. He published 155 titles during his lifetime, but his life's work was so prodigious that his *Collected Works* appeared during the period 1863 to 1933.

The law of quadratic reciprocity is not only a beautiful and subtle theoretical statement about numbers, it is also a practical tool for determining whether a number is a quadratic residue. Essentially, it lets us flip the Legendre symbol $\left(\frac{q}{p}\right)$ and replace it by $\pm\left(\frac{p}{q}\right)$. Then we can reduce p modulo q and repeat the process. This leads to Legendre symbols with smaller and smaller entries, so eventually we arrive at Legendre symbols that we can compute. Here's an example with detailed justification for each step.

$$\begin{aligned}
\left(\frac{14}{137}\right) &= \left(\frac{2}{137}\right)\left(\frac{7}{137}\right) && \text{Quadratic Residue Multiplication Rule,} \\
&= \left(\frac{7}{137}\right) && \text{Quadratic Reciprocity says } \left(\frac{2}{137}\right) = 1, \\
& && \text{since } 137 \equiv 1 \pmod{8}, \\
&= \left(\frac{137}{7}\right) && \text{Quadratic Reciprocity and } 137 \equiv 1 \pmod{4}, \\
&= \left(\frac{4}{7}\right) && \text{reducing 137 modulo 7,} \\
&= 1 && \text{since } 4 = 2^2 \text{ is certainly a square.}
\end{aligned}$$

Thus, 14 is a quadratic residue modulo 137. In fact, the solutions to the congruence $x^2 \equiv 14 \pmod{137}$ are $x \equiv 39 \pmod{137}$ and $x \equiv 98 \pmod{137}$.

Here's a second example that illustrates how the sign can change back and forth a number of times.

$$\begin{aligned}
\left(\frac{55}{179}\right) &= \left(\frac{5}{179}\right)\left(\frac{11}{179}\right) \\
&= \left(\frac{179}{5}\right) \times (-1) \times \left(\frac{179}{11}\right) && \text{since } 5 \equiv 1 \pmod 4 \text{ and} \\
& && \quad 11 \equiv 179 \equiv 3 \pmod 4, \\
&= \left(\frac{4}{5}\right) \times (-1) \times \left(\frac{3}{11}\right) && \text{since } 179 \equiv 4 \pmod 5 \text{ and} \\
& && \quad 179 \equiv 3 \pmod{11} \\
&= 1 \times (-1) \times \left(\frac{3}{11}\right) && \text{since } 4 = 2^2 \text{ is a square} \\
&= 1 \times (-1) \times (-1) \times \left(\frac{11}{3}\right) && \text{since } 3 \equiv 11 \equiv 3 \pmod 4, \\
&= 1 \times (-1) \times (-1) \times \left(\frac{2}{3}\right) && \text{since } 11 \equiv 2 \pmod 3, \\
&= 1 \times (-1) \times (-1) \times (-1) && \text{since 2 is a nonresidue mod 3,} \\
&= -1.
\end{aligned}$$

So 55 is a nonresidue modulo 179.

There is often more than one way to use quadratic reciprocity to evaluate a given Legendre symbol $\left(\frac{a}{p}\right)$, for example, by using the equality $\left(\frac{p}{q}\right) = \left(\frac{p-q}{q}\right)$. Thus we can compute $\left(\frac{299}{397}\right)$ as

$$\begin{aligned}
\left(\frac{299}{397}\right) &= \left(\frac{13}{397}\right)\left(\frac{23}{397}\right) = \left(\frac{397}{13}\right)\left(\frac{397}{23}\right) = \left(\frac{7}{13}\right)\left(\frac{6}{23}\right) \\
&= \left(\frac{13}{7}\right)\left(\frac{2}{23}\right)\left(\frac{3}{23}\right) = \left(\frac{-1}{7}\right) \times 1 \times -\left(\frac{23}{3}\right) = -1 \times -\left(\frac{2}{3}\right) = -1,
\end{aligned}$$

or we can compute it as

$$\left(\frac{299}{397}\right) = \left(\frac{-98}{397}\right) = \left(\frac{-1}{397}\right)\left(\frac{2}{397}\right)\left(\frac{7}{397}\right)^2 = 1 \times (-1) \times (\pm 1)^2 = -1.$$

Of course, regardless of the path taken, the final destination is always the same.

The Law of Quadratic Reciprocity furnishes an extremely efficient way to compute the Legendre symbol $\left(\frac{a}{p}\right)$, even for very large values of a and p. In fact, the number of steps to compute $\left(\frac{a}{p}\right)$ is more or less equal to the number of digits in p, so it is possible to evaluate Legendre symbols for numbers with hundreds of digits.

We won't spend the time to do an example that is that large, but we are content with the following more modest example.

$$\begin{aligned}\left(\frac{37603}{48611}\right) &= \left(\frac{31}{48611}\right)\left(\frac{1213}{48611}\right) = -\left(\frac{48611}{31}\right)\left(\frac{48611}{1213}\right)\\ &= -\left(\frac{3}{31}\right)\left(\frac{91}{1213}\right) = \left(\frac{31}{3}\right)\left(\frac{7}{1213}\right)\left(\frac{13}{1213}\right)\\ &= \left(\frac{1}{3}\right)\left(\frac{1213}{7}\right)\left(\frac{1213}{13}\right) = \left(\frac{2}{7}\right)\left(\frac{4}{13}\right) = 1\end{aligned}$$

Hence, 37603 is a quadratic residue modulo 48611.

The hardest part of computing $\left(\frac{a}{p}\right)$ lies not in the use of the Law of Quadratic Reciprocity, but rather in the necessity of factoring the number a before applying the law. Thus, in our example, it takes some work to recognize that 37603 factors as $31 \cdot 1213$, and if a has hundreds of digits, it may be virtually impossible to factor a. Surprisingly, it is possible to evaluate $\left(\frac{a}{p}\right)$ without doing any difficult factorizations. The idea is to use the Law of Quadratic Reciprocity to flip the Legendre symbol $\left(\frac{a}{p}\right)$ for any positive odd value of a, completely ignoring the question of whether a is prime. As usual, if both a and p are congruent to 3 modulo 4, then you must put in a minus sign. More generally, we can assign a value to the Legendre symbol $\left(\frac{a}{b}\right)$ for any integers a and b provided that b is positive and odd. (This generalized Legendre symbol is often called a Jacobi symbol.) We can evaluate the Legendre or Jacobi symbol by repeatedly applying the following generalized law of quadratic reciprocity.

Theorem 25.2 (Generalized Law of Quadratic Reciprocity). *Let a and b be odd numbers.*

$$\left(\frac{-1}{b}\right) = \begin{cases} 1 & \text{if } b \equiv 1 \pmod{4}, \\ -1 & \text{if } b \equiv 3 \pmod{4} \end{cases}$$

$$\left(\frac{2}{b}\right) = \begin{cases} 1 & \text{if } b \equiv 1 \text{ or } 7 \pmod{8}, \\ -1 & \text{if } b \equiv 3 \text{ or } 5 \pmod{8} \end{cases}$$

$$\left(\frac{a}{b}\right) = \begin{cases} \left(\frac{b}{a}\right) & \text{if } a \equiv 1 \pmod{4} \text{ or } b \equiv 1 \pmod{4}, \\ -\left(\frac{b}{a}\right) & \text{if } a \equiv b \equiv 3 \pmod{4} \end{cases}$$

Amazingly enough, if you use these rules, the multiplication formula $\left(\frac{a_1 a_2}{b}\right) = \left(\frac{a_1}{b}\right)\left(\frac{a_2}{b}\right)$, and the fact that $\left(\frac{a}{b}\right)$ only depends on the value of a modulo b, you'll end up with the correct value for the Legendre symbol. The only caveat, and it is extremely important, is that you're only allowed to flip $\left(\frac{a}{b}\right)$ for *odd positive* values of a. If a is even, then you must first factor off a power of $\left(\frac{2}{b}\right)$, and if it is negative, then you must factor off the $\left(\frac{-1}{b}\right)$.

We'll illustrate this new and improved quadratic reciprocity law by recomputing our earlier example.

$$\begin{aligned}\left(\frac{37603}{48611}\right) &= -\left(\frac{48611}{37603}\right) = -\left(\frac{11008}{37603}\right) = -\left(\frac{2^8\cdot 43}{37603}\right) = -\left(\frac{43}{37603}\right)\\ &= \left(\frac{37603}{43}\right) = \left(\frac{21}{43}\right) = \left(\frac{43}{21}\right) = \left(\frac{1}{21}\right) = 1\end{aligned}$$

Although this may not look much shorter than before, it actually required much less work, because we didn't need to find the prime factorization of 37603.

We have just verified that 37603 is a quadratic residue modulo 48611, so the congruence

$$x^2 \equiv 37603 \pmod{48611}$$

has a solution (in fact, two solutions). Unfortunately, nothing we have done helps us to find the solutions, which turn out to be $x \equiv 17173 \pmod{48611}$ and $x \equiv 31438 \pmod{48611}$. However, there do exist more advanced methods that actually solve the congruence $x^2 \equiv a \pmod{p}$. And for certain special sorts of primes, it is possible to write down the solutions explicitly, see Exercises 25.6 and 25.7.

Exercises

25.1. Use the Law of Quadratic Reciprocity to compute the following Legendre symbols.

(a) $\left(\frac{85}{101}\right)$ (b) $\left(\frac{29}{541}\right)$ (c) $\left(\frac{101}{1987}\right)$ (d) $\left(\frac{31706}{43789}\right)$

25.2. Does the congruence

$$x^2 - 3x - 1 \equiv 0 \pmod{31957}$$

have any solutions? (*Hint.* Use the quadratic formula to find out what number you need to take the square root of modulo the prime 31957.)

25.3. Show that there are infinitely many primes congruent to 1 modulo 3. [*Hint.* See the proof of the "1 (Modulo 4) Theorem" in Chapter 24, use $A = (2p_1p_2\cdots p_r)^2 + 3$, and try to pick out a good prime dividing A.]

25.4. Let p be a prime number ($p \neq 2$ and $p \neq 5$), and let A be some given number. Suppose that p divides the number $A^2 - 5$. Show that p must be congruent to either 1 or 4 modulo 5.

25.5. Write a program that uses quadratic reciprocity to compute the Legendre symbol $\left(\frac{a}{p}\right)$ or, more generally, the Jacobi symbol $\left(\frac{a}{b}\right)$.

25.6. Let p be a prime satisfying $p \equiv 3 \pmod 4$ and suppose that a is a quadratic residue modulo p.

(a) Show that $x = a^{(p+1)/4}$ is a solution to the congruence

$$x^2 \equiv a \pmod p.$$

This gives an explicit way to find square roots modulo p for primes congruent to 3 modulo 4.

(b) Find a solution to the congruence $x^2 \equiv 7 \pmod{787}$. (Your answer should lie between 1 and 786.)

25.7. Let p be a prime satisfying $p \equiv 5 \pmod 8$ and suppose that a is a quadratic residue modulo p.

(a) Show that one of the values

$$x = a^{(p+3)/8} \qquad \text{or} \qquad x = 2a \cdot (4a)^{(p-5)/8}$$

is a solution to the congruence

$$x^2 \equiv a \pmod p.$$

This gives an explicit way to find square roots modulo p for primes congruent to 5 modulo 8.

(b) Find a solution to the congruence $x^2 \equiv 5 \pmod{541}$. (Give an answer lying between 1 and 540.)

(c) Find a solution to the congruence $x^2 \equiv 13 \pmod{653}$. (Give an answer lying between 1 and 652.)

25.8. Let p be a prime that is congruent to 5 modulo 8. Write a program to solve the congruence

$$x^2 \equiv a \pmod p$$

using the method described in the previous exercise and successive squaring. The output should be a solution satisfying $0 \le x < p$. Be sure to check that a is a quadratic residue, and return an error message if it is not. Use your program to solve the congruences

$$x^2 \equiv 17 \pmod{1021}, \qquad x^2 \equiv 23 \pmod{1021}, \qquad x^2 \equiv 31 \pmod{1021}.$$

25.9. If $a^{m-1} \not\equiv 1 \pmod m$, then Fermat's Little Theorem tells us that m is composite. On the other hand, even if

$$a^{m-1} \equiv 1 \pmod m$$

for some (or all) a's satisfying $\gcd(a, m) = 1$, we cannot conclude that m is prime. This exercise describes a way to use Quadratic Reciprocity to check if a number is probably prime. (You might compare this method with the Rabin–Miller test described in Chapter 19.)

(a) Euler's criterion says that if p is prime then

$$a^{(p-1)/2} \equiv \left(\frac{a}{p}\right) \pmod{p}.$$

Use successive squaring to compute $11^{864} \pmod{1729}$ and use Quadratic Reciprocity to compute $\left(\frac{11}{1729}\right)$. Do they agree? What can you conclude concerning the possible primality of 1729?

(b) Use successive squaring to compute the quantities

$$2^{(1293337-1)/2} \pmod{1293337} \qquad \text{and} \qquad 2^{1293336} \pmod{1293337}.$$

What can you conclude concerning the possible primality of 1293337?

Chapter 26

Which Primes Are Sums of Two Squares?

Although our exploration of congruences has been interesting and fun, there is no doubt that the fundamental questions in number theory are questions about actual natural numbers. A congruence

$$A \equiv B \pmod{M}$$

is all well and good; it tells you that the difference $A - B$ is a multiple of M, but it can't compare to an actual equality

$$A = B.$$

One way to think of congruences is that they are approximations to true equalities. Such approximations are not to be despised. They have a certain intrinsic interest of their own and, furthermore, they can often be used as tools to construct true equalities. This is the path we take in this chapter, where we use the Law of Quadratic Reciprocity, which is a theorem about congruences, as a tool to construct equalities between whole numbers.

The question we address is as follows:

Which numbers can be written as a sum of two squares?

For example, 5, 10, and 65 are sums of two squares, since

$$5 = 2^2 + 1^2, \qquad 10 = 3^2 + 1^2, \qquad \text{and} \qquad 65 = 7^2 + 4^2.$$

On the other hand, the numbers 3, 19, and 154 cannot be written as a sum of two squares. To see this for 19, for example, we just need to check that none of the differences

$$19 - 1^2 = 18, \quad 19 - 2^2 = 15, \quad 19 - 3^2 = 10, \quad \text{or} \quad 19 - 4^2 = 3$$

is a square. In general, to check if a given number m is a sum of two squares, list the numbers

$$m - 1^2, \quad m - 2^2, \quad m - 3^2, \quad m - 4^2, \ldots$$

until either you get a square or the numbers become negative.[1]

As usual, we begin with a short table and look for patterns.

$1 = 1^2 + 0^2$	11 NO	21 NO	31 NO	$41 = 4^2 + 5^2$
$2 = 1^2 + 1^2$	12 NO	22 NO	$32 = 4^2 + 4^2$	42 NO
3 NO	$13 = 2^2 + 3^2$	23 NO	33 NO	43 NO
$4 = 0^2 + 2^2$	14 NO	24 NO	$34 = 3^2 + 5^2$	44 NO
$5 = 1^2 + 2^2$	15 NO	$25 = 3^2 + 4^2$	35 NO	$45 = 3^2 + 6^2$
6 NO	$16 = 0^2 + 4^2$	$26 = 1^2 + 5^2$	$36 = 0^2 + 6^2$	46 NO
7 NO	$17 = 1^2 + 4^2$	27 NO	$37 = 1^2 + 6^2$	47 NO
$8 = 2^2 + 2^2$	$18 = 3^2 + 3^2$	28 NO	38 NO	48 NO
$9 = 0^2 + 3^2$	19 NO	$29 = 2^2 + 5^2$	39 NO	$49 = 0^2 + 7^2$
$10 = 1^2 + 3^2$	$20 = 2^2 + 4^2$	30 NO	$40 = 2^2 + 6^2$	$50 = 5^2 + 5^2$

Numbers That Are Sums of Two Squares

From the table, we make a list of the numbers that are and are not a sum of two squares.

Numbers that are a sum of two squares	1, 2, 4, 5, 8, 9, 10, 13, 16, 17, 18, 20, 25, 26, 29, 32, 34, 36, 37, 40, 41, 45, 49, 50
Numbers that are not a sum of two squares	3, 6, 7, 11, 12, 14, 15, 19, 21, 22, 23, 24, 27, 28, 30, 31, 33, 35, 38, 39, 42, 43, 44, 46, 47, 48

Can you spot any patterns?

One immediate observation is that no number that is congruent to 3 modulo 4 can be written as a sum of two squares. Looking back at the first two columns of the table, we might have also guessed that if $m \equiv 1 \pmod 4$ then m is a sum of two squares. But this guess is not correct, since 21 is not a sum of two squares. Another exception is 33. However, both 21 and 33 are composite numbers, $21 = 3 \cdot 7$ and $33 = 3 \cdot 11$. If we only look at prime numbers, we see that every prime in our table satisfying

$$p \equiv 1 \pmod 4$$

[1]Actually, it's only necessary to check if $m - a^2$ is a square for all a's between 1 and $\sqrt{m/2}$. Do you see why this is enough?

is indeed a sum of two squares. This observation reminds us of the "prime directive" in number theoretic investigations: always start by investigating prime numbers. There are two reasons to do this. First, patterns are usually easier to spot for primes. Second, patterns for primes can often be used to deduce patterns for all numbers, since the Fundamental Theorem of Arithmetic (Chapter 7) says that the primes are the basic building blocks of all numbers.

Now that we've decided to concentrate on primes, let's compile a more extensive list of primes and see which can be written as a sum of two squares.

$2 = 1^2 + 1^2$	31 NO	$73 = 3^2 + 8^2$	127 NO	179 NO
3 NO	$37 = 1^2 + 6^2$	79 NO	131 NO	$181 = 9^2 + 10^2$
$5 = 1^2 + 2^2$	$41 = 4^2 + 5^2$	83 NO	$137 = 4^2 + 11^2$	191 NO
7 NO	43 NO	$89 = 5^2 + 8^2$	139 NO	$193 = 7^2 + 12^2$
11 NO	47 NO	$97 = 4^2 + 9^2$	$149 = 7^2 + 10^2$	$197 = 1^2 + 14^2$
$13 = 2^2 + 3^2$	$53 = 2^2 + 7^2$	$101 = 1^2 + 10^2$	151 NO	199 NO
$17 = 1^2 + 4^2$	59 NO	103 NO	$157 = 6^2 + 11^2$	211 NO
19 NO	$61 = 5^2 + 6^2$	107 NO	163 NO	223 NO
23 NO	67 NO	$109 = 3^2 + 10^2$	167 NO	227 NO
$29 = 2^2 + 5^2$	71 NO	$113 = 7^2 + 8^2$	$173 = 2^2 + 13^2$	$229 = 2^2 + 15^2$

Primes That Are Sums of Two Squares

This gives the following two lists.

Primes that are a sum of two squares	2, 5, 13, 17, 29, 37, 41, 53, 61, 73, 89, 97, 101, 109, 113, 137, 149, 157, 173, 181, 193, 197, 229
Primes that are not a sum of two squares	3, 7, 11, 19, 23, 31, 43, 47, 59, 67, 71, 79, 83, 103, 107, 127, 131, 139, 151, 163, 167, 179, 191, 199, 211, 223, 227

The right conjecture is obvious. Primes that are congruent to 1 modulo 4 seem to be a sum of two squares, and primes that are congruent to 3 modulo 4 seem not to be. (We're ignoring 2, which is a sum of two squares, but occupies a somewhat anomalous position.) The rest of this chapter is devoted to a discussion and verification of this conjecture.

Theorem 26.1 (Sum of Two Squares Theorem For Primes). *Let p be a prime. Then p is a sum of two squares exactly when*

$$p \equiv 1 \pmod 4 \qquad (\text{or } p = 2).$$

The Sum of Two Squares Theorem really consists of two statements.

Statement 1. If p is a sum of two squares, then $p \equiv 1 \pmod 4$.

Statement 2. If $p \equiv 1 \pmod 4$, then p is a sum of two squares.

One of these statements is fairly easy to verify, while the other is quite difficult. Can you guess which is which without actually trying to prove either of them? This is not an idle or frivolous question. Before trying to verify a mathematical statement, it helps to have some idea of how difficult the proof is likely to be, or, as a mathematician would say, to know the *depth* of the statement. The proof of a deep theorem is likely to require stronger tools and more effort than the proof of a "shallower" theorem, just as it requires specialized machinery and great effort to build a skyscraper, while hammer and nails suffice to construct a birdhouse.

So my question to you is "Which of the statements 1 and 2 is deeper?" Intuitively, a statement is deep if it starts with an easy assertion and uses it to prove a difficult assertion. Statements 1 and 2 deal with the following two assertions:

Assertion A. p is a sum of two squares.

Assertion B. $p \equiv 1 \pmod 4$.

Clearly, B is an easy assertion since for any given prime number p, it is easy to check whether it is true. Assertion A, on the other hand, is more difficult, since it can take a lot of work to check whether a given prime p is a sum of two squares. Thus, statement 1 says that if the deep assertion A is true, then so is the easy assertion B. This suggests that statement 1 won't be too difficult to prove. Statement 2 says that if the easy assertion B is true, then the deep assertion A is also true. This suggests that a proof of statement 2 is likely to be difficult.

Now that we know that statement 1 should be easy to prove, let's prove it. We are told that the prime p is a sum of two squares, say

$$p = a^2 + b^2.$$

We also know that p is odd, so one of a and b must be odd and the other one must be even. Switching them if necessary, we may assume that a is odd and b is even, say

$$a = 2n + 1 \qquad \text{and} \qquad b = 2m.$$

Then

$$p = a^2 + b^2 = (2n+1)^2 + (2m)^2 = 4n^2 + 4n + 1 + 4m^2 \equiv 1 \pmod 4,$$

which is exactly what we were trying to prove.

Having given this very easy proof of statement 1, I want to show you a more complicated proof. Why would we ever want to use a complicated proof in place of an easy one? One answer is that frequently the more complicated argument can be applied in situations where the simple ideas do not work.

Our easy proof was to take the given formula $p = a^2 + b^2$, reduce it modulo 4, and deduce something about p modulo 4. That's a very natural way to proceed. For our new proof, we reduce the formula modulo p. This gives

$$0 \equiv a^2 + b^2 \pmod{p}, \qquad \text{so} \qquad -a^2 \equiv b^2 \pmod{p}.$$

Next we take the Legendre symbol of both sides.

$$\begin{aligned}
\left(\frac{-a^2}{p}\right) &= \left(\frac{b^2}{p}\right)\\
\left(\frac{-1}{p}\right)\left(\frac{a}{p}\right)^2 &= \left(\frac{b}{p}\right)^2\\
\left(\frac{-1}{p}\right) &= 1
\end{aligned}$$

Thus, -1 is a quadratic residue modulo p, so the Law of Quadratic Reciprocity (Chapter 25) tells us that $p \equiv 1 \pmod{4}$. This second proof is especially amusing because we reduce modulo p to get information modulo 4.

The verification of statement 2, that every prime $p \equiv 1 \pmod{4}$ can be written as a sum of two squares, is more difficult. The proof we give is based on Fermat's famous *Method of Descent* and in this form is essentially due to Euler. We start by describing the basic idea of Fermat's descent method, since once you understand the concept, the details become much less fearsome.

We assume that $p \equiv 1 \pmod{4}$, and we want to write p as a sum of two squares. Rather than immediately trying to write $p = a^2 + b^2$, let's tackle the less onerous task of writing some multiple of p as a sum of two squares. For example, Quadratic Reciprocity tells us that $x^2 \equiv -1 \pmod{p}$ has a solution, say $x = A$, and then $A^2 + 1^2$ is a multiple of p. So we begin with the knowledge that

$$A^2 + B^2 = Mp$$

for some integers A, B, and M. If $M = 1$, then we're done, so we suppose that $M \geq 2$.

Fermat's brilliant idea is to use the numbers A, B, and M to find new integers a, b, and m with

$$a^2 + b^2 = mp \qquad \text{and} \qquad m \leq M - 1.$$

Of course, if $m = 1$, then we're done. And if $m \geq 2$, then we can apply Fermat's descent procedure again starting with a, b, and m to find a yet smaller multiple of p that is a sum of two squares. Continuing repeatedly in this fashion, we must eventually end up with p itself written as a sum of two squares.

This description has omitted one "minor" detail: how to use the known numbers A, B, and M to produce the new numbers a, b, and m. Before describing this crucial piece of the proof, we briefly digress to look at a beautiful (and useful) identity.

The identity says that if two numbers that are sums of two squares are multiplied together, then the product is also a sum of two squares.

$$(u^2 + v^2)(A^2 + B^2) = (uA + vB)^2 + (vA - uB)^2.$$

There is no difficulty in verifying that this identity is correct once it has been written down. (Discovering it in the first place is another matter, which we discuss at the end of this chapter.) Thus, multiplying out the right-hand side, we find that

$$\begin{aligned}(uA + vB)^2 &+ (vA - uB)^2 \\ &= (u^2A^2 + 2uAvB + v^2B^2) + (v^2A^2 - 2vAuB + u^2B^2) \\ &= u^2A^2 + v^2B^2 + v^2A^2 + u^2B^2 \\ &= (u^2 + v^2)(A^2 + B^2).\end{aligned}$$

We are now ready to describe Fermat's Descent Procedure for writing any prime

$$p \equiv 1 \pmod 4$$

as a sum of two squares. As explained above, the idea is to begin with some multiple Mp that is a sum of two squares and, by some clever manipulations, find a smaller multiple that is also a sum of two squares. To help you understand the various steps, we do the example

$$a^2 + b^2 = 881$$

side by side with the general procedure. The Descent Procedure, in all its glory, is on display in the table on page 192. Be sure to go over the procedure step by step before proceeding with the text.

The Descent Procedure described on page 192 reduced the initial equation

$$387^2 + 1^2 = 170 \cdot 881$$

to the smaller multiple

$$107^2 + 2^2 = 13 \cdot 881$$

of 881. To complete the task of writing 881 as a sum of two squares, we repeat the

Descent Procedure

<table>
<tr><th>$p = 881$</th><th>p any prime $\equiv 1 \pmod 4$</th></tr>
<tr><td>Write
$387^2 + 1^2 = 170 \cdot 881$
with $170 < 881$</td><td>Write
$A^2 + B^2 = Mp$
with $M < p$</td></tr>
<tr><td>Choose numbers with
$47 \equiv 387 \pmod{170}$
$1 \equiv 1 \pmod{170}$
$-\frac{170}{2} \le 47, 1 \le \frac{170}{2}$</td><td>Choose numbers u and v with
$u \equiv A \pmod M$
$v \equiv B \pmod M$
$-\frac{1}{2}M \le u, v \le \frac{1}{2}M$</td></tr>
<tr><td>Observe that
$47^2 + 1^2 \equiv 387^2 + 1^2$
$\equiv 0 \pmod{170}$</td><td>Observe that
$u^2 + v^2 \equiv A^2 + B^2$
$\equiv 0 \pmod M$</td></tr>
<tr><td>So we can write
$47^2 + 1^2 = 170 \cdot 13$
$387^2 + 1^2 = 170 \cdot 881$</td><td>So we can write
$u^2 + v^2 = Mr$
$A^2 + B^2 = Mp$
(for some $1 \le r < M$)</td></tr>
<tr><td>Multiply to get
$(47^2 + 1^2)(387^2 + 1^2)$
$= 170^2 \cdot 13 \cdot 881$</td><td>Multiply to get
$(u^2 + v^2)(A^2 + B^2) = M^2 rp$</td></tr>
<tr><td colspan="2">Use the identity $(u^2 + v^2)(A^2 + B^2) = (uA + vB)^2 + (vA - uB)^2$.</td></tr>
<tr><td>$(47 \cdot 387 + 1 \cdot 1)^2 + (1 \cdot 387 - 47 \cdot 1)^2$
$= 170^2 \cdot 13 \cdot 881$
$\underbrace{18190^2 + 340^2}_{\text{each divisible by } 170} = 170^2 \cdot 13 \cdot 881$</td><td>$\underbrace{(uA + vB)^2 + (vA - uB)^2}_{\text{each divisible by } M} = M^2 rp$</td></tr>
<tr><td>Divide by 170^2.
$\left(\frac{18190}{170}\right)^2 + \left(\frac{340}{170}\right)^2 = 13 \cdot 881$
$107^2 + 2^2 = 13 \cdot 881$
This gives a smaller multiple of 881 written as a sum of two squares.</td><td>Divide by M^2.
$\left(\frac{uA + vB}{M}\right)^2 + \left(\frac{vA - uB}{M}\right)^2 = rp$
This gives a smaller multiple of p written as a sum of two squares.</td></tr>
<tr><td colspan="2">Repeat the process until p itself is written as a sum of two squares.</td></tr>
</table>

Descent Procedure starting with the equation $107^2 + 2^2 = 13 \cdot 881$. This gives

$p = 881$	p any prime $\equiv 1 \pmod 4$
$107^2 + 2^2 = 13 \cdot 881$ $3 \equiv 107 \pmod{13}$ $2 \equiv 2 \pmod{13}$	$A^2 + B^2 = Mp$ $u \equiv A \pmod M$ $v \equiv B \pmod M$
$3^2 + 2^2 = 13 \cdot 1$	$u^2 + v^2 = Mr$
$(3^2 + 2^2)(107^2 + 2^2) = 13^2 \cdot 1 \cdot 881$	$(u^2 + v^2)(A^2 + B^2) = M^2 rp$
Use the identity $(u^2 + v^2)(A^2 + B^2) = (uA + vB)^2 + (vA - uB)^2$.	
$(3 \cdot 107 + 2 \cdot 2)^2 + (2 \cdot 107 - 3 \cdot 2)^2$ $= 13^2 \cdot 881$ $325^2 + 208^2 = 13^2 \cdot 881$	$(uA + vB)^2 + (vA - uB)^2 = M^2 rp$
Divide by 13^2. $25^2 + 16^2 = 881$	Divide by M^2. $\left(\frac{uA + vB}{M}\right)^2 + \left(\frac{vA - uB}{M}\right)^2 = rp$

This second application of the Descent Procedure has given us the solution to our original problem,

$$881 = 25^2 + 16^2.$$

Of course, for a small number such as 881 it might have been easier to solve $881 = a^2 + b^2$ by trial and error, but as soon as p becomes large, the Descent Procedure is definitely more efficient. In fact, each time the Descent Procedure is applied, the multiple of p is at least cut in half.

To show that the Descent Procedure actually works, we need to verify five assertions. At the first step we need to find numbers A and B with

(i) $A^2 + B^2 = Mp$ and $M < p$.

To do this, we take a solution to the congruence

$$x^2 \equiv -1 \pmod p$$

with $1 \le x < p$. Quadratic Reciprocity tells us that there is a solution,[2] since we are assuming that $p \equiv 1 \pmod 4$, and then $A = x$ and $B = 1$ have the property that $A^2 + B^2$ is divisible by p. Furthermore,

$$M = \frac{A^2 + B^2}{p} \le \frac{(p-1)^2 + 1^2}{p} = p - \frac{2p - 2}{p} < p.$$

[2]In practice, an easy way to solve $x^2 \equiv -1 \pmod p$ is to compute $b \equiv a^{(p-1)/4} \pmod p$ for some randomly chosen values of a. Euler's formula (Chapter 24) tells us that $b^2 \equiv \left(\frac{a}{p}\right) \pmod p$, so each choice of a gives us a 50% chance of winning.

In the second step of the descent procedure we chose numbers u and v satisfying

$$u \equiv A \pmod{M}, \quad v \equiv B \pmod{M}, \quad \text{and} \quad -\frac{1}{2}M \le u, v \le \frac{1}{2}M.$$

We then observed that

$$u^2 + v^2 \equiv A^2 + B^2 \equiv 0 \pmod{M},$$

so $u^2 + v^2$ is divisible by M, say $u^2 + v^2 = Mr$. The remaining four statements we need to check are as follows:

(ii) $r \ge 1$.

(iii) $r < M$.

(iv) $uA + vB$ is divisible by M.

(v) $vA - uB$ is divisible by M.

We check them in reverse order. To verify (v) we compute

$$vA - uB \equiv B \cdot A - A \cdot B \equiv 0 \pmod{M}.$$

Similarly, for (iv) we have

$$uA + vB \equiv A \cdot A + B \cdot B \equiv Mp \equiv 0 \pmod{M}.$$

For (iii) we use the fact that u and v are between $-M/2$ and $M/2$ to estimate

$$r = \frac{u^2 + v^2}{M} \le \frac{(M/2)^2 + (M/2)^2}{M} = \frac{M}{2} < M.$$

Notice that this actually shows that $r \le M/2$, so every time the Descent Procedure is used, the multiple of p is at least cut in half.

Finally, to show that (ii) is true, we need to check that $r \ne 0$. So we assume that $r = 0$ and see what happens. Well, if $r = 0$, then $u^2+v^2 = 0$, so we must have $u = v = 0$. But $u \equiv A \pmod{M}$ and $v \equiv B \pmod{M}$, so A and B are divisible by M. This implies that $A^2 + B^2$ is divisible by M^2. But $A^2 + B^2 = Mp$, so we see that M must divide the prime p. We also know that $M < p$, so it must be true that $M = 1$. This means that $A^2 + B^2 = p$ and we're already done writing p as a sum of two squares! Thus, either (ii) is true, or else we already had $A^2 + B^2 = p$ and there was no reason to use the Descent Procedure in the first place.

This completes the verification that the Descent Procedure always works, so we have now finished proving both parts of the Sum of Two Squares Theorem For Primes.

Digression on Sums of Squares and Complex Numbers

The identity

$$(u^2 + v^2)(A^2 + B^2) = (uA + vB)^2 + (vA - uB)^2,$$

which expresses the product of sums of two squares as a sum of two squares, has been very useful, and we will find further uses for it in the next chapter. You may well have wondered whence this identity comes. The answer lies in the realm of complex numbers, that is, numbers of the form

$$z = x + iy,$$

where i is a square root of -1. Two complex numbers can be multiplied together in the usual way as long as you remember to replace i^2 by -1. Thus,

$$\begin{aligned}(x_1 + iy_1)(x_2 + iy_2) &= x_1x_2 + ix_1y_2 + iy_1x_2 + i^2y_1y_2 \\ &= (x_1x_2 - y_1y_2) + i(x_1y_2 + y_1x_2).\end{aligned}$$

Complex numbers also have absolute values,

$$|z| = |x + iy| = \sqrt{x^2 + y^2}.$$

This idea is to imagine the number $z = x + iy$ as corresponding to the point (x, y) in the plane, and then $|z|$ is just the distance from z to the origin $(0, 0)$. Now our identity comes from the following fact:

The absolute value of a product is the product of the absolute values.

In other words, $|z_1z_2| = |z_1| \cdot |z_2|$. Writing this out in terms of x's and y's gives

$$\begin{aligned}\big|(x_1 + iy_1)(x_2 + iy_2)\big| &= |x_1 + iy_1| \cdot |x_2 + iy_2| \\ \big|(x_1x_2 - y_1y_2) + i(x_1y_2 + y_1x_2)\big| &= |x_1 + iy_1| \cdot |x_2 + iy_2| \\ \sqrt{(x_1x_2 - y_1y_2)^2 + (x_1y_2 + y_1x_2)^2} &= \sqrt{x_1^2 + y_1^2}\sqrt{x_2^2 + y_2^2}.\end{aligned}$$

If we square both sides of this last equation, we get exactly our identity (where $x_1 = u$, $y_1 = v$, $x_2 = A$, and $y_2 = -B$).

There is a similar identity involving sums of four squares, that is due to Euler:

$$\begin{aligned}&(a^2 + b^2 + c^2 + d^2)(A^2 + B^2 + C^2 + D^2) \\ &\qquad = (aA + bB + cC + dD)^2 + (aB - bA - cD + dC)^2 \\ &\qquad\quad + (aC + bD - cA - dB)^2 + (aD - bC + cB - dA)^2.\end{aligned}$$

This complicated identity is related to the theory of quaternions[3] in the same way that our identity is related to complex numbers. It is an unfortunate fact that there is no analogous identity for sums of three squares, and indeed the question of writing numbers as a sum of three squares is much more difficult than the same problem for either two or four squares.

Exercises

26.1. (a) Make a list of all primes $p < 50$ that can be written in the form $p = a^2 + ab + b^2$. For example, $p = 7$ has this form with $a = 2$ and $b = 1$, while $p = 11$ cannot be written in this form. Try to find a pattern and make a guess as to exactly which primes have this form. (Can you prove that at least part of your guess is correct?)
(b) Same question for primes p that can be written in the form[4] $p = a^2 + 2b^2$.

26.2. If the prime p can be written in the form $p = a^2 + 5b^2$, show that

$$p \equiv 1 \text{ or } 9 \pmod{20}.$$

(Of course, we are ignoring $5 = 0^2 + 5 \cdot 1^2$.)

26.3. Use the Descent Procedure twice, starting from the equation

$$557^2 + 55^2 = 26 \cdot 12049,$$

to write the prime 12049 as a sum of two squares.

26.4. (a) Start from $259^2 + 1^2 = 34 \cdot 1973$ and use the Descent Procedure to write the prime 1973 as a sum of two squares.
(b) Start from $261^2 + 947^2 = 10 \cdot 96493$ and use the Descent Procedure to write the prime 96493 as a sum of two squares.

26.5. (a) Which primes $p < 100$ can be written as a sum of three squares,

$$p = a^2 + b^2 + c^2?$$

(We allow one of a, b, c to equal 0, so, for example, $5 = 2^2 + 1^2 + 0^2$ is a sum of three squares.)
(b) Based on the data you collected in (a), try to make a conjecture describing which primes can be written as a sum of three squares. Your conjecture should consist of the following two statements, where you are to fill in the blanks:

---

[3]Quaternions are numbers of the form $a + ib + jc + kd$, where i, j, and k are three different square roots of -1 satisfying strange multiplication rules such as $ij = k = -ji$.

[4]The question of which primes p can be written in the form $p = a^2 + nb^2$ has been extensively studied and has connections with many branches of mathematics. There is even an entire book on the subject, *Primes of the Form $x^2 + ny^2$*, by David Cox (New York: John Wiley & Sons, 1989).

(i) If p satisfies ________________, then p is a sum of three squares.

(ii) If p satisfies ________________, then p is not a sum of three squares.

(c) Prove part (ii) of your conjecture in (b). [You might also try to prove part (i), but be warned, it is quite difficult.]

26.6. Write a program that solves $x^2 + y^2 = n$ by trying $x = 0, 1, 2, 3, \ldots$ and checking if $n - x^2$ is a perfect square. Your program should return all solutions with $x \leq y$ if any exist and should return an appropriate message if there are no solutions.

26.7. (a) Write a program that solves $x^2 + y^2 = p$ for primes $p \equiv 1 \pmod 4$ using Fermat's Descent Procedure. The input should consist of the prime p and a pair of numbers (A, B) satisfying

$$A^2 + B^2 \equiv 0 \pmod p.$$

(b) In the case that $p \equiv 5 \pmod 8$, modify your program as follows so that the user doesn't have to input (A, B). First, use successive squaring to compute the number $A \equiv -2 \cdot (-4)^{(p-5)/8} \pmod p$. Then $A^2 + 1 \equiv 0 \pmod p$ (see Exercise 25.7), so you can use $(A, 1)$ as your starting value to perform the descent.

Chapter 27

Which Numbers Are Sums of Two Squares?

In the last chapter we gave a definitive answer to the question of which primes can be written as a sum of two squares. We now take up the same question for arbitrary numbers. Part of our strategy, which can be summed up in three words, has a long and glorious history:

Divide and Conquer!

Of course, "Divide" doesn't mean division per se. Rather, it means to break up the problem into pieces of manageable size, and then "Conquer" means we need to solve each piece. But these two steps, which may suffice for warfare, have to be followed by a third step: fitting the pieces back together. This unification step uses the identity from the last chapter that expresses a product of sums of squares as a sum of squares:

$$(u^2 + v^2)(A^2 + B^2) = (uA + vB)^2 + (vA - uB)^2. \qquad (*)$$

Here, then, is our step by step strategy for expressing a number m as a sum of two squares.

Divide: Factor m into a product of primes $p_1 p_2 \cdots p_r$.
Conquer: Write each prime p_i as a sum of two squares.
Unify: Use the identity $(*)$ repeatedly to write m as a sum of two squares.

We know from the previous chapter exactly when the Conquer step works, since we know that a prime p is a sum of two squares if and only if either $p = 2$ or $p \equiv 1 \pmod 4$. For example, to write 10 as a sum of two squares, we factor

$10 = 2 \cdot 5$, write 2 and 5 as sums of two squares,

$$2 = 1^2 + 1^2 \qquad \text{and} \qquad 5 = 2^2 + 1^2,$$

and use the identity to recombine

$$10 = 2 \cdot 5 = (1^2 + 1^2)(2^2 + 1^2) = (2+1)^2 + (2-1)^2 = 3^2 + 1^2.$$

Here's a more complicated example. We'll write $m = 1105$ as a sum of two squares.

Divide: Factor $m = 1105 = 5 \cdot 13 \cdot 17$.

Conquer: Write each prime p as a sum of two squares.
$5 = 2^2 + 1^2, \qquad 13 = 3^2 + 2^2, \qquad 17 = 4^2 + 1^2$

Unify: Use the identity $(*)$ repeatedly to write m as a sum of two squares.

$$\begin{aligned} m = 1105 &= 5 \cdot 13 \cdot 17 \\ &= (2^2 + 1^2)(3^2 + 2^2)(4^2 + 1^2) \\ &= \big((6+2)^2 + (3-4)^2\big)(4^2 + 1^2) \\ &= (8^2 + 1^2)(4^2 + 1^2) \\ &= (32+1)^2 + (4-8)^2 \\ &= 33^2 + 4^2 \end{aligned}$$

Our Divide, Conquer, and Unify strategy is successful for the number m provided that each prime factor of m is itself a sum of two squares. We know which primes can be written as a sum of two squares, so we now have a method for writing m as a sum of two squares if m factors as

$$m = p_1^{k_1} p_2^{k_2} p_3^{k_3} \cdots p_r^{k_r},$$

where every prime in the factorization is either 2 or is congruent to 1 modulo 4.

However, if you look back at the list in the last chapter, you'll see that there are other m's that are a sum of two squares. For example,

$$9 = 3^2 + 0^2, \qquad 18 = 3^2 + 3^2, \qquad \text{and} \qquad 45 = 6^2 + 3^2.$$

What's going on? Notice that in each case m is divisible by 3^2 and $m = a^2 + b^2$ with both a and b divisible by 3. If we divide these three examples by 3^2, we get

$$\begin{aligned} 1 &= \frac{9}{3^2} = \frac{3^2 + 0^2}{3^2} = 1^2 + 0^2, \\ 2 &= \frac{18}{3^2} = \frac{3^2 + 3^2}{3^2} = 1^2 + 1^2, \\ 5 &= \frac{45}{3^2} = \frac{6^2 + 3^2}{3^2} = 2^2 + 1^2. \end{aligned}$$

In other words, these three examples were created by taking the equations

$$1 = 1^2 + 0^2, \qquad 2 = 1^2 + 1^2, \qquad \text{and} \qquad 5 = 2^2 + 1^2$$

and multiplying both sides by 3^2.

We can do this in general. Given any $m = a^2 + b^2$, we can multiply by d^2 to get

$$d^2 m = (da)^2 + (db)^2.$$

Thus, if m is a sum of two squares, then so is $d^2 m$ for any d. On the other hand, if $m = a^2 + b^2$ is a sum of two squares and if a and b have a common factor, say $a = dA$ and $b = dB$, then we can factor out d^2 to get

$$m = d^2(A^2 + B^2).$$

Thus, m is divisible by d^2, and m/d^2 is a sum of two squares.

The moral is that squares dividing m don't count when we're trying to write m as a sum of two squares. In other words, take m and factor it as

$$m = p_1 p_2 \cdots p_r M^2,$$

where the prime factors $p_1, p_2, \ldots, p_r$ are all different. Then m can be written as a sum of two squares provided that each of $p_1, p_2, \ldots, p_r$ can be written as a sum of two squares. For example, consider $m = 252000$. We factor m as

$$m = 252000 = 2^5 \cdot 3^2 \cdot 5^3 \cdot 7 = 2 \cdot 5 \cdot 7 \cdot (2^2 \cdot 3 \cdot 5)^2 = 2 \cdot 5 \cdot 7 \cdot 60^2.$$

The prime 7 is not a sum of two squares, so m is not a sum of two squares.

As another example, take $m = 25798500$. Then

$$m = 25798500 = 2^2 \cdot 3^4 \cdot 5^3 \cdot 7^2 \cdot 13 = 5 \cdot 13 \cdot (2 \cdot 3^2 \cdot 5 \cdot 7)^2 = 5 \cdot 13 \cdot 630^2.$$

In this case, 5 and 13 are sums of squares, and we easily find that $65 = 5 \cdot 13 = 8^2 + 1^2$. Multiplying both sides by 630^2 gives

$$m = 65 \cdot 630^2 = (8 \cdot 630)^2 + (1 \cdot 630)^2 = 5040^2 + 630^2.$$

In this chapter we have given a definitive answer to the question of which numbers are sums of two squares. We summarize our result in the following theorem, which also includes further interesting facts whose proof we leave as exercises.

Theorem 27.1 (Sum of Two Squares Theorem). *Let m be a positive integer.*

(a) *Factor m as*

$$m = p_1 p_2 \cdots p_r M^2$$

with distinct prime factors $p_1, p_2, \ldots, p_r$. Then m can be written as a sum of two squares exactly when every p_i is either 2 or is congruent to 1 modulo 4.

(b) *The number m can be written as a sum of two squares $m = a^2 + b^2$ with* $\gcd(a, b) = 1$ *if and only if it satisfies one of the following two conditions:*

  (i) *m is odd and every prime divisor of m is congruent to 1 modulo 4.*

  (ii) *m is even, $m/2$ is odd, and every prime divisor of $m/2$ is congruent to 1 modulo 4.*

The Return of the Pythagorean Triples

Recall that[1] a Pythagorean triple is a triple of positive integers (a, b, c) satisfying the equation

$$a^2 + b^2 = c^2,$$

and the triple is called primitive if $\gcd(a, b) = 1$. We are now in a position to completely describe all numbers that can appear as the hypotenuse c in a primitive Pythagorean triple.

The Pythagorean Triples Theorem says that every primitive Pythagorean triple can be obtained by choosing relatively prime odd integers $s > t \geq 1$ and setting

$$a = st, \qquad b = \frac{s^2 - t^2}{2}, \qquad c = \frac{s^2 + t^2}{2}.$$

So we are asking for a description of all numbers c for which we can find an s and a t, such that $c = (s^2 + t^2)/2$. In other words, c is the hypotenuse of a primitive Pythagorean triple exactly when the equation

$$2c = s^2 + t^2$$

has a solution in relatively prime odd integers s and t.

Note first that c must be odd. (We checked this in Chapter 2.) So we are asking which numbers $2c$ with c odd can be written as the sum of the squares of two relatively prime integers. The Sum of Two Squares Theorem says that this can be done if and only if every prime dividing c is congruent to 1 modulo 4. The following proposition records what we have proved.

[1] In this context, the phrase "Recall that..." is a polite way of saying "Now might be a good time to reread Chapter 2 and review the Pythagorean Triples Theorem in that chapter."

Theorem 27.2 (Pythagorean Hypotenuse Proposition). *A number c appears as the hypotenuse of a primitive Pythagorean triple (a, b, c) if and only if c is a product of primes each of which is congruent to* 1 *modulo* 4.

For example, the number $c = 1479$ cannot be the hypotenuse of a primitive Pythagorean triple, since $1479 = 3 \cdot 17 \cdot 29$. On the other hand, $c = 1105$ can be a hypotenuse, since $1105 = 5 \cdot 13 \cdot 17$. Furthermore, we can solve $s^2 + t^2 = 2c$ to find the values of s and t and then use these to find the corresponding a and b. Thus, $1105 = 33^2 + 4^2$ from earlier in this chapter, and then

$$2c = 2 \cdot 1105 = (1^2 + 1^2)(33^2 + 4^2) = 37^2 + 29^2.$$

Now $s = 37$ and $t = 29$, so $a = st = 1073$ and $b = (s^2 - t^2)/2 = 264$. This gives the desired primitive Pythagorean triple $(1073, 264, 1105)$ with hypotenuse 1105.

Exercises

27.1. For each of the following numbers m, either write m as a sum of two squares or explain why it is not possible to do so.

(a) 4370 (b) 1885 (c) 1189 (d) 3185

27.2. For each of the following numbers c, either find a primitive Pythagorean triple with hypotenuse c or explain why it is not possible to do so.

(a) 4370 (b) 1885 (c) 1189 (d) 3185

27.3. Find two pairs of positive integers (a, c) so that $a^2 + 5929 = c^2$.

27.4. In this exercise you will complete the proof of the first part of the Sum of Two Squares Theorem. Let m be a positive integer and factor m as $m = p_1 p_2 \cdots p_r M^2$ with distinct prime factors $p_1, p_2, \ldots, p_r$. If some p_i is congruent to 3 modulo 4, prove that m cannot be written as a sum of two squares.

27.5. In this exercise you will prove the second part of the Sum of Two Squares Theorem. Let m be a positive integer.

(a) If m is odd and if every prime dividing m is congruent to 1 modulo 4, prove that m can be written as a sum of two squares $m = a^2 + b^2$ with $\gcd(a, b) = 1$.

(b) If m is even and $m/2$ is odd and if every prime dividing $m/2$ is congruent to 1 modulo 4, prove that m can be written as a sum of two squares $m = a^2 + b^2$ with $\gcd(a, b) = 1$.

(c) If m can be written as a sum of two squares $m = a^2 + b^2$ with $\gcd(a, b) = 1$, prove that m is one of the numbers described in (a) or (b).

27.6. For any positive integer m, let

$$S(m) = (\text{\# of ways to write } m = a^2 + b^2 \text{ with } a \geq b \geq 0).$$

For example,

$$S(5) = 1, \quad \text{since} \quad 5 = 2^2 + 1^2,$$
$$S(65) = 2, \quad \text{since } 65 = 8^2 + 1^2 = 7^2 + 4^2,$$

while $S(15) = 0$.

(a) Compute the following values:

(i) $S(10)$ (ii) $S(70)$, (iii) $S(130)$ (iv) $S(1105)$

(b) If p is a prime and $p \equiv 1 \pmod 4$, what is the value of $S(p)$? Prove that your answer is correct.

(c) Let p and q be two different primes, both congruent to 1 modulo 4. What is the value of $S(pq)$? Prove that your answer is correct.

(d) More generally, if $p_1, \ldots, p_r$ are distinct primes, all congruent to 1 modulo 4, what is the value of $S(p_1 p_2 \ldots p_r)$? Prove that your answer is correct.

27.7. 💻 Write a program that solves $x^2 + y^2 = n$ by factoring n into a product of primes, first solving each $u^2 + v^2 = p$ using descent (Exercise 26.7) and then combining the solutions to find (x, y).

Chapter 28

The Equation $X^4 + Y^4 = Z^4$

Fermat's Last Theorem, scribbled by Fermat in a margin in the middle of the seventeenth century and finally proved by Andrew Wiles at the end of the twentieth, says that if $n \geq 3$ then the equation

$$a^n + b^n = c^n$$

has no solutions in positive integers a, b, and c. In this chapter we give Fermat's proof of this assertion for the particular exponent $n = 4$. In fact, we prove the following stronger statement.

Theorem 28.1 (Fermat's Last Theorem for Exponent 4). *The equation*

$$x^4 + y^4 = z^2$$

has no solutions in positive integers x, y, and z.

Verification. We use Fermat's method of descent to prove this theorem. Recall that the idea of "descent," as used in Chapter 26 to write a prime as a sum of two squares, is to descend from a large solution to a small solution. How does that help us in this instance, since we're trying to show that there aren't any solutions at all?

What we do is to suppose that there is a solution (x, y, z) in positive integers, and we use this supposed solution to produce a new solution (X, Y, Z) in positive integers with $Z < z$. Repeating this process, we would end up with a never-ending list of solutions

$$(x_1, y_1, z_1), (x_2, y_2, z_2), (x_3, y_3, z_3), \ldots \qquad \text{with } z_1 > z_2 > z_3 > \cdots.$$

This is, of course, completely absurd, since a decreasing list of positive integers can't continue indefinitely. The only escape from this absurdity lies in our original assumption that there is a solution. In other words, this contradiction shows that no solutions exist.

Now for the nitty-gritty details. We assume that we are given a solution (x, y, z) to the equation

$$x^4 + y^4 = z^2,$$

and we want to find a new smaller solution. If x, y, and z have a common factor, then we can factor it out and cancel it, so we may as well assume that they are relatively prime. Next, we observe that if we let $a = x^2$, $b = y^2$ and $c = z$ then (a, b, c) is a primitive Pythagorean triple,

$$a^2 + b^2 = c^2.$$

We know from Chapter 2 what all primitive Pythagorean triples look like. Possibly after switching x and y, there are odd integers s and t so that

$$x^2 = a = st, \qquad y^2 = b = \frac{s^2 - t^2}{2}, \qquad z = c = \frac{s^2 + t^2}{2}.$$

Notice that the product st is odd and equal to a square and that the only squares modulo 4 are 0 and 1, so we must have

$$st \equiv 1 \pmod{4}.$$

This means that s and t are either both 1 modulo 4 or both 3 modulo 4. In any case, we see that

$$s \equiv t \pmod{4}.$$

Next we look at the equation

$$2y^2 = s^2 - t^2 = (s - t)(s + t).$$

The fact that s and t are odd and relatively prime means that the only common factor of $s - t$ and $s + t$ is 2. We also know that $s - t$ is divisible by 4, so $s + t$ must be twice an odd number. Furthermore, we know that the product $(s - t)(s + t)$ is twice a square. The only way this can happen is if we have

$$s + t = 2u^2 \qquad \text{and} \qquad s - t = 4v^2$$

for some integers with u and $2v$ relatively prime.

We solve for s and t in terms of u and v,

$$s = u^2 + 2v^2 \qquad \text{and} \qquad t = u^2 - 2v^2,$$

and substitute into the formula $x^2 = st$ to get

$$x^2 = u^4 - 4v^4.$$

This can be rearranged to read

$$x^2 + 4v^4 = u^4.$$

Unfortunately, this isn't quite the equation we're looking for, so we repeat the process. If we let $A = x$, $B = 2v^2$, and $C = u^2$, then

$$A^2 + B^2 = C^2,$$

so (A, B, C) is a primitive Pythagorean triple. Again referring to Chapter 2, we can find odd relatively prime integers S and T so that

$$x = A = ST, \qquad 2v^2 = B = \frac{S^2 - T^2}{2}, \qquad u^2 = C = \frac{S^2 + T^2}{2}.$$

The middle formula says that

$$4v^2 = S^2 - T^2 = (S - T)(S + T).$$

Now S and T are odd and relatively prime, so the greatest common divisor of $S - T$ and $S + T$ is 2. Furthermore, their product is a square, so it must be true that

$$S + T = 2X^2 \qquad \text{and} \qquad S - T = 2Y^2$$

for some numbers X and Y. Solving for S and T in terms of X and Y gives

$$S = X^2 + Y^2 \qquad \text{and} \qquad T = X^2 - Y^2,$$

and then substituting into the formula for u^2 yields

$$u^2 = \frac{S^2 + T^2}{2} = \frac{(X^2 + Y^2)^2 + (X^2 - Y^2)^2}{2} = X^4 + Y^4.$$

Voilà! We have a new solution (X, Y, u) to our original equation

$$x^4 + y^4 = z^2.$$

It only remains to verify that the new solution is smaller than the original one. Using various formulas from above, we find that

$$z = \frac{s^2 + t^2}{2} = \frac{(u^2 + 2v^2)^2 + (u^2 - 2v^2)^2}{2} = u^4 + 4v^4.$$

This makes it clear that u is smaller than z. □

Exercises

28.1. Show that the equation $y^2 = x^3 + xz^4$ has no solutions in nonzero integers x, y, z.

Chapter 29

Square–Triangular Numbers Revisited

Some numbers are "shapely" in that they can be laid out in some sort of regular shape. For example, a *square number* n^2 can be arranged in the shape of an n-by-n square. Similarly, a *triangular number* is a number that can be arranged in the shape of a triangle. The following picture illustrates the first few triangular and square numbers (other than 1).

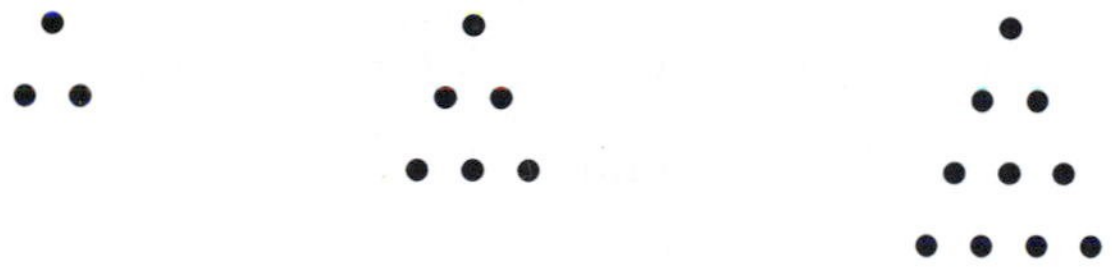

$1 + 2 = 3$ $\quad$ $1 + 2 + 3 = 6$ $\quad$ $1 + 2 + 3 + 4 = 10$

Triangular Numbers

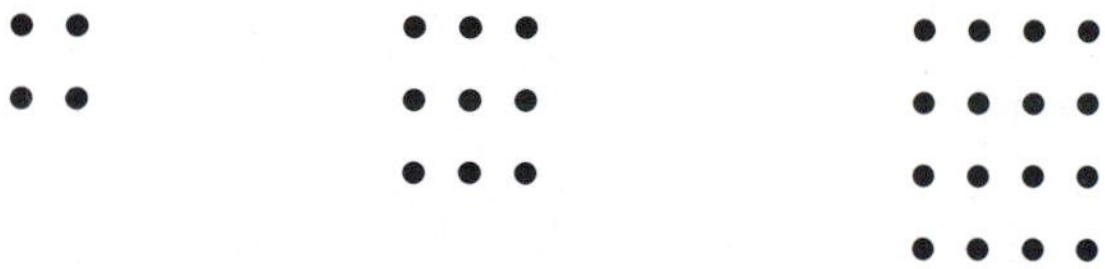

$2^2 = 4$ $\quad$ $3^2 = 9$ $\quad$ $4^2 = 16$

Square Numbers

Triangular numbers are thus formed by adding

$$1 + 2 + 3 + \cdots + m$$

for different values of m. We found a formula for the mth triangular number in Chapter 1,

$$1 + 2 + 3 + \cdots + m = \frac{m(m+1)}{2}.$$

Here's a list of the first few triangular and square numbers.

Triangular Numbers	$1, 3, 6, 10, 15, 21, 28, 36, 45, 55, 66, 78, 91, 105$
Square Numbers	$1, 4, 9, 16, 25, 36, 49, 64, 81, 100, 121, 144, 169$

In Chapter 1 we posed the question of "Squaring the Triangle," that is, finding square numbers that are also triangular numbers. Even our short list reveals two examples, 1 (which isn't very interesting) and 36. This means that 36 pebbles can be arranged in the shape of a 6-by-6 square, and they can also be arranged in the shape of a triangle with 8 rows. An exercise in Chapter 1 asked you to find one or two more examples of these square–triangular numbers and to think about the question of how many there are. Using the mathematical sophistication we've gained in the subsequent 20+ chapters, we're now going to develop a method for finding all square–triangular numbers.

Triangular numbers look like $m(m+1)/2$ and square numbers look like n^2, so square–triangular numbers are solutions to the equation

$$n^2 = \frac{m(m+1)}{2}$$

with positive integers n and m. If we multiply both sides by 8, we can do a little algebra to get

$$8n^2 = 4m^2 + 4m = (2m+1)^2 - 1.$$

This suggests that we make the substitution

$$x = 2m+1 \qquad \text{and} \qquad y = 2n$$

to get the equation

$$2y^2 = x^2 - 1,$$

which we rearrange into the form

$$x^2 - 2y^2 = 1.$$

Solutions to this equation give square–triangular numbers with

$$m = \frac{x-1}{2} \qquad \text{and} \qquad n = \frac{y}{2}.$$

By trial and error we notice one solution, $(x, y) = (3, 2)$, which gives the square–triangular number $(m, n) = (1, 1)$. With a little more experimentation (or using the fact that 36 is square–triangular), we find another solution $(x, y) = (17, 12)$ corresponding to $(m, n) = (8, 6)$. Using a computer, we can search for more solutions by substituting $y = 1, 2, 3, \ldots$ and checking if $1 + 2y^2$ is a square. The next solution found is $(x, y) = (99, 70)$, which gives us a new square–triangular number with $(m, n) = (49, 35)$. In other words, 1225 is a square–triangular number, since

$$35^2 = 1225 = 1 + 2 + 3 + \cdots + 48 + 49.$$

What tools can we use to solve the equation

$$x^2 - 2y^2 = 1?$$

One method we've used repeatedly in the past is factorization. Unfortunately, $x^2 - 2y^2$ does not factor if we stay within the realm of whole numbers; but if we expand our horizons a little, it does factor as

$$x^2 - 2y^2 = \left(x + y\sqrt{2}\right)\left(x - y\sqrt{2}\right).$$

For example, our solution $(x, y) = (3, 2)$ can be written as

$$1 = 3^2 - 2 \cdot 2^2 = \left(3 + 2\sqrt{2}\right)\left(3 - 2\sqrt{2}\right).$$

Now see what happens if we square the left- and right-hand sides of this equation.

$$\begin{aligned} 1 = 1^2 &= \left(3 + 2\sqrt{2}\right)^2 \left(3 - 2\sqrt{2}\right)^2 \\ &= \left(17 + 12\sqrt{2}\right)\left(17 - 12\sqrt{2}\right) \\ &= 17^2 - 2 \cdot 12^2 \end{aligned}$$

So by "squaring" the solution $(x, y) = (3, 2)$, we have constructed the next solution $(x, y) = (17, 12)$.

This process can be repeated to find more solutions. Thus, cubing the $(x, y) = (3, 2)$ solution gives

$$\begin{aligned} 1 = 1^3 &= \left(3 + 2\sqrt{2}\right)^3 \left(3 - 2\sqrt{2}\right)^3 \\ &= \left(99 + 70\sqrt{2}\right)\left(99 - 70\sqrt{2}\right) \\ &= 99^2 - 2 \cdot 70^2, \end{aligned}$$

and taking the fourth power gives

$$\begin{aligned} 1 = 1^4 &= \left(3 + 2\sqrt{2}\right)^4 \left(3 - 2\sqrt{2}\right)^4 \\ &= \left(577 + 408\sqrt{2}\right)\left(577 - 408\sqrt{2}\right) \\ &= 577^2 - 2 \cdot 408^2. \end{aligned}$$

Notice that the fourth power gives us a new square–triangular number, $(m, n) = (288, 204)$. When doing computations of this sort, it's not necessary to raise the original solution to a large power. Instead, we can just multiply the original solution by the current one to get the next one. Thus, to find the 5th-power solution, we multiply the original solution $3 + 2\sqrt{2}$ by the 4th-power solution $577 + 408\sqrt{2}$. This gives

$$\left(3 + 2\sqrt{2}\right)\left(577 + 408\sqrt{2}\right) = 3363 + 2378\sqrt{2},$$

and from this we read off the 5th-power solution $(x, y) = (3363, 2378)$. Continuing in this fashion, we can construct a list of square–triangular numbers.

x	y	m	n	$n^2 = \dfrac{m(m+1)}{2}$
3	2	1	1	1
17	12	8	6	36
99	70	49	35	1225
577	408	288	204	41616
3363	2378	1681	1189	1413721
19601	13860	9800	6930	48024900
114243	80782	57121	40391	1631432881
665857	470832	332928	235416	55420693056

As you see, these square–triangular numbers get quite large.

By raising $3 + 2\sqrt{2}$ to higher and higher powers, we can find more and more solutions to the equation

$$x^2 - 2y^2 = 1,$$

which gives us an inexhaustible supply of square–triangular numbers. Thus, there are infinitely many square–triangular numbers, which answers our original question, but now we ask if this procedure actually produces all of them. The answer is that it does, and you won't be surprised to learn that we use a descent argument to verify this fact.

Theorem 29.1 (Square–Triangular Number Theorem). (a) *Every solution in positive integers to the equation*

$$x^2 - 2y^2 = 1$$

is obtained by raising $3+2\sqrt{2}$ to powers. That is, the solutions (x_k, y_k) can all be found by multiplying out

$$x_k + y_k\sqrt{2} = \left(3+2\sqrt{2}\right)^k \qquad \textit{for } k = 1, 2, 3, \ldots.$$

(b) *Every square–triangular number $n^2 = \frac{1}{2}m(m+1)$ is given by*

$$m = \frac{x_k - 1}{2} \qquad n = \frac{y_k}{2} \qquad \textit{for } k = 1, 2, 3, \ldots,$$

where the (x_k, y_k)'s are the solutions from (a).

Verification. The only thing we have left to check is that, if (u, v) is any solution to $x^2 - 2y^2 = 1$, then it comes from a power of the solution $(3, 2)$. In other words, we must show that

$$u + v\sqrt{2} = \left(3+2\sqrt{2}\right)^k \qquad \text{for some } k.$$

We prove this by the method of descent. Here's the plan. If $u = 3$, then we must have $v = 2$, so there's really nothing to check. So we suppose that $u > 3$, and we show that there is then another solution (s, t) in positive integers so that

$$u + v\sqrt{2} = \left(3+2\sqrt{2}\right)\left(s+t\sqrt{2}\right) \qquad \text{and} \qquad s < u.$$

Why does this help? Well, if $(s, t) = (3, 2)$, then we're done; otherwise, s must be larger than 3, so we can do the same thing starting from (s, t) to find a new solution (q, r) with

$$s + t\sqrt{2} = \left(3+2\sqrt{2}\right)\left(q+r\sqrt{2}\right) \qquad \text{and} \qquad q < s.$$

This means that

$$u + v\sqrt{2} = \left(3+2\sqrt{2}\right)^2\left(q+r\sqrt{2}\right).$$

Now if $(q, r) = (3, 2)$, we're done, and if not, then we apply the procedure yet again. Continuing in this fashion, we observe that this process cannot go on forever, since each time we get a new solution, the value of x is smaller. But these values are all positive integers, so they cannot keep getting smaller indefinitely. Therefore, eventually we get $(3, 2)$ as a solution, which means that eventually we end up with $u + v\sqrt{2}$ written as a power of $3 + 2\sqrt{2}$.

So now we begin with a solution (u, v) with $u > 3$, and we are looking for a solution (s, t) with the property

$$u + v\sqrt{2} = \left(3 + 2\sqrt{2}\right)\left(s + t\sqrt{2}\right) \qquad \text{and} \qquad s < u.$$

Multiplying out the right-hand side of the equation, we need to solve

$$u + v\sqrt{2} = (3s + 4t) + (2s + 3t)\sqrt{2}$$

for s and t. In other words, we need to solve

$$u = 3s + 4t \qquad \text{and} \qquad v = 2s + 3t.$$

This is done easily, the answer being

$$s = 3u - 4v \qquad \text{and} \qquad t = -2u + 3v.$$

Let's check that this (s, t) really gives a solution.

$$\begin{aligned} s^2 - 2t^2 &= (3u - 4v)^2 - 2(-2u + 3v)^2 \\ &= (9u^2 - 24uv + 16v^2) - 2(4u^2 - 12uv + 9v^2) \\ &= u^2 - 2v^2 \\ &= 1 \end{aligned}$$

since we know that (u, v) is a solution. So that's fine. There are two more things we need to check. First, we need to check that s and t are both positive. Second, we must verify that $s < u$, since we want the new solution to be "smaller" than the original solution.

It's easy to see that s is positive using the fact that

$$u^2 = 1 + 2v^2 > 2v^2, \quad \text{which tells us that} \quad u > \sqrt{2}\,v.$$

Then

$$s = 3u - 4v > 3\sqrt{2}\,v - 4v = \left(3\sqrt{2} - 4\right)v > 0,$$

since $3\sqrt{2} \approx 4.242$ is greater than 4.

The verification that t is positive is a little trickier. Here's one way to do it:

$u > 3$	We assumed this.
$u^2 > 9$	Square both sides.
$9u^2 > 9 + 8u^2$	Add $8u^2$ to both sides.
$9u^2 - 9 > 8u^2$	Move the 9 to the other side.
$u^2 - 1 > \frac{8}{9}u^2$	Divide both sides by 9.
$2v^2 > \frac{8}{9}u^2$	Since we know that $u^2 - 2v^2 = 1$.
$v > \frac{2}{3}u$	Divide by 2 and take square roots.

Using this last inequality, it is now easy to check that t is positive.

$$t = -2u + 3v > -2u + 3 \cdot \frac{2}{3}u = 0.$$

We now know that s and t are positive, from which it follows that $s < u$, since $u = 3s + 4t$. This completes our verification that the descent process works and thus completes our proof of the Square–Triangular Number Theorem. □

The Square–Triangular Number Theorem says that every solution (x_k, y_k) in positive integers to the equation

$$x^2 - 2y^2 = 1$$

can be obtained by multiplying out

$$x_k + y_k\sqrt{2} = \left(3 + 2\sqrt{2}\right)^k \qquad \text{for } k = 1, 2, 3, \ldots.$$

The table at the beginning of this chapter makes it clear that the size of the solutions grows very rapidly as k increases. We'd like to get a more precise idea of just how large the kth solution is. To do this, we note that the preceding formula is still correct if we replace $\sqrt{2}$ by $-\sqrt{2}$. In other words, it's also true that

$$x_k - y_k\sqrt{2} = \left(3 - 2\sqrt{2}\right)^k \qquad \text{for } k = 1, 2, 3, \ldots.$$

Now if we add these two formulas together and divide by 2, we obtain a formula for x_k:

$$x_k = \frac{(3 + 2\sqrt{2})^k + (3 - 2\sqrt{2})^k}{2}.$$

Similarly, if we subtract the second formula from the first and divide by $2\sqrt{2}$, we get a formula for y_k:

$$y_k = \frac{\left(3+2\sqrt{2}\right)^k - \left(3-2\sqrt{2}\right)^k}{2\sqrt{2}}.$$

These formulas for x_k and y_k are useful because

$$3+2\sqrt{2} \approx 5.82843 \qquad \text{and} \qquad 3-2\sqrt{2} \approx 0.17157.$$

The fact that $3 - 2\sqrt{2}$ is less than 1 means that, when we take a large power of $3 - 2\sqrt{2}$, we'll get a very tiny number. For example,

$$\left(3-2\sqrt{2}\right)^{10} \approx 0.0000000221,$$

so

$$x_{10} \approx \frac{\left(3+2\sqrt{2}\right)^{10}}{2} \approx 22619536.99999998895 \qquad \text{and}$$

$$y_{10} \approx \frac{\left(3+2\sqrt{2}\right)^{10}}{2\sqrt{2}} \approx 15994428.000000007815.$$

But we know that x_{10} and y_{10} are integers, so the 10th solution is

$$(x_{10}, y_{10}) = (22619537, 15994428).$$

Using this we find that the 10th square–triangular number $n^2 = m(m+1)/2$ is given by

$$n = 7997214 \qquad \text{and} \qquad m = 11309768.$$

It's also apparent from the formulas for x_k and y_k why the solutions grow so rapidly, since

$$x_k \approx \frac{1}{2}(5.82843)^k \qquad \text{and} \qquad y_k \approx \frac{1}{2\sqrt{2}}(5.82843)^k.$$

Thus, each successive solution is more than five times as large as the previous one. Mathematically, we say that the size of the solutions grows *exponentially*. Later, when we study elliptic curves in Chapter 43, we'll see some equations whose solutions grow even faster than this!

Exercises

29.1. Find four solutions in positive integers to the equation

$$x^2 - 5y^2 = 1.$$

[*Hint*. Use trial and error to find a small solution (a, b) and then take powers of $a + b\sqrt{5}$.]

29.2. (a) In Chapters 26 and 27 we studied which numbers can be written as a sum of two squares. Compile some data and try to make a conjecture as to which numbers can be written as a sum of (one or) two triangular numbers. For example, $7 = 1 + 6$ and $25 = 10 + 15$ are sums of two triangular numbers, while 19 is not.

(b) Prove that your conjecture in (a) is correct.

(c) Which numbers do you think can be written as the sum of one, two, or three triangular numbers?

29.3. (a) Fill in the blanks with positive numbers so that the following statement is true: If (m, n) gives a square–triangular number, that is, if the pair (m, n) satisfies the formula $n^2 = m(m+1)/2$, then

$$(1 + \_\_m + \_\_n, 1 + \_\_m + \_\_n)$$

also gives a square–triangular number.

(b) If L is a square–triangular number, explain why $1 + 17L + 6\sqrt{L + 8L^2}$ is the next largest square–triangular number.

29.4. A number n is called a *pentagonal number* if n pebbles can be arranged in the shape of a (filled in) pentagon. The first four pentagonal numbers are 1, 5, 12, and 22, as illustrated in Figure 29.1. You should visualize each pentagon as sitting inside the next larger pentagon. The nth pentagonal number is formed using an outer pentagon whose sides have n pebbles.

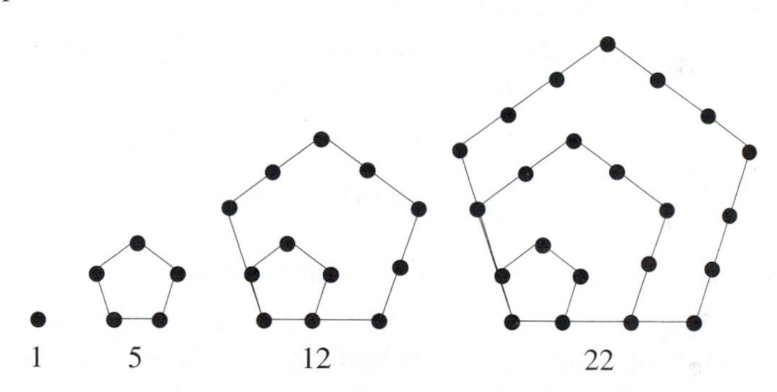

Figure 29.1: The First Four Pentagonal Numbers

(a) Draw a picture for the 5th pentagonal number.

(b) Figure out the pattern and find a simple formula for the nth pentagonal number.

(c) What is the 10th pentagonal number? What is the 100th pentagonal number?

Chapter 30

Pell's Equation

In the last chapter we gave a complete description of the solutions to the equation

$$x^2 - 2y^2 = 1 \qquad \text{in positive integers } x \text{ and } y.$$

This is an example of what is called a *Pell equation*, which is an equation of the form

$$x^2 - Dy^2 = 1,$$

where D is a fixed positive integer that is not a perfect square.

Pell's equation has a long and fascinating history. Its first recorded appearance is in the "Cattle problem of Archimedes." This problem involves eight different kinds of cattle and asks the reader to determine how many there are of each kind. Various linear relations are given, together with two conditions specifying that certain quantities are perfect squares. After a lot of algebra, the problem finally reduces to solving the Pell equation

$$x^2 - 4729494y^2 = 1.$$

The y coordinate of the smallest solution, which was first determined by Amthor in 1880, has 41 digits, and then the answer to the original cattle problem has hundreds of thousands of digits! It seems unlikely that Archimedes or his contemporaries could have determined the solution, but it is fascinating that they even thought to pose such a problem.

Fast-forwarding through the centuries, the first significant progress in solving Pell's equation was made in India. As early as AD 628, Brahmagupta described how to use known solutions to Pell's equation to create new solutions, and in AD 1150 Bhaskaracharya gave an ingenious method, with a surprisingly modern flavor, for finding an initial solution. Unfortunately, this groundbreaking work remained unknown in Europe until long after it had been rediscovered and superceded during the seventeenth century.

> **Brahmagupta** (598–670) One of India's most famous mathematicians of his era, Brahmagupta's best known work is the *Brahmasphutasiddhanta* (*The Opening of the Universe*) written in AD 628. This extraordinary book includes a discussion of equations of the form $x^2 - Dy^2 = A$, and in particular "Pell's" equation $x^2 - Dy^2 = 1$. Brahmagupta describes a composition method for creating new solutions from old ones, which he calls *samasa*, and he gives an algorithm that (sometimes) produces an initial solution.
>
> Approximately 500 years later the Indian mathematician Bhaskaracharya (AD 1114–1185) extended Brahmagupta's work on "Pell's" equation by describing a method that uses an initial approximate solution to find a true solution via repeated reductions. Bhaskaracharya called his method *chakravala*; today arguments of this type go by the name "Fermat descent." We saw examples of Fermat descent in Chapters 26 and 28. Bhaskara illustrates his method by solving $x^2 - 61y^2 = 1$ more than 500 years before Fermat used this equation to issue a challenge.

The modern European history of Pell's equation begins in 1657 when Fermat challenged his fellow mathematicians to solve the equation $x^2 - 61y^2 = 1$. Several of them found the smallest solution, which is

$$(x, y) = (1766319049, 226153980),$$

and in 1657 William Brouncker described a general method for solving Pell's equation. Brouncker demonstrated the efficiency of his method by finding, in just a couple of hours, the solution

$$(32188120829134849, 1819380158564160)$$

to the equation

$$x^2 - 313y^2 = 1$$

J. Wallis described Brouncker's method in a book on algebra and number theory, and Wallis and Fermat both asserted that Pell's equation always has a solution. Euler mistakenly thought that the method in Wallis's book was due to John Pell, another English mathematician, and it is Euler who assigned the equation the name by which it has since been known. Of such misapprehensions is mathematical immortality attained![1]

[1] Some are born great, some achieve greatness, and some have mathematical greatness thrust upon them. With the benefit of historical hindsight, a better name for "Pell's equation" might be the "B^3 equation," in honor of the three mathematicians Brahmagupta, Bhaskaracharya, and Brouncker.

Suppose that we are able to find a solution (x_1, y_1) to the Pell equation

$$x^2 - Dy^2 = 1.$$

Then we can produce new solutions using the same method described in the last chapter for $D = 2$. Factoring the known solution as

$$1 = x_1^2 - Dy_1^2 = \left(x_1 + y_1\sqrt{D}\right)\left(x_1 - y_1\sqrt{D}\right),$$

we square both sides to get a new solution

$$\begin{aligned} 1 = 1^2 &= \left(x_1 + y_1\sqrt{D}\right)^2 \left(x_1 - y_1\sqrt{D}\right)^2 \\ &= \left((x_1^2 + y_1^2 D) + 2x_1y_1\sqrt{D}\right)\left((x_1^2 + y_1^2 D) - 2x_1y_1\sqrt{D}\right) \\ &= (x_1^2 + y_1^2 D)^2 - (2x_1y_1)^2 D. \end{aligned}$$

In other words, $(x_1^2 + y_1^2 D, 2x_1y_1)$ is a new solution. Taking the third power, fourth power, and so on, we can continue to find as many more additional solutions as we desire.

This leaves two vexing questions. First, does every Pell equation have a solution? Note that this question didn't arise when we studied the Pell equation $x^2 - 2y^2 = 1$, since for this specific equation it was easy to find the solution $(3, 2)$. Second, assuming that a given Pell equation does have a solution, is it true that every solution can be found by taking powers of the smallest solution? For the equation $x^2 - 2y^2 = 1$ we showed that this is true; every solution comes from powers of $3 + 2\sqrt{2}$. The answers to both of these questions are given in the following theorem.

Theorem 30.1 (Pell's Equation Theorem). *Let D be a positive integer that is not a perfect square. Then Pell's equation*

$$x^2 - Dy^2 = 1$$

always has solutions in positive integers. If (x_1, y_1) is the solution with smallest x_1, then every solution (x_k, y_k) can be obtained by taking powers

$$x_k + y_k\sqrt{D} = \left(x_1 + y_1\sqrt{D}\right)^k \qquad \textit{for } k = 1, 2, 3, \ldots.$$

For example, the smallest solution to the Pell equation

$$x^2 - 47y^2 = 1$$

is $(x, y) = (48, 7)$. Then all solutions can be obtained by taking powers of $48 + 7\sqrt{47}$. The second and third smallest solutions are

$$\left(48 + 7\sqrt{47}\right)^2 = 4607 + 672\sqrt{47} \quad \text{and}$$
$$\left(48 + 7\sqrt{47}\right)^3 = 442224 + 64505\sqrt{47}.$$

The second part of Pell's Equation Theorem, which says that every solution to Pell's equation is a power of the smallest solution, is actually not too difficult to verify. It can be proved for arbitrary values of D in much the same way that we proved it for $D = 2$ in the previous chapter. The first part, however, which asserts that there is always at least one solution, is somewhat more difficult. We postpone the proof of both parts until Chapter 32.

Table 30.1 lists the smallest solution to Pell's equation for all D up to 75. As you can see, sometimes the smallest solution is quite small. For example, the equation $x^2 - 72y^2 = 1$ has the comparatively tiny solution $(17, 2)$, as does the equation $x^2 - 75y^2 = 1$ with small solution $(26, 3)$. On the other hand, sometimes the smallest solution is huge. Striking examples in the table include

$$x^2 - 61y^2 = 1 \quad \text{with smallest solution} \quad (1766319049, 226153980),$$

and

$$x^2 - 73y^2 = 1 \quad \text{with smallest solution} \quad (2281249, 267000).$$

Another example of this phenomenon is given by

$$x^2 - 97y^2 = 1 \quad \text{with smallest solution} \quad (62809633, 6377352),$$

and of course there's the equation $x^2 - 313y^2 = 1$, already mentioned, which has a similarly spectacular smallest solution.

There is no known pattern as to when the smallest solution is actually small and when it is large. It is known that the smallest solution (x, y) to $x^2 - Dy^2 = 1$ is no larger than $x < 2^D$, but obviously this is not a very good estimate.[2] Maybe you'll be able to discern a pattern that no one else has noticed and use it to prove hitherto unknown properties of the solutions to Pell's equation.

[2]There is a more precise bound for the smallest solution (x, y) that is due to C.L. Siegel. He showed that for each D there is a positive integer h so that the number $h \cdot \log(x + y\sqrt{D})$ has the same order of magnitude as $\sqrt{D}$. In particular, $\log(x)$ and $\log(y)$ won't be much larger than some multiple of $\sqrt{D}$. So, for x and y to be small, this mysterious number h (which is called the *class number* for D) needs to be large. There are many unsolved problems concerning the class number, including the famous conjecture that there are infinitely many D's whose class number equals 1.

D	x	y	D	x	y	D	x	y
1	–	–	26	51	10	51	50	7
2	3	2	27	26	5	52	649	90
3	2	1	28	127	24	53	66249	9100
4	–	–	29	9801	1820	54	485	66
5	9	4	30	11	2	55	89	12
6	5	2	31	1520	273	56	15	2
7	8	3	32	17	3	57	151	20
8	3	1	33	23	4	58	19603	2574
9	–	–	34	35	6	59	530	69
10	19	6	35	6	1	60	31	4
11	10	3	36	–	–	61	1766319049	226153980
12	7	2	37	73	12	62	63	8
13	649	180	38	37	6	63	8	1
14	15	4	39	25	4	64	–	–
15	4	1	40	19	3	65	129	16
16	–	–	41	2049	320	66	65	8
17	33	8	42	13	2	67	48842	5967
18	17	4	43	3482	531	68	33	4
19	170	39	44	199	30	69	7775	936
20	9	2	45	161	24	70	251	30
21	55	12	46	24335	3588	71	3480	413
22	197	42	47	48	7	72	17	2
23	24	5	48	7	1	73	2281249	267000
24	5	1	49	–	–	74	3699	430
25	–	–	50	99	14	75	26	3

Table 30.1: The Smallest Solution to the Pell Equation $x^2 - Dy^2 = 1$

Exercises

30.1. A Pell equation is an equation $x^2 - Dy^2 = 1$, where D is a positive integer that is not a perfect square. Can you figure out why we do not want D to be a perfect square? Suppose that D is a perfect square, say $D = A^2$. Can you describe the integer solutions of the equation $x^2 - A^2y^2 = 1$?

30.2. Find a solution to the Pell equation $x^2 - 22y^2 = 1$ whose x is larger than 10^6.

30.3. Prove that every solution to the Pell equation $x^2 - 11y^2 = 1$ is obtained by taking powers of $10 + 3\sqrt{11}$. (Do not just quote the Pell Equation Theorem. I want you to give a proof for this equation using the same ideas that we used to handle the equation $x^2 - 2y^2 = 1$ in Chapter 29.)

30.4. We continue our study of the pentagonal numbers described in Exercise 29.4.

(a) Are there any pentagonal numbers (aside from 1) that are also triangular numbers? Are there infinitely many?

(b) Are there any pentagonal numbers (aside from 1) that are also square numbers? Are there infinitely many?

(c) Are there any numbers, aside from 1, that are simultaneously triangular, square, and pentagonal? Are there infinitely many?

Chapter 31

Diophantine Approximation

How might we go about finding a solution to Pell's equation

$$x^2 - Dy^2 = 1$$

in positive integers x and y? The factorization

$$\left(x - y\sqrt{D}\right)\left(x + y\sqrt{D}\right) = 1$$

expresses the number 1 as the product of two numbers, one of which is fairly large. More precisely, the number $x + y\sqrt{D}$ is large, especially if x and y are large, so the other factor

$$x - y\sqrt{D} = \frac{1}{x + y\sqrt{D}}$$

must be rather small.

We capitalize on this observation by investigating the following question:

$$\text{How small can we make } x - y\sqrt{D}?$$

If we can find integers x and y that make $x - y\sqrt{D}$ very small, we might hope that x and y are a solution to Pell's equation.[1] For the remainder of this chapter we concentrate on giving Lejeune Dirichlet's beautiful solution to this problem. We return to Pell's equation in the next chapter.

[1]Unfortunately, as happens so often in life, our hopes are dashed when it turns out that x and y are only solutions to a "Pell-like" equation $x^2 - Dy^2 = M$. Don't despair. Turning sorrow into joy, we will be able to take two carefully chosen solutions to $x^2 - Dy^2 = M$ and transform them miraculously into the sought after solution to $x^2 - Dy^2 = 1$.

Let's begin with the easiest answer to our question. For any positive integer y, if we take x to be the integer closest to the number $y\sqrt{D}$, then the difference

$$\left|x - y\sqrt{D}\right| \text{ is at most } \frac{1}{2}.$$

This is true because any real number lies between two integers, so its distance to the nearest integer is at most $\frac{1}{2}$.

Can we do better? Here is a brief table for $\sqrt{13}$. For each integer y from 1 to 40, we have listed the integer x that is closest to $y\sqrt{13}$, together with the values of $\left|x - y\sqrt{13}\right|$ and $x^2 - 13y^2$.

x	y	$\lvert x - y\sqrt{13}\rvert$	$x^2 - 13y^2$
4	1	0.394449	3.000
7	2	0.211103	–3.000
11	3	0.183346	4.000
14	4	0.422205	–12.000
18	5	0.027756	–1.000
22	6	0.366692	16.000
25	7	0.238859	–12.000
29	8	0.155590	9.000
32	9	0.449961	–29.000
36	10	0.055513	–4.000
40	11	0.338936	27.000
43	12	0.266615	–23.000
47	13	0.127833	12.000
50	14	0.477718	–48.000
54	15	0.083269	–9.000
58	16	0.311180	36.000
61	17	0.294372	–36.000
65	18	0.100077	13.000
69	19	0.494526	68.000
72	20	0.111026	–16.000

x	y	$\lvert x - y\sqrt{13}\rvert$	$x^2 - 13y^2$
76	21	0.283423	43.000
79	22	0.322128	–51.000
83	23	0.072321	12.000
87	24	0.466769	81.000
90	25	0.138782	–25.000
94	26	0.255667	48.000
97	27	0.349884	–68.000
101	28	0.044564	9.000
105	29	0.439013	92.000
108	30	0.166538	–36.000
112	31	0.227910	51.000
115	32	0.377641	–87.000
119	33	0.016808	4.000
123	34	0.411257	101.000
126	35	0.194295	–49.000
130	36	0.200154	52.000
133	37	0.405397	–108.000
137	38	0.010948	–3.000
141	39	0.383500	108.000
144	40	0.222051	–64.000

Notice that $\left|x - y\sqrt{13}\right|$ is always less than $1/2$, just as we predicted. Sometimes it is close to $1/2$, as happens for $y = 19$ and $y = 24$, but sometimes it is much smaller. For example, there are four instances in the table for which it is smaller than 0.05:

$$
\begin{aligned}
(x,y) &= (18,5), & \left|x - y\sqrt{13}\right| &= 0.027756, & x^2 - 13y^2 &= -1,\\
(x,y) &= (101,28), & \left|x - y\sqrt{13}\right| &= 0.044564, & x^2 - 13y^2 &= 9,\\
(x,y) &= (119,33), & \left|x - y\sqrt{13}\right| &= 0.016808, & x^2 - 13y^2 &= 4,\\
(x,y) &= (137,38), & \left|x - y\sqrt{13}\right| &= 0.010948, & x^2 - 13y^2 &= -3.
\end{aligned}
$$

If we extend the table up to $y = 200$, we find that all the following pairs (x,y) satisfy $\left|x - y\sqrt{13}\right| < 0.05$:

$$
\begin{gathered}
(18,5), (101,28), (119,33), (137,38), (155,43), (238,66), (256,71),\\
(274,76), (292,81), (375,104), (393,109), (411,114), (494,137),\\
(512,142), (530,147), (548,152), (631,175), (649,180), (667,185).
\end{gathered}
$$

Do you see a pattern? Well, I don't either.

Since there doesn't seem to be any obvious pattern, we take a different approach to making $\left|x - y\sqrt{13}\right|$ small. The method that we use is called

The Pigeonhole Principle

This marvelous principle says that if you have more pigeons than pigeonholes, then at least one of the pigeonholes contains more than one pigeon![2] Although seemingly obvious and trivial, the proper application of this principle yields a bountiful mathematical harvest.

What we are going to do is look for two different multiples $y_1\sqrt{D}$ and $y_2\sqrt{D}$ whose difference is very close to a whole number. To do this, we pick some large number Y and consider all the multiples

$$0\sqrt{D}, 1\sqrt{D}, 2\sqrt{D}, 3\sqrt{D}, \ldots, Y\sqrt{D}.$$

We write each of these multiples as the sum of a whole number and a decimal

[2]The Pigeonhole Principle, so called while residing in town, often Bunburies* in the country under the name of the Box Principle or the Schubfachschluß. The Box Principle asserts that if there are more objects than boxes, then some box contains at least two objects. Many consider Boxes to be dull when compared to Pigeonholes, while the Germanic Schubfachschluß sounds thoroughly respectable and, indeed, I believe is so.**

* The art and artifice of Bunburying is fully explained by Algernon Montcrieff in Act I of Oscar Wilde's *The Importance of Being Earnest*.

** See Algernon's Aunt Augusta (*ibid.*) for more on the merits of the German language.

between 0 and 1,

$$\begin{aligned}
0\sqrt{D} &= N_0 + F_0 && \text{with } N_0 = 0 \text{ and } F_0 = 0.\\
1\sqrt{D} &= N_1 + F_1 && \text{with } N_1 \text{ an integer and } 0 \le F_1 < 1.\\
2\sqrt{D} &= N_2 + F_2 && \text{with } N_2 \text{ an integer and } 0 \le F_2 < 1.\\
3\sqrt{D} &= N_3 + F_3 && \text{with } N_3 \text{ an integer and } 0 \le F_3 < 1.\\
&\vdots\\
Y\sqrt{D} &= N_Y + F_Y && \text{with } N_Y \text{ an integer and } 0 \le F_Y < 1.
\end{aligned}$$

Our pigeons are the $Y + 1$ numbers $F_0, F_1, \ldots, F_Y$. All the pigeons are between 0 and 1, so they are all sitting in the interval $0 \le t < 1$. We form Y pigeonholes by dividing up this interval into Y pieces of equal length. In other words, we take as pigeonholes the intervals

$$\begin{aligned}
&\text{Pigeonhole 1:} && 0/Y \le t < 1/Y.\\
&\text{Pigeonhole 2:} && 1/Y \le t < 2/Y.\\
&\text{Pigeonhole 3:} && 2/Y \le t < 3/Y.\\
& && \vdots\\
&\text{Pigeonhole Y:} && (Y-1)/Y \le t < Y/Y.
\end{aligned}$$

Each pigeon is roosting in one pigeonhole, and there are more pigeons than holes, so the Pigeonhole Principle assures us that some hole contains at least two pigeons. Figure 31.1 illustrates the pigeons and pigeonholes for $D = 13$ and $Y = 5$, where we see that Pigeon 0 and Pigeon 5 are both nesting in Pigeonhole 1.

We now know that there are two pigeons, say pigeons F_m and F_n, that are in the same pigeonhole. We label these two pigeons so that $m < n$. Notice that the pigeonholes are quite narrow, only measuring $1/Y$ from side to side, so the distance between F_m and F_n is less than $1/Y$. In mathematical terms,

$$|F_m - F_n| < 1/Y.$$

Next we use the fact that

$$m\sqrt{D} = N_m + F_m \qquad \text{and} \qquad n\sqrt{D} = N_n + F_n$$

to rewrite this inequality as

$$\left|\left(m\sqrt{D} - N_m\right) - \left(n\sqrt{D} - N_n\right)\right| < 1/Y.$$

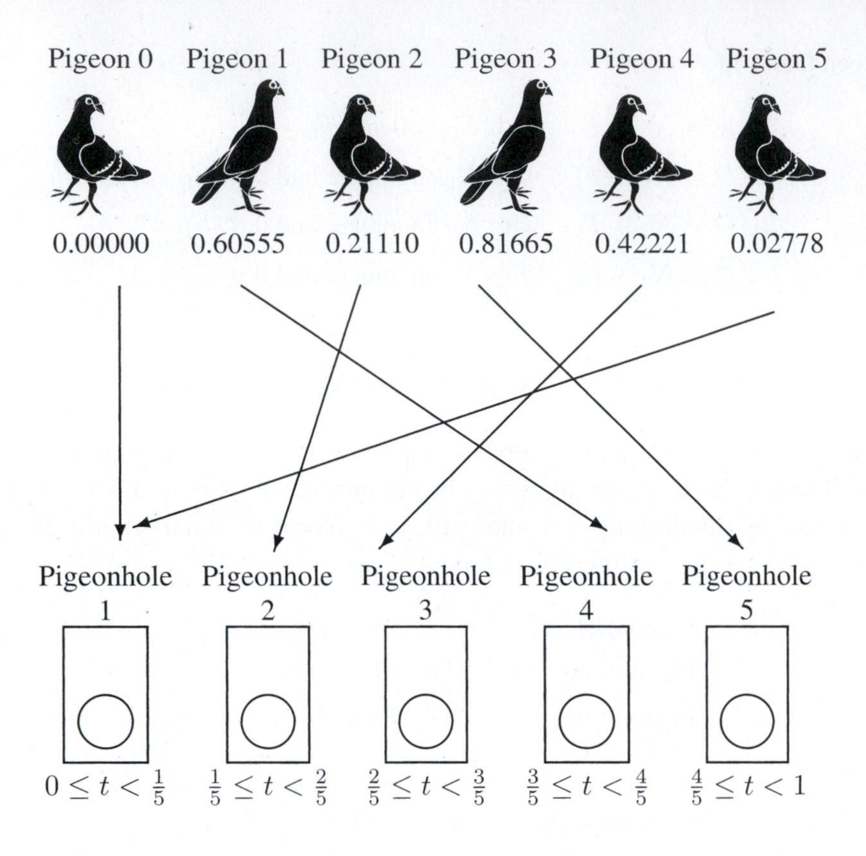

Figure 31.1: Pigeons and Pigeonholes for $D = 13$ and $Y = 5$

Rearranging the terms on the left-hand side gives

$$\left|(N_n - N_m) - (n - m)\sqrt{D}\right| < 1/Y.$$

Note that the quantities $N_n - N_m$ and $n - m$ are (positive) integers. If we call them x and y, respectively, then we have accomplished our aim of making the quantity $|x - y\sqrt{D}|$ quite small.

Our final task is to estimate the size of the integer $y = n - m$. The numbers m and n were chosen so that among the pigeons $F_0, F_1, \ldots, F_Y$, the pigeons F_m and F_n are in the same hole. In particular, m and n are between 0 and Y, and since we chose them with $n > m$, we have $0 < m < n \le Y$. It follows that y satisfies

$$0 < y \le Y.$$

In summary, we have shown that for any integer Y, we can find integers x and y so that

$$0 < y \le Y \qquad \text{and} \qquad \left| x - y\sqrt{D} \right| < 1/Y.$$

Furthermore, by taking Y larger and larger, we automatically get new x's and y's. This is true since for any fixed x and y the inequality

$$\left| x - y\sqrt{D} \right| < 1/Y$$

is false when Y is large enough.[3] Finally, we make the trivial observation that $1/Y \le 1/y$, which completes the verification of the following theorem of Dirichlet.

Theorem 31.1 (Dirichlet's Diophantine Approximation Theorem). (Version 1) *Suppose that D is a positive integer that is not a perfect square. Then there are infinitely many pairs of positive integers (x, y) such that*

$$\left| x - y\sqrt{D} \right| < 1/y.$$

We can use our table with $D = 13$ to illustrate Dirichlet's Diophantine Approximation Theorem. There are seven pairs of numbers (x, y) in the table satisfying the inequality

$$\left| x - y\sqrt{D} \right| < 1/y$$

They are

$$(4, 1),\ (7, 2),\ (11, 3),\ (18, 5),\ (36, 10),\ (119, 33),\ (137, 38).$$

This looks like a lot, but such pairs are actually rather rare.[4] If we were to extend the table up to $y = 1000$, we would find four more pairs:

$$(256, 71),\ (393, 109),\ (649, 180),\ (1298, 360);$$

and even if we go up to $y = 5000$, we only find an additional four pairs:

$$(4287, 1189),\ (4936, 1369),\ (9223, 2558),\ (14159, 3927).$$

---

[3]We are implicitly using the fact that D is not a perfect square, since otherwise the quantity $|x - y\sqrt{D}|$ could equal 0.

[4]Of course, in some sense such pairs are not rare, since there are infinitely many of them, and at first it seems nonsensical to call an infinitely available resource "rare." However, such pairs are extremely rare among the set of all pairs of whole numbers. This is similar to our observation in Chapter 13 that "most" numbers are composite numbers, despite the fact that there are also infinitely many prime numbers.

There is an entire subject, called the theory of *Diophantine Approximation*, which deals with the approximation of irrational quantities by rational numbers. In Dirichlet's Diophantine Approximation Theorem, the irrational number $\sqrt{D}$ is being approximated by the rational number x/y, since dividing both sides of Dirichlet's inequality by y gives

$$\left|\frac{x}{y} - \sqrt{D}\right| < \frac{1}{y^2}.$$

This shows clearly that if y is large, then x/y is extremely close to $\sqrt{D}$.

If you look back at our proof of Dirichlet's Diophantine Approximation Theorem, you will see that we never really used the fact that $\sqrt{D}$ is the square root of D. All we really needed to know was that $\sqrt{D}$ is not itself a rational number. So what we really proved is the following much more general result.

Theorem 31.2 (Dirichlet's Diophantine Approximation Theorem). (Version 2) *Suppose that $\alpha > 0$ is an irrational number. That is, α is a real number that is not a fraction a/b. Then there are infinitely many pairs of positive integers (x, y) such that*

$$|x - y\alpha| < \frac{1}{y}.$$

For example, we could take α to be

$$\pi = 3.14159265358979323846264338 3\ldots.$$

The following table lists all the (x, y)'s with $y < 500$ such that

$$|x - y\pi| < 1/y,$$

together with the values of $|x - y\pi| \cdot y$ and x/y. More precisely, since we're mainly interested in the ratio x/y, the table only lists pairs with $\gcd(x, y) = 1$.

x	y	$\lvert x - y\pi\rvert \cdot y$	x/y
3	1	0.141593	3.0000000000
19	6	0.902664	3.1666666667
22	7	0.061960	3.1428571429
333	106	0.935056	3.1415094340
355	113	0.003406	3.1415929204

Notice that the fractions $22/7$ and $355/113$ are especially close to π. They have been widely used in the past as approximations for π. We would have to extend our search considerably to find a better approximation, since it turns out that the next rational number close to π is

$$\frac{103993}{33102} = 3.141592653011903\ldots.$$

We have been using the brute force approach for finding rational approximations to irrational numbers. For each y, we chose the integer x closest to $y\alpha$ and then checked to see how close x/y comes to α. There is a more systematic method for finding the best x/y's based on the theory of continued fractions. We will study continued fractions in Chapters 39 and 40, and you can read further about them in Davenport's *The Higher Arithmetic* or any standard text on elementary number theory. To illustrate the power of continued fractions, we mention that they can be used to find the rational numbers

$$\frac{5419351}{1725033} = 3.14159265358981538\ldots$$

and

$$\frac{21053343141}{6701487259} = 3.1415926535897932384623817\ldots,$$

which approximate π to 13 and 21 decimal places, respectively. Clearly, we would not want to look for such examples using the brute force approach! Continued fractions also provide an efficient method for solving Pell's equation, even when the solution is extremely large. In the exercises you will see how the continued fraction method is used to find close rational approximations to a certain number called the Golden Ratio.

Exercises

31.1. Prove Version 2 of Dirichlet's Diophantine Approximation Theorem.

31.2. The number

$$\gamma = \frac{1+\sqrt{5}}{2} = 1.61803398874989\ldots$$

was called the *Golden Ratio* by the Greeks.

(a) For each $y \le 20$, find the integer x making $|x - y\gamma|$ as small as possible. Which rational number x/y with $y \le 20$ most closely approximates γ?

(b) If you have access to a computer, find all pairs (x, y) satisfying

$$21 < y \le 1000, \quad \gcd(x, y) = 1, \quad \text{and} \quad |x - y\gamma| < 1/y.$$

Compare the values of x/y and γ.

(c) Find out why the Greeks called γ the Golden Ratio, and write a paragraph or two explaining the mathematical significance of γ and how it appears in art and architecture.

31.3. Consider the following rules for producing a list of rational numbers.

- The first number is $r_1 = 1$.
- The second number is $r_2 = 1 + 1/r_1 = 1 + 1/1 = 2$.
- The third number is $r_3 = 1 + 1/r_2 = 1 + 1/2 = 3/2$.
- The fourth number is $r_4 = 1 + 1/r_3 = 1 + 2/3 = 5/3$.

In general, the nth number in the list is given by $r_n = 1 + 1/r_{n-1}$.

(a) Compute the values of $r_1, r_2, \ldots, r_{10}$. (You should get $r_{10} = 89/55$.)

(b) Let $\gamma = \frac{1}{2}(1 + \sqrt{5}\,)$ be the Golden Ratio. Compute the differences

$$|r_1 - \gamma|, \quad |r_2 - \gamma|, \quad \ldots \quad |r_{10} - \gamma|$$

as decimals. Do you notice anything?

(c) If you have a computer or programmable calculator, compute r_{20}, r_{30}, and r_{40} and compare them with γ.

(d) Suppose that the numbers in the list $r_1, r_2, r_3, \ldots$ get closer and closer to some number r. (In calculus notation, $r = \lim_{n\to\infty} r_n$.) Use the fact that $r_n = 1 + 1/r_{n-1}$ to explain why r should satisfy the relation $r = 1 + 1/r$. Use this to show that $r = \gamma$, thereby explaining your observations in (b) and (c).

(e) Look again at the numerators and denominators of the fractions $r_1, r_2, r_3, \ldots$. Do you recognize these numbers? If you do, prove that they have the value that you claim.

31.4. Dirichlet's Diophantine Approximation Theorem tells us that there are infinitely many pairs of positive integers (x, y) with $\left|x - y\sqrt{2}\right| < 1/y$. This exercise asks you to see if we can do better.

(a) For each of the following y's, find an x so that $\left|x - y\sqrt{2}\right| < 1/y$:

$$y = 12, 17, 29, 41, 70, 99, 169, 239, 408, 577, 985, 1393, 2378, 3363.$$

(This list gives all the y's between 10 and 5000 for which this is possible.) Is the value of $\left|x - y\sqrt{2}\right|$ ever much less than $1/y$? Is it ever as small as $1/y^2$? A good way to compare the value of $\left|x - y\sqrt{2}\right|$ with $1/y$ and $1/y^2$ is to compute the quantities $y\left|x - y\sqrt{2}\right|$ and $y^2\left|x - y\sqrt{2}\right|$. Can you make a guess as to the smallest possible value of $y\left|x - y\sqrt{2}\right|$?

(b) Prove that the following two statements are true for every pair of positive integers (x, y):

(i) $|x^2 - 2y^2| \geq 1$.

(ii) If $\left|x - y\sqrt{2}\right| < 1/y$, then $x + y\sqrt{2} < 2y\sqrt{2} + 1/y$.

Now use (i) and (ii) to show that

$$\left|x - y\sqrt{2}\right| > \frac{1}{2y\sqrt{2} + 1/y} \qquad \text{for all pairs of positive integers } (x, y).$$

Does this explain your computations in (a)?

Chapter 32

Diophantine Approximation and Pell's Equation

We now return to the problem of finding solutions to Pell's equation

$$x^2 - Dy^2 = 1.$$

As we observed in the last chapter, we should look for solutions among those pairs (x, y) making $\left|x - y\sqrt{D}\right|$ small, since any solution to Pell's equation satisfies

$$\left|x - y\sqrt{D}\right| = \frac{1}{\left|x + y\sqrt{D}\right|} < \frac{1}{y}.$$

The idea we use is to take two pairs for which $x^2 - Dy^2$ has the same value and "divide them."

An example helps illustrate what we mean. We take $D = 13$. Looking at the table in Chapter 31, we see that the pairs $(x_1, y_1) = (11, 3)$ and $(x_2, y_2) = (119, 33)$ are both solutions to the equation $x^2 - 13y^2 = 4$. We "divide" these two solutions as follows:

$$\begin{aligned}\frac{119 - 33\sqrt{13}}{11 - 3\sqrt{13}} &= \left(\frac{119 - 33\sqrt{13}}{11 - 3\sqrt{13}}\right)\left(\frac{11 + 3\sqrt{13}}{11 + 3\sqrt{13}}\right) \\ &= \frac{22 - 6\sqrt{13}}{4} \\ &= \frac{11}{2} - \frac{3}{2}\sqrt{13}.\end{aligned}$$

Voilà! The pair $(11/2, 3/2)$ is a solution to Pell's equation $x^2 - 13y^2 = 1$. Unfortunately, as you may already have noticed, it is not a solution in integers. The difficulty is the appearance of that pesky 2 in the denominator. More precisely, notice that there was a 4 in the denominator coming from the fact that $11^2 - 3^2 \cdot 13 = 4$; and we were only able to cancel 2 out of the denominator.

Maybe if we look for more solutions to $x^2 - 13y^2 = 4$, we'll find one that allows us to cancel the entire 4 in the denominator. Searching for additional solutions, we eventually find $(14159, 3927)$ and, using this solution as our (x_2, y_2), we calculate

$$\frac{14159 - 3927\sqrt{13}}{11 - 3\sqrt{13}} = \frac{2596 - 720\sqrt{13}}{4} = 649 - 180\sqrt{13}.$$

Eureka! The Pell equation $x^2 - 13y^2 = 1$ has the solution in integers $(x, y) = (649, 180)$.

Why did the pairs $(11, 3)$ and $(14159, 3927)$ successfully lead to a solution in integers? It turns out that these pairs got rid of the 4 in the denominator because

$$11 \equiv 14159 \pmod 4 \qquad \text{and} \qquad 3 \equiv 3927 \pmod 4.$$

Armed with this crucial observation, we are finally ready to verify Pell's Equation Theorem as stated in Chapter 30. For your convenience, we restate it here.

Theorem 32.1 (Pell's Equation Theorem). *Let D be a positive integer that is not a perfect square. Then Pell's equation*

$$x^2 - Dy^2 = 1$$

always has solutions in positive integers. If (x_1, y_1) is the solution with smallest x_1, then every solution (x_k, y_k) can be obtained by taking powers

$$x_k + y_k\sqrt{D} = \left(x_1 + y_1\sqrt{D}\right)^k \qquad \textit{for } k = 1, 2, 3, \ldots.$$

Verification. Our first goal is to show that Pell's equation has at least one solution. Dirichlet's Diophantine Approximation Theorem 31.1 tells us that there are infinitely many pairs of positive integers (x, y) that satisfy the inequality

$$\left|x - y\sqrt{D}\right| < \frac{1}{y}.$$

Suppose that (x, y) is such a pair. We want to estimate the size of

$$\left|x^2 - Dy^2\right| = \left|x - y\sqrt{D}\right| \cdot \left|x + y\sqrt{D}\right|.$$

The first factor on the right is less than $1/y$. What can we say about the second factor?

Using the fact that $|x - y\sqrt{D}| < 1/y$, we see that x is bounded by

$$x < y\sqrt{D} + 1/y,$$

and so

$$x + y\sqrt{D} < \left(y\sqrt{D} + 1/y\right) + y\sqrt{D} < 2y\sqrt{D} + 1/y < 3y\sqrt{D}.$$

Multiplying both sides of this last inequality by $|x - y\sqrt{D}|$ gives

$$\left|x^2 - Dy^2\right| < \left|x - y\sqrt{D}\right| \cdot 3y\sqrt{D} < (1/y) \cdot \left(3y\sqrt{D}\right) = 3\sqrt{D}.$$

To recapitulate, we have shown that every solution in positive integers (x, y) to the inequality

$$\left|x - y\sqrt{D}\right| < 1/y$$

also satisfies the estimate

$$\left|x^2 - Dy^2\right| < 3\sqrt{D}.$$

We now use a variant of the Pigeonhole Principle introduced in Chapter 31. Our pigeons are the positive integer solutions (x, y) to the inequality $|x - y\sqrt{D}| < 1/y$. Dirichlet's Diophantine Approximation Theorem 31.1 tells us that there are infinitely many pigeons.[1] For pigeonholes we take the integers

$$-T, -T+1, -T+2, \ldots, -3, -2, -1, 0, 1, 2, 3, \ldots, T-2, T-1, T,$$

where T is the largest integer less than $3\sqrt{D}$. We know that if (x, y) is a pigeon, then the quantity $x^2 - Dy^2$ is between $-T$ and T, so we can assign the pigeon (x, y) to the pigeonhole numbered $x^2 - Dy^2$.

We've now taken infinitely many pigeons and stuffed them into a finite collection of pigeonholes![2] Clearly, there must be some pigeonhole that contains infinitely many pigeons. Say pigeonhole M contains infinitely many pigeons. (To simplify the exposition, we assume that M is positive. The argument for negative M is very similar and is left for you to do.) In mathematical terms, this means that the "Pell-like" equation

$$x^2 - Dy^2 = M$$

[1]Don't worry, you won't be assigned the job of feeding the pigeons, nor will you have to clean out the pigeonholes.

[2]This is a task akin to, but messier than, that of getting infinitely many angels to dance on the head of a pin. Which brings up a question you may care to ponder: "To what extent is the Pigeonhole Principle an Obvious Truth, and to what extent is it an Act of Faith?"

has infinitely many solutions in positive integers (x, y). We write the list of solutions as

$$(X_1, Y_1), (X_2, Y_2), (X_3, Y_3), (X_4, Y_4), \ldots.$$

Keep firmly in mind that this list continues indefinitely.

Following the path suggested by the example at the beginning of this chapter, we look for two solutions (X_j, Y_j) and (X_k, Y_k) that also satisfy

$$X_j \equiv X_k \pmod{M} \qquad \text{and} \qquad Y_j \equiv Y_k \pmod{M}.$$

We'll find them by once again using the Pigeonhole Principle. This time our pigeons are the solutions $(X_1, Y_1), (X_2, Y_2), \ldots$, so we have infinitely many pigeons. The pigeonholes are the pairs

$$(A, B) \qquad \text{with } 0 \le A < M \text{ and } 0 \le B < M,$$

so there are M^2 pigeonholes. We assign each pigeon (X_i, Y_i) to a pigeonhole by reducing the numbers X_i and Y_i modulo M. In other words, the pigeon (X_i, Y_i) is assigned to the pigeonhole (A, B) by choosing A and B to satisfy

$$X_i \equiv A \pmod{M}, \quad Y_i \equiv B \pmod{M}, \quad 0 \le A, B < M.$$

We have again managed to stuff infinitely many pigeons into a finite number of pigeonholes, so again there must be some pigeonhole containing infinitely many pigeons. In particular, we can find two different pigeons (X_j, Y_j) and (X_k, Y_k) nesting in the same hole. Mathematically, we have produced two pairs of positive integers (X_j, Y_j) and (X_k, Y_k) with the following properties:

$$\begin{aligned} X_j &\equiv X_k \pmod{M}, \qquad & X_j^2 - DY_j^2 &= M, \\ Y_j &\equiv Y_k \pmod{M}, \qquad & X_k^2 - DY_k^2 &= M. \end{aligned}$$

As described earlier in this chapter, we now expect to get a solution (x, y) to Pell's equation $x^2 - Dy^2 = 1$ by setting

$$x + y\sqrt{D} = \frac{X_j - Y_j\sqrt{D}}{X_k - Y_k\sqrt{D}} = \frac{(X_jX_k - Y_jY_kD) + (X_jY_k - X_kY_j)\sqrt{D}}{X_k^2 - DY_k^2}.$$

In other words, we claim that the formulas

$$x = \frac{X_jX_k - Y_jY_kD}{M} \qquad \text{and} \qquad y = \frac{X_jY_k - X_kY_j}{M}$$

give a solution to $x^2 - Dy^2 = 1$ in integers.

First we check that (x, y) satisfies Pell's equation.

$$\begin{aligned} x^2 - Dy^2 &= \left(\frac{X_jX_k - Y_jY_kD}{M}\right)^2 - D\left(\frac{X_jY_k - X_kY_j}{M}\right)^2 \\ &= \frac{(X_j^2 - DY_j^2)(X_k^2 - DY_j^2)}{M^2} \\ &= 1. \end{aligned}$$

Second, we must verify that x and y are integers. Using the congruences

$$X_j \equiv X_k \pmod{M} \qquad \text{and} \qquad Y_j \equiv Y_k \pmod{M},$$

we find that the "numerators" of x and y satisfy

$$\begin{aligned} X_jX_k - Y_jY_kD \equiv X_j^2 - Y_j^2D = M \equiv 0 \pmod{M}, \\ X_jY_k - X_kY_j \equiv X_jY_j - X_jY_j \equiv 0 \pmod{M}. \end{aligned}$$

Thus the numerators are divisible by M, so the M's in the denominators can be canceled. This shows that x and y are indeed integers, and replacing them by their negatives if necessary, we have found a solution to Pell's equation $x^2 - Dy^2 = 1$ in integers $x, y \geq 0$.

Clearly, $x \geq 1$. It remains to show that $y \neq 0$. But if $y = 0$, then $X_jY_k = X_kY_j$, so we find that

$$\begin{aligned} Y_k^2M = Y_k^2(X_j^2 - DY_j^2) = (X_jY_k)^2 - D(Y_jY_k)^2 \\ = (X_kY_j)^2 - D(Y_jY_k)^2 = Y_j^2(X_k^2 - DY_k^2) = Y_j^2M. \end{aligned}$$

However, we chose Y_j and Y_k to be positive and unequal, so this cannot happen. Therefore, $y \neq 0$, and we have found a solution (x, y) to Pell's equation in positive integers. This completes the verification of the first half of Pell's Equation Theorem.

For the second half, we let (x_1, y_1) be the solution in positive integers with smallest x_1, and we need to show that every solution is obtained by taking powers of $x_1 + y_1\sqrt{D}$. We could reprise the proof that we gave in Chapter 29 when $D = 2$, but instead we present a different and interesting proof that is useful in more general situations.

Suppose that (u, v) is any solution to $x^2 - Dy^2 = 1$ in positive integers. We consider the two real numbers

$$z = x_1 + y_1\sqrt{D} \qquad \text{and} \qquad r = u + v\sqrt{D}.$$

The number z satisfies $z > 1$, so the number r lies between two powers of z, say

$$z^k \leq r < z^{k+1}.$$

[To be precise, take k to be the greatest integer less than $\log(r)/\log(z)$.] Dividing by z^k gives

$$1 \leq z^{-k} \cdot r < z.$$

We observe that $z^k = x_k + y_k\sqrt{D}$, since that is how we defined x_k and y_k, and hence $z^{-k} = x_k - y_k\sqrt{D}$, since we know that

$$\left(x_k + y_k\sqrt{D}\right)\left(x_k - y_k\sqrt{D}\right) = x_k^2 - Dy_k^2 = 1.$$

Thus

$$z^{-k} \cdot r = \left(x_k - y_k\sqrt{D}\right)\left(u + v\sqrt{D}\right) = \underbrace{(x_k u - y_k v D)}_{\text{call this } s} + \underbrace{(x_k v - y_k u)}_{\text{call this } t}\sqrt{D}.$$

We know the following three facts about s and t:

(1) $s^2 - Dt^2 = 1$.
(2) $s + t\sqrt{D} \geq 1$.
(3) $s + t\sqrt{D} < z$.

We claim that $s \geq 0$ and $t \geq 0$. To verify this claim, we eliminate the other possibilities. Fact (1) shows immediately that s and t cannot both be negative. Suppose that $s \geq 0$ and $t < 0$. Then Fact (2) tells us that $s - t\sqrt{D} > s + t\sqrt{D} \geq 1$, so using Fact (1) gives

$$1 = s^2 - Dt^2 = \left(s - t\sqrt{D}\right)\left(s + t\sqrt{D}\right) > 1.$$

This is impossible, so we cannot have $s \geq 0$ and $t < 0$. Similarly, if $s < 0$ and $t \geq 0$, then $-s + t\sqrt{D} > s + \sqrt{D} \geq 1$, so

$$-1 = -s^2 + Dt^2 = \left(-s + t\sqrt{D}\right)\left(s + t\sqrt{D}\right) > 1,$$

which is also impossible. We have eliminated every possibility except $s \geq 0$ and $t \geq 0$, which completes the proof of the claim.

We now know that (s, t) is a solution to $x^2 - Dy^2 = 1$ in nonnegative integers. If s and t were both positive, then the assumption that (x_1, y_1) is the smallest such solution would imply that $s \geq x_1$. Furthermore,

$$t^2 = \frac{s^2 - 1}{D} \geq \frac{x_1^2 - 1}{D} = y_1^2,$$

so we also find that $t \geq y_1$, and hence

$$s + t\sqrt{D} \geq x_1 + y_1\sqrt{D} = z.$$

This contradicts the inequality $s + t\sqrt{D} < z$ in Fact (3). Thus, although s and t are both nonnegative, they are not both positive. So one of them is zero, and from $s^2 - Dt^2 = 1$, we must have $t = 0$ and $s = 1$.

To recapitulate, we have shown that $z^{-k} \cdot r$ is equal to 1, which is the same as saying that $r = z^k$. In other words, we have shown that if (u, v) is any solution to $x^2 - Dy^2 = 1$ in positive integers then there is some exponent $k \geq 1$ so that

$$r = u + v\sqrt{D} \quad \text{is equal to} \quad z^k = \left(x_1 + y_1\sqrt{D}\right)^k = x_k + y_k\sqrt{D}.$$

This shows that $u + v\sqrt{D}$ is a power of $x_1 + y_1\sqrt{D}$, which completes the verification of Pell's Equation Theorem. □

Exercises

32.1. In this chapter we have shown that Pell's equation $x^2 - Dy^2 = 1$ always has a solution in positive integers. This exercise explores what happens if the 1 on the right-hand side is replaced by some other number.

(a) For each $2 \leq D \leq 15$ that is not a perfect square, determine whether or not the equation $x^2 - Dy^2 = -1$ has a solution in positive integers. Can you determine a pattern that lets you predict for which D's it has a solution?

(b) If (x_0, y_0) is a solution to $x^2 - Dy^2 = -1$ in positive integers, show that $(x_0^2 + Dy_0^2, 2x_0y_0)$ is a solution to Pell's equation $x^2 - Dy^2 = 1$.

(c) Find a solution to $x^2 - 41y^2 = -1$ by plugging in $y = 1, 2, 3, \ldots$ until you find a value for which $41y^2 - 1$ is a perfect square. (You won't need to go very far.) Use your answer and (b) to find a solution to Pell's equation $x^2 - 41y^2 = 1$ in positive integers.

(d) If (x_0, y_0) is a solution to the equation $x^2 - Dy^2 = M$, and if (x_1, y_1) is a solution to Pell's equation $x^2 - Dy^2 = 1$, show that $(x_0x_1 + Dy_0y_1, x_0y_1 + y_0x_1)$ is also a solution to the equation $x^2 - Dy^2 = M$. Use this to find five different solutions in positive integers to the equation $x^2 - 2y^2 = 7$.

32.2. For each of the following equations, either find a solution (x, y) in positive integers, or explain why no such solution can exist.

(a) $x^2 - 11y^2 = 7$ (b) $x^2 - 11y^2 = 433$ (c) $x^2 - 11y^2 = 3$

Chapter 33

Number Theory and Imaginary Numbers

Most everyone these days is familiar with the "number"

$$i = \sqrt{-1}.$$

The use of "i" to denote the square root of negative 1 dates back to the days when people viewed such numbers with great suspicion and, indeed, felt that they were so far from being real numbers that they deserved to be called *imaginary*. In these more enlightened times we recognize that all[1] numbers are, to some extent, abstractions that can be used to solve certain sorts of problems. For example, negative numbers (which were not used by European mathematicians even in the fourteenth century, although they were in use in India as early as AD 600) are not needed for counting cattle, but they are useful in keeping track of who owes how many cattle to whom. Fractions arise naturally when people start dealing with objects that can be subdivided, such as bushels of wheat or fields of corn. Irrational numbers—that is, numbers that are not fractions—appear in even the simplest sorts of measurements, as the Pythagoreans discovered when they found that the diagonal of certain geometric figures may be incommensurable with their sides. This discovery upset the conventional mathematical wisdom of the time, and the penalty was death for those who revealed the secret. The introduction of imaginary numbers caused similar consternation in nineteenth-century Europe, although thankfully the sanctions imposed on those using imaginary numbers were less severe than in earlier times.

Imaginary numbers and more generally *complex numbers*,

$$z = x + iy,$$

[1]Or at least, almost all. As Leopold Kronecker (1823–1891) so eloquently put it, "God made the integers, all else is the work of man."

were introduced into mathematics for a definite purpose, the solution of equations. Thus negative numbers are needed to solve the equation $x + 3 = 0$, while fractions are needed for $3x - 7 = 0$. For the equation $x^2 - 5 = 0$, we need the irrational quantity $\sqrt{5}$, but even if we introduce more general irrational numbers, we still won't be able to solve the very simple equation $x^2 + 1 = 0$. Since this equation doesn't have any solutions in "real numbers," there's nothing to stop us from creating a new sort of number to be a solution and giving that new number the name i. This is no different from observing that, since the equation $x^2 - 5 = 0$ has no solution in fractions, we are free to create a solution and call it $\sqrt{5}$. In fact, we're even doing the same thing when we observe that $3x - 7 = 0$ has no solutions in whole numbers, so we create a solution and call it $\frac{7}{3}$.

Complex numbers were thus invented[2] to solve equations such as $x^2 + 1 = 0$, but why stop there? Now that we know about complex numbers, we can try to solve more complicated equations, even equations with complex coefficients such as

$$(3 + 2i)x^3 - (\sqrt{3} - 5i)x^2 - (\sqrt[7]{5} + \sqrt[3]{14}i)x + 17 - 8i = 0.$$

If this equation had no solutions, then we would be forced to invent even more numbers. Amazingly, it turns out that this equation does have solutions in complex numbers. The solutions (accurate to five decimal places) are

$$1.27609 + 0.72035i, \quad 0.03296 - 2.11802i, \quad -1.67858 - 0.02264i.$$

In fact, there are enough complex numbers to solve every equation of this sort, a statement that has a long history and an impressive name.

Theorem 33.1 (The Fundamental Theorem of Algebra). *If $a_0, a_1, a_2, \ldots, a_d$ are complex numbers with $a_0 \neq 0$ and $d \geq 1$, then the equation*

$$a_0x^d + a_1x^{d-1} + a_2x^{d-2} + \cdots + a_{d-1}x + a_d = 0$$

has a solution in complex numbers.

This theorem was formulated (and used) by many mathematicians during the eighteenth century, but the first satisfactory proofs weren't discovered until the early part of the nineteenth century. Many proofs are now known, some using mostly algebra, some using analysis (calculus), and some using geometric ideas. Unfortunately, none of the proofs is easy, so we won't give one here.

[2] Some would say that complex numbers already existed and were merely discovered, while others believe just as strongly that mathematical entities such as the complex numbers are abstractions that were created by human imagination. This question of whether mathematics is discovered or created is a fascinating philosophical conundrum for which (as with most good philosophical questions) there is unlikely ever to be a definitive answer.

Instead, we investigate the number theory of the complex numbers. You undoubtedly recall the simple rules for adding and subtracting complex numbers, and multiplication is just as easy,

$$(a+bi)(c+di) = ac + adi + bci + bdi^2 = (ac - bd) + (ad + bc)i.$$

For division we use the old trick of rationalizing the denominator,

$$\frac{a+bi}{c+di} = \frac{a+bi}{c+di} \cdot \frac{c-di}{c-di} = \frac{(ac+bd) + (-ad+bc)i}{c^2+d^2}.$$

We do number theory with a certain subset of the complex numbers called the *Gaussian integers*. These are the complex numbers that look like

$$a + bi \qquad \text{with } a \text{ and } b \text{ both integers.}$$

The Gaussian integers have many properties in common with the ordinary integers. For example, if α and β are Gaussian integers, then so are their sum $\alpha + \beta$, their difference $\alpha - \beta$, and their product $\alpha\beta$. However, the quotient of two Gaussian integers need not be a Gaussian integer (just as the quotient of two ordinary integers need not be an integer). For example,

$$\frac{3+2i}{1-6i} = \frac{-9+20i}{37}$$

is not a Gaussian integer, while

$$\frac{16-11i}{3+2i} = \frac{26-65i}{13} = 2 - 5i$$

is a Gaussian integer.

This suggests that we define a notion of divisibility for Gaussian integers just as we did for ordinary integers. So we say that the Gaussian integer $a + bi$ *divides* the Gaussian integer $c + di$ if we can find a Gaussian integer $e + fi$ so that

$$c + di = (a + bi)(e + fi).$$

Of course, this is the same as saying that the quotient $\dfrac{c+di}{a+bi}$ is a Gaussian integer. For example, we saw that $3 + 2i$ divides $16 - 11i$, but $1 - 6i$ does not divide $3 + 2i$.

Now that we know how to talk about divisibility, we can talk about factorization. For example, the number $1238 - 1484i$ factors as

$$1238 - 1484i = (2+3i)^3 \cdot (-1+4i) \cdot (3+i)^2.$$

And even ordinary integers such as $600 = 2^3 \cdot 3 \cdot 5^2$, which we think we already know how to factor, can be factored further using Gaussian integers:

$$600 = -i \cdot (1+i)^6 \cdot 3 \cdot (2+i)^2 \cdot (2-i)^2.$$

For the ordinary integers, the primes are the basic building blocks because they cannot be factored any further. Of course, technically this isn't quite true, since the prime 7 can be "factored" as

$$7 = 1 \cdot 7, \quad \text{or as} \quad 7 = (-1) \cdot (-1) \cdot 7, \quad \text{or even as}$$
$$7 = (-1) \cdot (-1) \cdot (-1) \cdot (-1) \cdot 1 \cdot 1 \cdot 1 \cdot 7.$$

However, we recognize that these aren't really different factorizations, because we can always put in more 1's and pairs of -1's. What is it about 1 and -1 that makes them special? The answer is that they are the only two integers that have integer (multiplicative) inverses:

$$1 \cdot 1 = 1 \qquad \text{and} \qquad (-1) \cdot (-1) = 1.$$

(In fact, they are their own inverses, but this turns out to be less important.) Notice that if a is any integer other than 1 and -1 then a does not have an integer multiplicative inverse, since the equation $ab = 1$ does not have a solution b that is an integer. We say that 1 and -1 are the only *units* in the ordinary integers.

The Gaussian integers are blessed with more units than the ordinary integers. For example, i itself is a unit, since

$$i \cdot (-i) = 1.$$

This equation also shows that $-i$ is a unit, so we see that the Gaussian integers have at least four units: 1, -1, i, and $-i$. Are there any others?

To answer this question, we suppose that $a + bi$ is a unit in the Gaussian integers. This means that it has a multiplicative inverse, so there is another Gaussian integer $c + di$ such that

$$(a + bi)(c + di) = 1.$$

Multiplying everything out, we find that

$$ac - bd = 1 \qquad \text{and} \qquad ad + bc = 0,$$

so we are looking for solutions (a, b, c, d) to these equations in ordinary integers. Later we will see a fancier (and more geometric) way to solve this problem but, for now, let's just use a little algebra and a little common sense.

We need to consider several cases. First, if $a = 0$ then $-bd = 1$, so $b = \pm 1$ and $a + bi = \pm i$. Second, if $b = 0$, then $ac = 1$, so $a = \pm 1$ and $a + bi = \pm 1$. These first two cases lead to the four units that we already know.

For our third and final case, suppose that $a \neq 0$ and $b \neq 0$. Then we can solve the first equation for c and substitute it into the second equation:

$$c = \frac{1+bd}{a} \quad \Longrightarrow \quad ad + b\left(\frac{1+bd}{a}\right) = 0 \quad \Longrightarrow \quad \frac{a^2d + b + b^2d}{a} = 0.$$

Thus any solution with $a \neq 0$ must satisfy

$$(a^2 + b^2)d = -b.$$

This means that $a^2 + b^2$ divides b, which is absurd, since $a^2 + b^2$ is larger than b (remember neither a nor b is 0). This means that Case 3 yields no new units, so we have completed the proof of our first theorem about the Gaussian integers.

Theorem 33.2 (Gaussian Unit Theorem). *The only units in the Gaussian integers are* 1, -1, i, *and* $-i$. *That is, these are the only Gaussian integers that have Gaussian integer multiplicative inverses.*

One thing that makes the Gaussian integers an interesting subset of the complex numbers is that the sum, difference, and product of any two Gaussian integers is again a Gaussian integer. Notice that the ordinary integers also have this property. A subset of the complex numbers that has this property (and also contains the numbers 0 and 1) is called a *ring*, so the ordinary integers and the Gaussian integers are examples of rings. Many other interesting rings lurk within the complex numbers, some of which you will have an opportunity to study in the Exercises for this chapter.

Returning to our discussion of factorization in the Gaussian integers, we might say that a Gaussian integer α is prime if it is only divisible by ± 1 and itself, but this is clearly the wrong thing to do. For example, we can always write

$$\alpha = i \cdot (-i) \cdot \alpha,$$

so any α is divisible by i and by $-i$ and by $i\alpha$ and by $-i\alpha$. This leads us to the correct definition. A Gaussian integer α is called a *Gaussian prime* if the only Gaussian integers that divide α are the eight numbers

$$1, \quad -1, \quad i, \quad -i, \quad \alpha, \quad -\alpha, \quad i\alpha, \quad \text{and} \quad -i\alpha.$$

In other words, the only numbers dividing α are units and α times a unit.

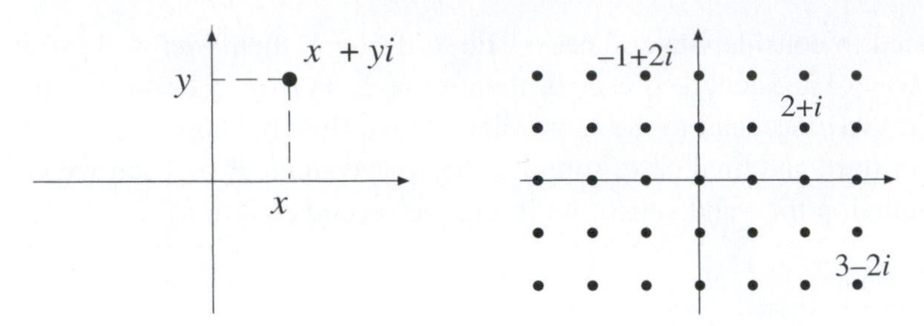

(a) Complex Numbers in the Plane (b) The Gaussian Integers

Figure 33.1: The Geometry of the Complex Numbers

Now that we know what Gaussian primes are, can we identify them? For example, which of the following do you think are Gaussian primes?

$$2, \qquad 3, \qquad 5, \qquad 1+i, \qquad 3+i, \qquad 2+3i.$$

We could answer this question using the algebraic ideas we employed earlier when we determined all the Gaussian units, but our task is easier if we use a soupçon of geometry.

We introduce geometry into our study of complex numbers by identifying each complex number $x+yi$ with the point (x,y) in the plane. This idea is illustrated in Figure 33.1(a). The Gaussian integers are then identified with the integer points (x,y), that is, the points with x and y both integers. Figure 33.1(b) shows that the Gaussian integers form a square-shaped lattice of points in the plane.

Having identified the complex number $x+yi$ with the point (x,y), we can talk about the distance between two complex numbers. In particular, the distance from $x+yi$ to 0 is $\sqrt{x^2+y^2}$. It is a little more convenient to work with the square of this distance, so we define the *norm of* $x+yi$ to be

$$\mathrm{N}(x+yi) = x^2+y^2.$$

Intuitively, the norm of a complex number α is a sort of measure of the size or magnitude of α. In this sense, the norm measures a geometric quantity. On the other hand, the norm also has a very important algebraic property: the norm of a product equals the product of the norms. It is the interplay between these geometric and algebraic properties that makes the norm such a useful tool for studying the Gaussian integers. We now verify the multiplication property.

Theorem 33.3 (Norm Multiplication Property). *Let α and β be any complex numbers. Then*

$$\mathrm{N}(\alpha\beta) = \mathrm{N}(\alpha)\,\mathrm{N}(\beta).$$

Proof. If we write $\alpha = a + bi$ and $\beta = c + di$, then

$$\alpha\beta = (ac - bd) + (ad + bc)i,$$

so we need to check that

$$(ac - bd)^2 + (ad + bc)^2 = (a^2 + b^2)(c^2 + d^2).$$

This is easily verified by multiplying out both sides, a task that we leave for you (or look back at Chapter 26). □

Before returning to the problem of factorization, it is instructive to see how the norm can be used to find the units. Thus, suppose that $\alpha = a + bi$ is a unit. This means that there is a $\beta = c + di$ so that $\alpha\beta = 1$. Taking norms of both sides and using the Norm Multiplication Property yields

$$\mathrm{N}(\alpha)\,\mathrm{N}(\beta) = \mathrm{N}(\alpha\beta) = \mathrm{N}(1) = 1,$$

so

$$(a^2 + b^2)(c^2 + d^2) = 1.$$

But a, b, c, d are all integers, so we must have $a^2 + b^2 = 1$. The only solutions to this equation are

$$(a, b) = (1, 0),\ (-1, 0),\ (0, 1),\ (0, -1),$$

which gives us a new proof that the only Gaussian units are 1, -1, i, and $-i$. We also obtain a useful characterization:

A Gaussian integer α is a unit if and only if $\mathrm{N}(\alpha) = 1$.

Now let's try to factor some numbers. We start with the number 2 and try to factor it as

$$(a + bi)(c + di) = 2.$$

Taking the norm of both sides yields

$$(a^2 + b^2)(c^2 + d^2) = 4.$$

[Note that $\mathrm{N}(2) = \mathrm{N}(2 + 0i) = 2^2 + 0^2 = 4$.] We don't want either of $a + bi$ or $c + di$ to be a unit, so neither $a^2 + b^2$ nor $c^2 + d^2$ is allowed to equal 1. Since their product is supposed to equal 4 and they're both positive integers, we must have

$$a^2 + b^2 = 2 \qquad \text{and} \qquad c^2 + d^2 = 2.$$

These equations certainly have solutions. For example, if we take $(a, b) = (1, 1)$ and divide 2 by $a + bi = 1 + i$, we get

$$c + di = \frac{2}{1+i} = \frac{2(1-i)}{2} = 1 - i.$$

Hence $2 = (1 + i)(1 - i)$, so 2 is not a Gaussian integer prime!

If we try to factor 3 in a similar fashion, we end up with the equations

$$a^2 + b^2 = 3 \qquad \text{and} \qquad c^2 + d^2 = 3.$$

These equations clearly have no solutions, so 3 is a Gaussian prime. On the other hand, if we start with 5, we end up with the factorization $5 = (2 + i)(2 - i)$.

We can use the same procedure to factor Gaussian integers that are not ordinary integers. The general method for factoring a Gaussian integer α is to set

$$(a + bi)(c + di) = \alpha$$

and take the norm of both sides to obtain

$$(a^2 + b^2)(c^2 + d^2) = \mathrm{N}(\alpha).$$

This is an equation in integers, and we want a nontrivial solution, by which we mean a solution where neither $a^2 + b^2$ nor $c^2 + d^2$ equals 1. So the first thing we need to do is factor the integer $\mathrm{N}(\alpha)$ into a product AB with $A \neq 1$ and $B \neq 1$. Then we need to solve

$$a^2 + b^2 = A \qquad \text{and} \qquad c^2 + d^2 = B.$$

Thus, factorization of Gaussian integers leads us back to the sums of two squares problem that we studied in Chapters 26 and 27.

To see how this works in practice, we factor $\alpha = 3 + i$. The norm of α is $\mathrm{N}(\alpha) = 10$, which factors as $2 \cdot 5$, so we solve $a^2 + b^2 = 2$ and $c^2 + d^2 = 5$. There are a number of solutions. For example, if we take $(a, b) = (1, 1)$, then we obtain the factorization

$$3 + i = (1 + i)(2 - i).$$

Do you understand why we get several solutions? It has to do with the fact that the factorization of $3 + i$ can always be changed by units. Thus, if we take $(a, b) = (-1, 1)$, we get $3 + i = (-1 + i)(-1 - 2i)$, which is really the same factorization, since $-1 + i = i(1 + i)$ and $-1 - 2i = -i(2 - i)$.

What happens when we try to factor $\alpha = 1 + i$? The norm of α is $\mathrm{N}(\alpha) = 2$, and 2 cannot be factored as $2 = AB$ with ordinary integers $A, B > 1$. This means that α has no nontrivial factorizations in the Gaussian integers, so it is prime. Similarly, we cannot factor $2+3i$ in the Gaussian integers, since $\mathrm{N}(2+3i)=13$ is a prime in the ordinary integers. [But note that 13 is not a Gaussian prime, since $13=(2+3i)(2-3i)$.] So $2+3i$ is a Gaussian prime. More generally, if $\mathrm{N}(\alpha)$ is an ordinary prime, then the same reasoning shows that α must be a Gaussian prime. It turns out that these are half the Gaussian primes, and the other half are numbers like 3 that are both ordinary primes and Gaussian primes. The following theorem gives a complete description of all Gaussian primes. Don't be fooled by the shortness of its proof, which merely reflects all our hard work in earlier chapters. It is a deep and beautiful result.

Theorem 33.4 (Gaussian Prime Theorem). *The Gaussian primes can be described as follows:*

(i) $1+i$ *is a Gaussian prime.*

(ii) *Let* p *be an ordinary prime with* $p \equiv 3 \pmod 4$. *Then* p *is a Gaussian prime.*

(iii) *Let* p *be an ordinary prime with* $p \equiv 1 \pmod 4$ *and write* p *as a sum of two squares* $p=u^2+v^2$ (see Chapter 26). Then $u+vi$ is a Gaussian prime.

Every Gaussian prime is equal to a unit (± 1 or $\pm i$) multiplied by a Gaussian prime of the form (i), (ii), or (iii).[3]

Proof. As we observed previously, if $\mathrm{N}(\alpha)$ is an ordinary prime, then α must be a Gaussian prime. The number $1+i$ in category (i) has norm 2, so it is a Gaussian prime. Similarly, the numbers $u+vi$ in category (iii) have norm $u^2+v^2=p$, so they are also Gaussian primes.

Next we check category (ii), so we let $\alpha = p$ be an ordinary prime with $p \equiv 3 \pmod 4$. If α were to have a factorization into Gaussian integers, say $(a+bi)(c+di)=\alpha$, then taking norms would yield

$$(a^2+b^2)(c^2+d^2)=\mathrm{N}(\alpha)=p^2.$$

To get a nontrivial factorization, we would need to solve

$$a^2+b^2=p \qquad \text{and} \qquad c^2+d^2=p.$$

But we know from the Sum of Two Squares Theorem (Chapter 26) that since $p \equiv 3 \pmod 4$, it cannot be written as a sum of two squares, so there are no solutions. Therefore, p cannot be factored, so it is a Gaussian prime.

[3]As you know, mathematicians love to give obscure names to the objects they study. In this instance, category (i) primes are called *ramified*, category (ii) primes are called *inert*, and category (iii) primes are called *split*. This means that if we cover a prime in category (iii) with ice cream and fruit, it becomes a *banana split prime*!

We've now shown that the numbers in categories (i), (ii), and (iii) are indeed Gaussian primes, so it remains to show that every Gaussian prime fits into one of these three categories. To do this, we use the following lemma.

Lemma 33.5 (Gaussian Divisibility Lemma). *Let $\alpha = a + bi$ be a Gaussian integer.*
(a) *If 2 divides $\mathrm{N}(\alpha)$, then $1 + i$ divides α.*
(b) *Let $\pi = p$ be a category (ii) prime, and suppose that p divides $\mathrm{N}(\alpha)$ as ordinary integers. Then π divides α as Gaussian integers.*
(c) *Let $\pi = u + vi$ be a Gaussian prime in category (iii), and let $\bar{\pi} = u - vi$. (This is a natural notation, since $\bar{\pi}$ is indeed the complex conjugate of the complex number π.) Suppose that $N(\pi) = p$ divides $\mathrm{N}(\alpha)$ as ordinary integers. Then at least one of π and $\bar{\pi}$ divides α as Gaussian integers.*

Proof of Lemma. (a) We are given that 2 divides $\mathrm{N}(\alpha) = a^2 + b^2$, so a and b are either both odd or both even. It follows that $a + b$ and $-a + b$ are both divisible by 2, so the quotient

$$\frac{a+bi}{1+i} = \frac{(a+b)+(-a+b)i}{2}$$

is a Gaussian integer. Hence $a + bi$ is divisible by $1 + i$.
(b) We are given that $p \equiv 3 \pmod 4$ and that p divides $a^2 + b^2$. This means that $a^2 \equiv -b^2 \pmod p$, so we can compute the Legendre symbols

$$\left(\frac{a}{p}\right)^2 = \left(\frac{a^2}{p}\right) = \left(\frac{-b^2}{p}\right) = \left(\frac{-1}{p}\right)\left(\frac{b}{p}\right)^2.$$

Since $p \equiv 3 \pmod 4$, the Law of Quadratic Reciprocity (Chapter 24) tells us that $\left(\frac{-1}{p}\right) = -1$, so we get

$$\left(\frac{a}{p}\right)^2 = -\left(\frac{b}{p}\right)^2.$$

But the value of a Legendre symbol is ± 1, so we seem to have ended up with $1 = -1$. What went wrong? Stop and try to figure it out for yourself before you read on.

The answer is that a Legendre symbol such as $\left(\frac{a}{p}\right)$ only makes sense if $a \not\equiv 0 \pmod p$; we never assigned a value to $\left(\frac{0}{p}\right)$. So the egress from our seeming contradiction is that a and b must both be divisible by p, say $a = pa'$ and $b = pb'$. Hence $\alpha = a + bi = p(a' + b'i)$ is divisible by $p = \pi$, which is what we were trying to prove.
(c) We are given that p divides $\mathrm{N}(\alpha)$, so we can write

$$\mathrm{N}(\alpha) = a^2 + b^2 = pK \qquad \text{for some integer } K \geq 1.$$

We need to show that at least one of the two numbers

$$\frac{\alpha}{\pi} = \frac{(au+bv)+(-av+bu)i}{p} \quad \text{and} \quad \frac{\alpha}{\bar{\pi}} = \frac{(au-bv)+(av+bu)i}{p}$$

is a Gaussian integer.

Our first observation is that

$$\begin{aligned}(au+bv)(au-bv) &= a^2u^2 - b^2v^2 \\ &= a^2u^2 - b^2(p-u^2) \\ &= (a^2+b^2)u^2 - pb^2 \\ &= pKu^2 - pb^2,\end{aligned}$$

so at least one of the two integers $au+bv$ and $au-bv$ is divisible by p. A similar calculation shows that

$$(-av+bu)(av+bu) = pKu^2 - pa^2,$$

so at least one of $-av+bu$ and $av+bu$ is divisible by p. There are thus four cases to consider:

Case 1. $au+bv$ and $-av+bu$ are divisible by p.

Case 2. $au+bv$ and $av+bu$ are divisible by p.

Case 3. $au-bv$ and $-av+bu$ are divisible by p.

Case 4. $au-bv$ and $av+bu$ are divisible by p.

Case 1 is easy, since it immediately implies that the quotient α/π is a Gaussian integer, hence π divides α. Similarly, for Case 4, the quotient $\alpha/\bar{\pi}$ is a Gaussian integer, so $\bar{\pi}$ divides α. This takes care of Cases 1 and 4.

Next consider Case 2, which is a little more complicated. We are given that p divides both $au+bv$ and $av+bu$, from which we deduce that p divides

$$(au+bv)b - (av+bu)a = (b^2-a^2)v.$$

(The idea here is that we "eliminated" u from the equation.) Since p clearly doesn't divide v (remember that $p = u^2+v^2$), we see that p divides b^2-a^2. However, we also know that p divides a^2+b^2, so we find that p divides both

$$2a^2 = (a^2+b^2) - (b^2-a^2) \quad \text{and} \quad 2b^2 = (a^2+b^2) + (b^2-a^2).$$

Since $p \neq 2$, we finally deduce that p divides both a and b, say $a = pa'$ and $b = pb'$. Then

$$\alpha = a + bi = p(a'+b'i) = (u^2+v^2)(a'+b'i) = \pi\bar{\pi}(a'+b'i),$$

so for Case 2 we find that α is actually divisible by both π and $\bar{\pi}$.

Finally, we observe that a similar argument for Case 3 leads to the same conclusion as in Case 2; we leave the details for you to complete. □

Resumption of Proof. After that (not so) brief interlude, we're ready to resume proving the Gaussian Prime Theorem. We suppose that $\alpha = a + bi$ is a Gaussian prime, and our aim is to show that α fits into one of the three categories of Gaussian primes. We know that $\mathrm{N}(\alpha) \neq 1$, since α is not a unit, so there is (at least) one prime p that divides $\mathrm{N}(\alpha)$.

Suppose first that $p = 2$. Then Part (a) of the Lemma tells us that $1 + i$ divides α. But α is supposed to be prime, so this means that α must equal $1 + i$ multiplied by a unit, so α fits into category (i).

Next suppose that $p \equiv 3 \pmod 4$. Then Part (b) of the Lemma tells us that p divides α, so again the primality of α implies that α equals a unit times p. Hence α is a category (ii) prime.

Finally, suppose that $p \equiv 1 \pmod 4$. The Sum of Two Squares Theorem in Chapter 26 tells us that p can be written as a sum of two squares, say $p = u^2 + v^2$, and then Part (c) of the Lemma says that α is divisible by either $u + iv$ or by $u - iv$. Hence α is a unit times one of $u + iv$ or $u - iv$. In particular, $a^2 + b^2 = u^2 + v^2 = p$, so α is a category (iii) prime. This completes our proof that every Gaussian prime fits into one of the three categories. □

Exercises

33.1. Write a short essay (one or two pages) on the following topics:
- (a) The introduction of complex numbers in nineteenth-century Europe
- (b) The discovery of irrational numbers in ancient Greece
- (c) The introduction of zero and negative numbers into Indian mathematics, Arabic mathematics, and European mathematics
- (d) The discovery of transcendental numbers in nineteenth-century Europe

33.2. (a) Choose one of the following two statements and write a one-page essay defending it. Be sure to give at least three specific reasons why your statement is true and the opposing statement is incorrect.

Statement 1. Mathematics already exists and is merely discovered by people (in the same sense that the planet Pluto existed before it was discovered in 1930).

Statement 2. Mathematics is an abstract creation invented by people to describe the world (and possibly even an abstract creation with no relation to the real world).

(b) Now switch your perspective and repeat part (a) using the other statement.

33.3. Write each of the following quantities as a complex number.

(a) $(3-2i)\cdot(1+4i)$ (b) $\dfrac{3-2i}{1+4i}$ (c) $\left(\dfrac{1+i}{\sqrt{2}}\right)^2$

33.4. (a) Solve the equation $x^2 = 95 - 168i$ using complex numbers. [*Hint.* Set $(u + vi)^2 = 95 - 168i$, square the left-hand side, and solve for u and v.]

(b) Solve the equation $x^2 = 1 + 2i$ using complex numbers.

33.5. For each part, check whether the Gaussian integer α divides the Gaussian integer β and, if it does, find the quotient.

(a) $\alpha = 3 + 5i$ and $\beta = 11 - 8i$

(b) $\alpha = 2 - 3i$ and $\beta = 4 + 7i$

(c) $\alpha = 3 - 39i$ and $\beta = 3 - 5i$

(d) $\alpha = 3 - 5i$ and $\beta = 3 - 39i$

33.6. (a) Show that the statement that $a + bi$ divides $c + di$ is equivalent to the statement that the ordinary integer $a^2 + b^2$ divides both of the integers $ac + bd$ and $-ad + bc$.

(b) Suppose that $a + bi$ divides $c + di$. Show that $a^2 + b^2$ divides $c^2 + d^2$.

33.7. Verify that each of the following subsets R_1, R_2, R_3, R_4 of the complex numbers is a ring. In other words, show that if α and β are in the set, then $\alpha + \beta$, $\alpha - \beta$, and $\alpha\beta$ are also in the set.

(a) $R_1 = \{a + bi\sqrt{2} \,:\, a \text{ and } b \text{ are ordinary integers}\}$.

(b) Let ρ be the complex number $\rho = -\frac{1}{2} + \frac{1}{2}i\sqrt{3}$.
$R_2 = \{a + b\rho \,:\, a \text{ and } b \text{ are ordinary integers}\}$.
(*Hint.* ρ satisfies the equation $\rho^2 + \rho + 1 = 0$.)

(c) Let p be a fixed prime number.
$R_3 = \{a/d \,:\, a \text{ and } d \text{ are ordinary integers such that } p \nmid d\}$.

(d) $R_4 = \{a + b\sqrt{3} \,:\, a \text{ and } b \text{ are ordinary integers}\}$.

33.8. An element α of a ring R is called a *unit* if there is an element $\beta \in R$ satisfying $\alpha\beta = 1$. In other words, $\alpha \in R$ is a unit if it has a multiplicative inverse in R. Describe all the units in each of the following rings.

(a) $R_1 = \{a + bi\sqrt{2} \,:\, a \text{ and } b \text{ are ordinary integers}\}$.
(*Hint.* Use the Norm Multiplication Property for numbers $a + bi\sqrt{2}$.)

(b) Let ρ be the complex number $\rho = -\frac{1}{2} + \frac{1}{2}i\sqrt{3}$.
$R_2 = \{a + b\rho \,:\, a \text{ and } b \text{ are ordinary integers}\}$.

(c) Let p be a fixed prime number.
$R_3 = \{a/d \,:\, a \text{ and } d \text{ are ordinary integers such that } p \nmid d\}$.

33.9. Let R be the ring $\{a + b\sqrt{3} \,:\, a \text{ and } b \text{ are ordinary integers}\}$. For any element $\alpha = a + b\sqrt{3}$ in R, define the" norm" of α to be $\mathrm{N}(\alpha) = a^2 - 3b^2$. (Note that R is a subset of the real numbers, and this "norm" is not the square of the distance from α to 0.)

(a) Show that $\mathrm{N}(\alpha\beta) = \mathrm{N}(\alpha)\,\mathrm{N}(\beta)$ for every α and β in R.

(b) If α is a unit in R, show that $\mathrm{N}(\alpha)$ equals 1. [*Hint.* First show that $\mathrm{N}(\alpha)$ must equal ± 1; then figure out why it can't equal -1.]

(c) If $\mathrm{N}(\alpha) = 1$, show that α is a unit in R.
(d) Find eight different units in R.
(e) Describe all the units in R. (*Hint*. See Chapter 32.)

33.10. Complete the proof of the Gaussian Divisibility Lemma Part (c) by proving that in Case 3 the Gaussian integer α is divisible by both π and $\bar{\pi}$.

33.11. Factor each of the following Gaussian integers into a product of Gaussian primes. (You may find the Gaussian Divisibility Lemma helpful in deciding which Gaussian primes to try as factors.)

(a) $91 + 63i$ (b) 975 (c) $53 + 62i$

Chapter 34

The Gaussian Integers and Unique Factorization

We saw in the last chapter that it can be as much fun doing number theory with the Gaussian integers as it is doing number theory with the ordinary integers. In fact, some might consider the Gaussian integers to be even more fun, since they contain even more prime numbers to play with. We saw long ago how the ordinary primes are the basic building blocks used to form all other integers, and we proved the fundamental result that each integer can be constructed from primes in exactly one way. Although this Unique Factorization Theorem studied in Chapter 7 seemed obvious at first, our trip to the "Even Number World" (the $\mathbb{E}$-Zone) convinced us that it is far more subtle than it initially appeared.

The question now arises as to whether every Gaussian integer can be written as a product of Gaussian primes in exactly one way. Of course, rearranging the order of the factors is not considered a different factorization, but there are other possible difficulties. For example, consider the two factorizations

$$11 - 10i = (3 + 2i)(1 - 4i) \qquad \text{and} \qquad 11 - 10i = (2 - 3i)(4 + i).$$

They look different, but if you remember our discussion of units, you'll notice that

$$3 + 2i = i \cdot (2 - 3i) \qquad \text{and} \qquad 1 - 4i = -i \cdot (4 + i).$$

So the two supposedly different factorizations of $11 - 10i$ arise from the relation $-i \cdot i = 1$.

We would have had the same problem with the ordinary integers if we had allowed both positive and negative prime numbers, since, for example, $6 = 2 \cdot 3 = (-2) \cdot (-3)$ has two seemingly "different" factorizations into primes. To avoid this difficulty, we selected the positive primes as our basic building blocks. This suggests that we do something similar for the Gaussian integers, but clearly we can't talk about positive complex numbers versus negative complex numbers.

If $\alpha = a + bi$ is any nonzero Gaussian integer, then we can multiply α by each of the units 1, -1, i, and $-i$ to obtain the numbers

$$\alpha = a + bi, \qquad i\alpha = -b + ai, \qquad -\alpha = -a - bi, \qquad -i\alpha = b - ai.$$

If you plot these four Gaussian integers in the complex plane, you will find that exactly one of them is in the first quadrant. More precisely, exactly one of them has its x-coordinate > 0 and its y-coordinate ≥ 0. We say that

$x + yi$ is *normalized* if $x > 0$ and $y \geq 0$.

These normalized Gaussian integers will play the same role as positive ordinary integers.

Theorem 34.1 (Unique Factorization of Gaussian Integers). *Every Gaussian integer $\alpha \neq 0$ can be factored into a unit u multiplied by a product of normalized Gaussian primes*

$$\alpha = u\pi_1\pi_2\cdots\pi_r$$

in exactly one way.

As usual, a few words of explanation are required. First, if α is itself a unit, we take $r = 0$ and $u = \alpha$ and let the factorization of α be simply $\alpha = u$. Second, the Gaussian primes $\pi_1, \ldots, \pi_r$ do not have to be different; an alternative description is to write the factorization of α as

$$\alpha = u\pi_1^{e_1}\pi_2^{e_2}\cdots\pi_r^{e_r}$$

using distinct Gaussian primes $\pi_1, \ldots, \pi_r$ and exponents $e_1, \ldots, e_r > 0$. Third, when we say that there is exactly one factorization, we obviously do not consider a rearrangement of the factors to be a new factorization.

If you review[1] the proof of the Fundamental Theorem of Arithmetic in Chapter 7, you will see that the decisive property of primes, from which all else naturally flows, is the following simple assertion:

If a prime divides a product of two numbers, then it divides (at least) one of the numbers.

[1]So what are you waiting for, go back and review!

Luckily for us, the Gaussian integers also have this property, but before giving the proof, we need to know that when we divide one Gaussian integer by another the remainder is smaller than the number with which we're dividing.

This is so obvious for ordinary integers that you probably wouldn't think it was worth mentioning. For example, if we divide 177 by 37, we get a quotient of 4 and a remainder of 29. In other words,

$$177 = 4 \cdot 37 + 29,$$

and the remainder 29 is smaller than the divisor 37.

However, matters are far less clear for the Gaussian integers. For example, if we divide $237 + 504i$ by $15 - 17i$, what are the quotient and remainder, and how can we even talk about the remainder being smaller than the divisor? The answer to the second question is easy; we measure the size of a Gaussian integer $a + bi$ by its norm $\mathrm{N}(a + bi) = a^2 + b^2$ so we can ask that the remainder have smaller norm than the divisor. But is it possible to divide $237 + 504i$ by $15 - 17i$ and get a remainder whose norm is smaller than $\mathrm{N}(15 - 17i) = 514$? The answer is Yes since

$$237 + 504i = (-10 + 23i)(15 - 17i) + (-4 - 11i).$$

This says that $237 + 504i$ divided by $-10 + 23i$ gives a quotient of $15 - 17i$ and a remainder of $-4 - 11i$, and clearly $\mathrm{N}(-4 - 11i) = 137$ is smaller than $\mathrm{N}(15 - 17i) = 514$.

We now prove that it is always possible to divide Gaussian integers and get a small remainder. The proof is a pleasing blend of algebra and geometry.

Theorem 34.2 (Gaussian Integer Division with Remainder). *Let α and β be Gaussian integers with $\beta \neq 0$. Then there are Gaussian integers γ and ρ so that*

$$\alpha = \beta\gamma + \rho \qquad \textit{and} \qquad \mathrm{N}(\rho) < \mathrm{N}(\beta).$$

Proof. If we divide the equation we're trying to prove by β, it becomes

$$\frac{\alpha}{\beta} = \gamma + \frac{\rho}{\beta} \qquad \text{with} \qquad \mathrm{N}\left(\frac{\rho}{\beta}\right) < 1.$$

This means that we should choose γ to be as close to α/β as possible, since we want the difference between γ and α/β to be small.

If the ratio α/β is itself a Gaussian integer, then we can take $\gamma = \alpha/\beta$ and $\rho = 0$; but in general α/β is not a Gaussian integer. However, it is certainly a complex number, so we can mark it in the complex plane as illustrated in Figure 34.1. We next tile the complex planes into square boxes by drawing vertical and horizontal lines through all the Gaussian integers. The complex number α/β lies in one of these squares, and we take γ to be the closest corner of the square that contains α/β. Note that γ is a Gaussian integer, since the corners of the squares are the Gaussian integers.

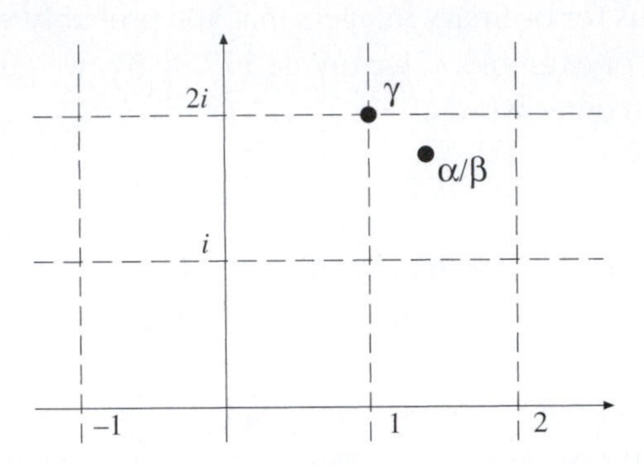

Figure 34.1: Closest Gaussian Integer γ to the Quantity α/β

The farthest that α/β can be from γ occurs if α/β is exactly in the middle of a square, so

$$\text{(Distance from } \alpha/\beta \text{ to } \gamma) \le \frac{\sqrt{2}}{2}.$$

(The diagonal of the square has length $\sqrt{2}$, so the middle of the square is half of $\sqrt{2}$ from the corners.) If we square both sides and use the fact that the norm is the square of the length, we obtain

$$\mathrm{N}\left(\frac{\alpha}{\beta} - \gamma\right) \le \frac{1}{2}.$$

Next we multiply both sides by $\mathrm{N}(\beta)$ and use the Norm Multiplication Property to obtain

$$\mathrm{N}(\alpha - \beta\gamma) \le \frac{1}{2}\mathrm{N}(\beta).$$

Finally, we simply choose ρ to be $\rho = \alpha - \beta\gamma$, and then we get the desired properties:

$$\alpha = \beta\gamma + \rho \qquad \text{and} \qquad \mathrm{N}(\rho) < \mathrm{N}(\beta).$$

[In fact, we get the stronger inequality $\mathrm{N}(\rho) \le \frac{1}{2}\mathrm{N}(\beta)$.] □

The next step is to use Gaussian Integer Division with Remainder to show that the "smallest" nonzero number of the form $A\alpha + B\beta$ divides both α and β. It is instructive to compare this with the analogous property of ordinary integers that we proved in Chapter 6.

Theorem 34.3 (Gaussian Integer Common Divisor Property). *Let α and β be Gaussian integers, and let S be the collection of Gaussian integers*

$$A\alpha + B\beta, \qquad \text{where } A \text{ and } B \text{ are any Gaussian integers.}$$

Among all the Gaussian integers in S, choose an element

$$g = a\alpha + b\beta$$

having the smallest nonzero *norm. In other words,*

$$0 < \mathrm{N}(g) \le \mathrm{N}(A\alpha + B\beta) \qquad \begin{array}{l}\text{for any Gaussian integers } A \\ \text{and } B \text{ with } A\alpha + B\beta \ne 0.\end{array}$$

Then g divides both α and β.

Proof. We use Gaussian Integer Division with Remainder to divide α by g,

$$\alpha = g\gamma + \rho \qquad \text{with} \qquad 0 \le \mathrm{N}(\rho) < \mathrm{N}(g).$$

Our goal is to show that the remainder ρ is zero.

Substituting $g = a\alpha + b\beta$ into $\alpha = g\gamma + \rho$ and doing a little algebra yields

$$(1 - a\gamma)\alpha - b\gamma\beta = \rho.$$

Thus ρ is in the set S, since it has the form

$$(\text{Gaussian integer times } \alpha) + (\text{Gaussian integer times } \beta).$$

On the other hand, $\mathrm{N}(\rho) < \mathrm{N}(g)$, and we chose g to have the smallest nonzero norm among the elements of S. Therefore, $\mathrm{N}(\rho)$ must equal 0, which means that $\rho = 0$. This shows that $\alpha = g\gamma$, so g divides α.

Finally, reversing the roles of α and β and repeating the argument shows that g also divides β. □

Now we're ready to show that if a Gaussian prime divides a product of two Gaussian integers, then it divides at least one of the two.

Theorem 34.4 (Gaussian Prime Divisibility Property). *Let π be a Gaussian prime, let α and β be Gaussian integers, and suppose the π divides the product $\alpha\beta$. Then either π divides α or π divides β (or both).*

More generally, if π divides a product $\alpha_1\alpha_2\cdots\alpha_n$ of Gaussian integers, then it divides at least one of the factors $\alpha_1, \alpha_2, \ldots, \alpha_n$.

Proof. We apply the Gaussian Integer Common Divisor Property to the two numbers α and π. This tells us that we can find Gaussian integers a and b so that the number

$$g = a\alpha + b\pi \quad \text{divides both } \alpha \text{ and } \pi.$$

But π is a prime, so the fact that g divides π means either that g is a unit or else g is equal to π times a unit. We consider these two cases separately.

First, suppose that $g = u\pi$ for some unit u (that is, u is one of the numbers 1, -1, i or $-i$). Since we also know that g divides α, it follows that π divides α, so we are done.

Second, suppose that g itself is a unit. We multiply the equation $g = a\alpha + b\pi$ by β to get

$$g\beta = a\alpha\beta + b\pi\beta.$$

We are told that π divides $\alpha\beta$, so this equation tells us that π divides $g\beta$. Since g is a unit, it follows that π divides β, so again we are done. This completes the proof that if a prime π divides a product $\alpha\beta$, then it divides at least one of the factors α or β.

This proves the first part of the Gaussian Prime Divisibility Property. For the second part we can use induction on the number n of factors. We have proved the case $n = 2$ (i.e., two factors $\alpha_1\alpha_2$), which is enough to get our induction started. Now suppose that we have proved the Gaussian Prime Divisibility Property for all products having fewer than n factors, and suppose that π divides a product $\alpha_1\alpha_2\cdots\alpha_n$ having n factors. If we let $\alpha = \alpha_1\cdots\alpha_{n-1}$ and $\beta = \alpha_n$, then π divides $\alpha\beta$, so we know from above that either π divides α or π divides β. If π divides β, then we're done, since $\beta = \alpha_n$. On the other hand, if π divides α, then π divides the product $\alpha_1\cdots\alpha_{n-1}$ consisting of $n-1$ factors, so by the induction hypothesis we know that π divides one of the factors $\alpha_1,\ldots,\alpha_{n-1}$. This completes the proof of the Gaussian Prime Divisibility Property. □

We are finally ready to prove that every nonzero Gaussian integer has a unique factorization into primes.

Proof of Unique Factorization of Gaussian Integers. We begin by demonstrating that every Gaussian integer has some factorization into primes. We could simply mimic the proof we gave in Chapter 7, but for the sake of variety and to introduce you to a new mathematical tool, we instead give a "Proof by Contradiction."[2] In

[2]The classical phrase for "proof by contradiction" is *reductio ad absurdum*, literally "reduction to an absurdity." As G.H. Hardy says in his monograph *A Mathematician's Apology*, proof by contradiction "is one of a mathematician's finest weapons. It is a far finer gambit than any chess gambit: a chess player may offer the sacrifice of a pawn or even a piece, but a mathematician offers *the game*."

a proof by contradiction, we begin by making a statement. We then use that statement to make deductions, eventually ending up with a conclusion that is clearly false. This allows us to deduce that the original statement was false, since it led to a false conclusion.

The particular statement with which we are going to begin is the following:

Statement: $\begin{cases}\text{There exists at least one nonzero Gaussian} \\ \text{integer that does not factor into primes.}\end{cases}$

Among the nonzero Gaussian integers with this property, we choose one (call it α) having smallest norm. We can do this, since the norms of Gaussian integers are positive integers, and any collection of positive integers has a smallest element. Notice that α cannot itself be prime, since otherwise $\alpha = \alpha$ is already a factorization of α into primes. Similarly, α cannot be a unit, since otherwise $\alpha = \alpha$ would again be a factorization into primes (in this case, into zero primes). But if α is neither prime nor a unit, then it must factor $\alpha = \beta\gamma$ into a product of two Gaussian integers, neither of which is a unit.

Now consider the norms of β and γ. Since β and γ are not units, we know that $\mathrm{N}(\beta) > 1$ and $\mathrm{N}(\gamma) > 1$. We also have the multiplication property $\mathrm{N}(\beta)\,\mathrm{N}(\gamma) = \mathrm{N}(\alpha)$, so

$$\mathrm{N}(\beta) = \frac{\mathrm{N}(\alpha)}{\mathrm{N}(\gamma)} < \mathrm{N}(\alpha) \qquad \text{and} \qquad \mathrm{N}(\gamma) = \frac{\mathrm{N}(\alpha)}{\mathrm{N}(\beta)} < \mathrm{N}(\alpha).$$

But we chose α to be the Gaussian integer of smallest norm that does not factor into primes, so both β and γ do factor into primes. In other words,

$$\beta = \pi_1\pi_2\cdots\pi_r \qquad \text{and} \qquad \gamma = \pi_1'\pi_2'\cdots\pi_s'$$

for certain Gaussian primes $\pi_1, \ldots, \pi_r, \pi_1', \ldots, \pi_s'$. But then

$$\alpha = \beta\gamma = \pi_1\pi_2\cdots\pi_r\pi_1'\pi_2'\cdots\pi_s'$$

is also a product of primes, which contradicts the choice of α as a number that cannot be written as a product of primes. This proves that our Statement must be false, since it leads to the absurdity that α both does and does not factor into primes. In other words, we have proved that the statement "there exist nonzero Gaussian integers that do not factor into primes" is false, so we have proved that every nonzero Gaussian integer does factor into primes.

The second part of the theorem requires us to show that the factorization into primes can be done in only one way, subject to the caveats already described. Again we could mimic the proof in Chapter 7, but instead we use a proof by contradiction. We start with the following statement:

Statement: $\begin{cases} \text{There exists at least one nonzero Gaussian} \\ \text{integer with two distinct factorizations into primes.} \end{cases}$

Assuming the truth of this statement, we look at the set of all Gaussian integers having two distinct factorizations into primes (the statement assures us this set is not empty), and we take α to be an element of the set having the smallest possible norm.

This means that α has two different factorizations

$$\alpha = u\pi_1\pi_2\cdots\pi_r = u'\pi_1'\pi_2'\cdots\pi_s',$$

where the primes are normalized as described at the beginning of this chapter. Clearly, α cannot be a unit, since otherwise we would have $\alpha = u = u'$, so the factorizations would not be different. This means that $r \geq 1$, so there is a prime π_1 in the first factorization. Then π_1 divides α, so

$$\pi_1 \quad \text{divides the product} \quad u'\pi_1'\pi_2'\cdots\pi_s'.$$

The Gaussian Prime Divisibility Property tells us that π_1 divides at least one of the number $u', \pi_1', \ldots, \pi_s'$. It certainly doesn't divide the unit u', so it divides one of the factors. Rearranging the order of these other factors, we may assume that π_1 divides π_1'. However, the number π_1' is a Gaussian integer prime, so its only divisors are units and itself times units. Since π_1 is not a unit, we deduce that

$$\pi_1 = (\text{unit}) \times \pi_1'.$$

Furthermore, both π_1 and π_1' are normalized, so the unit must equal 1 and $\pi_1 = \pi_1'$.

Let $\beta = \alpha/\pi_1 = \alpha/\pi_1'$. Canceling π_1 from the two factorizations of α yields

$$\beta = u\pi_2\cdots\pi_r = u'\pi_2'\cdots\pi_s'.$$

This number β has the following two properties:

- $\mathrm{N}(\beta) = \mathrm{N}(\alpha)/\mathrm{N}(\pi) < \mathrm{N}(\alpha)$.
- β has two distinct factorizations into primes (since α has this property, and we canceled the same factor from both sides of the two factorizations of α).

This contradicts the choice of α as the smallest number with two distinct factorizations into primes, and hence our original statement must be false. Thus, there do not exist any Gaussian integers with two distinct factorizations into primes, so every Gaussian integer has a unique such factorization. □

We use the Gaussian Integer Unique Factorization Theorem to count how many different ways a number can be written as a sum of two squares. For example, how many ways can the number 45 be written as a sum of two squares? A little experimentation quickly yields

$$45 = 3^2 + 6^2,$$

and this is the only way to write 45 as $a^2 + b^2$ with a and b positive and $a < b$. Of course, we could switch the two terms to get $45 = 6^2 + 3^2$, and we could also use negative numbers, for example,

$$45 = (-3)^2 + 6^2 \qquad \text{and} \qquad 45 = (-6)^2 + (-3)^2.$$

It is convenient to count all of these as different. So we say that 45 can be written as a sum of two squares in 8 different ways:

$$\begin{array}{ll} 45 = 3^2 + 6^2 & 45 = 6^2 + 3^2 \\ 45 = (-3)^2 + 6^2 & 45 = 6^2 + (-3)^2 \\ 45 = 3^2 + (-6)^2 & 45 = (-6)^2 + 3^2 \\ 45 = (-3)^2 + (-6)^2 & 45 = (-6)^2 + (-3)^2 \end{array}$$

In general, we write

$$R(N) = \text{number of ways to write } N \text{ as a sum of two squares.}$$

This is also known as the number of *representations* of N as a sum of two squares, which explains the nomenclature. Our example says that

$$R(45) = 8.$$

Similarly, $R(65) = 16$, since

$$\begin{array}{ll} 65 = 1^2 + 8^2 & 65 = 8^2 + 1^2 \\ 65 = (-1)^2 + 8^2 & 65 = 8^2 + (-1)^2 \\ 65 = 1^2 + (-8)^2 & 65 = (-8)^2 + 1^2 \\ 65 = (-1)^2 + (-8)^2 & 65 = (-8)^2 + (-1)^2 \\ 65 = 4^2 + 7^2 & 65 = 7^2 + 4^2 \\ 65 = (-4)^2 + 7^2 & 65 = 7^2 + (-4)^2 \\ 65 = 4^2 + (-7)^2 & 65 = (-7)^2 + 4^2 \\ 65 = (-4)^2 + (-7)^2 & 65 = (-7)^2 + (-4)^2. \end{array}$$

The following beautiful theorem gives a surprisingly simple formula for the number of representations of an integer N as a sum of two squares.

Theorem 34.5 (Sum of Two Squares Theorem (Legendre)). *For a given positive integer N, let*

$$D_1 = \big(\text{the number of positive integers } d \text{ dividing } N \text{ that satisfy } d \equiv 1 \pmod{4}\big),$$
$$D_3 = \big(\text{the number of positive integers } d \text{ dividing } N \text{ that satisfy } d \equiv 3 \pmod{4}\big).$$

Then N can be written as a sum of two squares in exactly

$$R(N) = 4(D_1 - D_3) \quad \text{ways.}$$

Before giving the proof of Legendre's formula, we illustrate the theorem with the number $N = 45$. The divisors of 45 are

$$1,\ 3,\ 5,\ 9,\ 15,\ 45.$$

Four of these divisors (1, 5, 9, 45) are congruent to 1 modulo 4, so $D_1 = 4$, while two of the divisors (3 and 15) are congruent to 3 modulo 4, so $D_3 = 2$. The theorem says that

$$R(45) = 4(D_1 - D_3) = 4(4 - 2) = 8,$$

which agrees with our earlier calculation. Similarly, the number 65 has the four divisors 1, 5, 13, and 65, all of which are congruent to 1 modulo 4. Thus the theorem predicts that

$$R(65) = 4(4 - 0) = 16,$$

again agreeing with the preceding calculation.

Proof of Legendre's Sum of Two Squares Theorem. The proof has two steps. First we find a formula for $R(N)$. Next we find a formula for $D_1 - D_3$. Comparing the two formulas completes the proof.

Although the proof is not very hard, it may seem complicated because of the notation. So we first explain how to use Gaussian integers to compute $R(N)$ for a particular number N. If you can follow the proof for this value of N, then you should have no trouble with the general proof.

We use the number $N = 28949649300$. We begin by factoring N into a product of ordinary primes and grouping together the primes that are congruent to 1 modulo 4 and the ones congruent to 3 modulo 4,

$$N = 28949649300 = 2^2 \cdot \underbrace{(5^2 \cdot 13^3)}_{\text{(1 mod 4 primes)}} \cdot \underbrace{(3^2 \cdot 11^4)}_{\text{(3 mod 4 primes)}}.$$

Next we factor N into a product of Gaussian primes. Using the fact that $2 = -i(1+i)^2$, the primes congruent to 1 modulo 4 factor into a product of conjugate

Gaussian primes, and the primes congruent to 3 modulo 4 are already Gaussian primes, which gives the factorization

$$N = -(1+i)^4 \cdot \left((2+i)^2(2-i)^2 \cdot (2+3i)^3(2-3i)^3\right) \cdot (3^2 \cdot 11^4).$$

Now suppose that we want to write N as a sum of two squares, say $N = A^2 + B^2$. This means that

$$N = (A + Bi)(A - Bi),$$

so by unique factorization of Gaussian integers, $A + Bi$ is a product of some of the primes dividing N, and $A - Bi$ is the product of the remaining ones.

However, we don't have complete freedom in distributing the primes dividing N, because $A + Bi$ and $A - Bi$ are complex conjugates of one another. That is, changing i to $-i$ changes one into the other. This means that, if some prime power $(a + bi)^e$ divides $A + Bi$, then the conjugate prime power $(a - bi)^e$ must divide $A - Bi$. So, for example, if $(2 + i)^2$ divides $A + Bi$, then $(2 - i)^2$ divides $A - Bi$, so there won't be any $2 - i$ factors left to divide $A + Bi$.

This reasoning also applies to the Gaussian primes congruent to 3 modulo 4. Thus we can't have 9 dividing $A + Bi$, since then there wouldn't be any factors of 3 left over to divide $A - Bi$. These observations show that the factors $A + Bi$ of $N = 28949649300$ must look like

$$A + Bi = \text{unit} \cdot (1+i)^2 \cdot (2+i)^n(2-i)^{2-n} \cdot (2+3i)^m(2-3i)^{3-m} \cdot 3 \cdot 11^2,$$

where we can take any $0 \le n \le 2$ and any $0 \le m \le 3$. There are thus 3 choices for n, there are 4 choices for m, and there are the usual 4 choices of the unit, so there are $4 \cdot 3 \cdot 4 = 48$ possibilities for $A + Bi$. The unique factorization property of Gaussian integers tells us that writing N as a sum of two squares is exactly the same problem as finding an $A + Bi$ dividing N, so we conclude that $R(N) = 48$. But it is important to keep in mind that this number 48 is really the product of the following three quantities:

- the number of units in the Gaussian integers
- one more than the exponent of $2 + i$
- one more than the exponent of $2 + 3i$

We now begin the proof of Legendre's Sum of Two Squares Theorem. We begin by factoring N into a product of ordinary primes

$$N = 2^t \underbrace{p_1^{e_1} p_2^{e_2} \cdots p_r^{e_r}}_{\text{(1 mod 4 primes)}} \cdot \underbrace{q_1^{f_1} q_2^{f_2} \cdots q_s^{f_s}}_{\text{(3 mod 4 primes)}},$$

where $p_1, \ldots, p_r$ are congruent to 1 modulo 4, and $q_1, \ldots, q_s$ are congruent to 3 modulo 4. We use Gaussian integers to find a formula for $R(N)$ in terms of the exponents $e_1, \ldots, e_r, f_1, \ldots, f_s$.

We factor N into a product of Gaussian primes. The integer 2 factors as $2 = -i(1+i)^2$, and each p_j factors as

$$p_j = (a_j + b_j i)(a_j - b_j i),$$

while the q_j are themselves Gaussian primes. This gives the factorization

$$N = (-i)^t(1+i)^{2t}\big((a_1+b_1 i)(a_1-b_1 i)\big)^{e_1}\big((a_2+b_2 i)(a_2-b_2 i)\big)^{e_2} \\ \cdots \big((a_r+b_r i)(a_r-b_r i)\big)^{e_r} q_1^{f_1} q_2^{f_2} \cdots q_s^{f_s}.$$

If any of the exponents $f_1, \ldots, f_s$ is odd, then we know that N cannot be written as a sum of two squares, so $R(N) = 0$. So we now suppose that all of $f_1, \ldots, f_s$ are even, and we suppose that N is written as a sum of two squares, say $N = A^2 + B^2$. This means that

$$N = (A + Bi)(A - Bi),$$

so $A + Bi$ and $A - Bi$ are composed of the prime factors of N. Furthermore, since $A + Bi$ and $A - Bi$ are complex conjugates of one another, each prime that appears in one of their factorizations must have its complex conjugate appearing in the other. This means that $A + Bi$ looks like

$$A + Bi = u(1+i)^t\big((a_1+b_1 i)^{x_1}(a_1-b_1 i)^{e_1-x_1}\big) \\ \cdots \big((a_r+b_r i)^{x_r}(a_r-b_r i)^{e_r-x_r}\big) q_1^{f_1/2} q_2^{f_2/2} \cdots q_s^{f_s/2},$$

where u is a unit and the exponents $x_1, \ldots, x_r$ satisfy

$$0 \le x_1 \le e_1, \quad 0 \le x_2 \le e_2, \ldots \quad 0 \le x_r \le e_r.$$

Taking the norm of both sides expresses N as a sum of two squares, so counting the number of choices for the exponents, we find that this gives

$$4(e_1+1)(e_2+1)\cdots(e_r+1)$$

ways to write N as a sum of two squares. (We leave it as an exercise for you to check that different choices of $u, x_1, \ldots, x_r$ yield different values of A and B.)

To recapitulate, we have proved that if the integer N is factored as

$$N = 2^t p_1^{e_1} \cdots p_r^{e_r} q_1^{f_1} \cdots q_s^{f_s}$$

with $p_1, \ldots, p_r$ all congruent to 1 modulo 4 and $q_1, \ldots, q_s$ all congruent to 3 modulo 4 then

$$R(N) = \begin{cases} 4(e_1+1)(e_2+1)\cdots(e_r+1) & \text{if } f_1, \ldots, f_s \text{ are all even,} \\ 0 & \text{if any of } f_1, \ldots, f_s \text{ is odd.} \end{cases}$$

The proof of Legendre's Sum of Two Squares Theorem will thus be complete once we prove that the difference $D_1 - D_3$ is given by the same formula.

Theorem 34.6 (Difference of $D_1 - D_3$ Theorem). *Factor the integer N into a product of ordinary primes as*

$$N = 2^t \underbrace{p_1^{e_1} p_2^{e_2} \cdots p_r^{e_r}}_{\text{(1 mod 4 primes)}} \cdot \underbrace{q_1^{f_1} q_2^{f_2} \cdots q_s^{f_s}}_{\text{(3 mod 4 primes)}}.$$

Let

$$D_1 = \bigl(\textit{the number of integers } d \textit{ dividing } N \textit{ that satisfy } d \equiv 1 \pmod 4\bigr),$$
$$D_3 = \bigl(\textit{the number of integers } d \textit{ dividing } N \textit{ that satisfy } d \equiv 3 \pmod 4\bigr).$$

Then the difference $D_1 - D_3$ is given by the rule

$$D_1 - D_3 = \begin{cases} (e_1+1)(e_2+1)\cdots(e_r+1) & \textit{if } f_1, \ldots, f_s \textit{ are all even,} \\ 0 & \textit{if any of } f_1, \ldots, f_s \textit{ is odd.} \end{cases}$$

Proof. We give a proof by induction on s. First, if $s = 0$, then $N = 2^t p_1^{e_1} \cdots p_r^{e_r}$, so every odd divisor of N is congruent to 1 modulo 4. In other words, $D_3 = 0$, and D_1 is the number of odd divisors of N. The odd divisors of N are the numbers $p_1^{u_1} \cdots p_r^{u_r}$ with each exponent u_i satisfying $0 \le u_i \le e_i$. There are thus $e_i + 1$ choices for u_i, which means that the total number of odd divisors is

$$D_1 - D_3 = D_1 = (e_1+1)(e_2+1)\cdots(e_r+1).$$

This completes the proof if $s = 0$, that is, if N is not divisible by any 3 modulo 4 primes.

Now let N be divisible by q for some prime $q \equiv 3 \pmod 4$, and assume that we have completed the proof for all numbers having fewer 3 modulo 4 prime divisors than N. Let q^f be the highest power of q dividing N, so $N = q^f n$ with $f \ge 1$ and $q \nmid n$. We consider two cases, depending on whether f is odd or even.

First, suppose that f is odd. The odd divisors of N are the numbers

$$q^i d \qquad \text{with } 0 \le i \le f \text{ and } d \text{ odd and dividing } n.$$

Thus each divisor d of n gives rise to exactly $f+1$ divisors of N, that is, to the divisors $q^i d$ with $0 \le i \le f$; and of these $f+1$ divisors of N, exactly half are congruent to 1 modulo 4 and exactly half are congruent to 3 modulo 4. Thus the divisors of N are equally split among D_1 and D_3, so we have $D_1 - D_3 = 0$. This completes the proof in the case that N is divisible by an odd power of a 3 modulo 4 prime.

Second, suppose that $N = q^f n$ with f even. Again the odd divisors of N look like $q^i d$ with $0 \le i \le f$ and d odd and dividing n. If we only consider divisors $q^i d$ with exponents $0 \le i \le f-1$, then the same reasoning as before shows that the number of 1 modulo 4 divisors is exactly the same as the number of 3 modulo 4 divisors, so they cancel out in the difference $D_1 - D_3$. So we are left to consider the divisors of N of the form $q^f d$. The exponent f is even, so $q^f \equiv 1 \pmod 4$. This means that $q^f d$ counts in D_1 if $d \equiv 1 \pmod 4$ and it counts in D_3 if $d \equiv 3 \pmod 4$. In other words,

$$(D_1 - D_3 \text{ for } N) = (D_1 - D_3 \text{ for } n).$$

Our induction hypothesis tells us that the theorem is true for n, so we deduce that the theorem is also true for N. This completes the proof of the $D_1 - D_3$ Theorem. □

Exercises

34.1. (a) Let $\alpha = 2 + 3i$. Plot the four points α, $i\alpha$, $-\alpha$, $-i\alpha$ in the complex plane. Connect the four points. What sort of figure do you get?

(b) Same question with $\alpha = -3 + 4i$.

(c) Let $\alpha = a + bi$ be any nonzero Gaussian integer. Let A be the point in the complex plane corresponding to α, let B be the point in the complex plane corresponding to $i\alpha$, and let $O = (0,0)$ be the point corresponding to 0. What is the measure of the angle $\angle AOB$? That is, what is the measure of the angle made by the rays $\overrightarrow{OA}$ and $\overrightarrow{OB}$?

(d) Again let $\alpha = a + bi$ be any nonzero Gaussian integer. What sort of shape is formed by connecting the four points α, $i\alpha$, $-\alpha$, and $-i\alpha$? Prove that your answer is correct.

34.2. For each of the following pairs of Gaussian integers α and β, find Gaussian integers γ and ρ satisfying

$$\alpha = \beta\gamma + \rho \qquad \text{and} \qquad \mathrm{N}(\rho) < \mathrm{N}(\beta).$$

(a) $\alpha = 11 + 17i$, $\beta = 5 + 3i$

(b) $\alpha = 12 - 23i$, $\beta = 7 - 5i$

(c) $\alpha = 21 - 20i$, $\beta = 3 - 7i$

34.3. Let α and β be Gaussian integers with $\beta \neq 0$. We proved that we can always find a pair of Gaussian integers (γ, ρ) that satisfy

$$\alpha = \beta\gamma + \rho \qquad \text{and} \qquad \mathrm{N}(\rho) < \mathrm{N}(\beta).$$

(a) Show that there are actually always at least two different pairs (γ, ρ) with the desired properties.
(b) Can you find an α and β with exactly three different pairs (γ, ρ) having the desired properties? Either give an example or prove that none exists.
(c) Same as (b), but with exactly four different pairs (γ, ρ).
(d) Same as (b), but with exactly five different pairs (γ, ρ).
(e) Illustrate your results in (a), (b), (c) and (d) geometrically by dividing a square into several different regions corresponding to the value of α/β.

34.4. Let α and β be Gaussian integers that are not both zero. We say that a Gaussian integer γ is a *greatest common divisor of* α *and* β if (i) γ divides both α and β, and (ii) among all common divisors of α and β, the quantity $\mathrm{N}(\gamma)$ is as large as possible.
(a) Suppose that γ and δ are both greatest common divisors of α and β. Prove that γ divides δ. Use this fact to deduce that $\delta = u\gamma$ for some unit u.
(b) Prove that the set

$$\{\alpha r + \beta s \ : \ r \text{ and } s \text{ are Gaussian integers}\}$$

contains a greatest common divisor of α and β. (*Hint.* Look at the element in the set that has smallest norm.)
(c) Let γ be a greatest common divisor of α and β. Prove that the set in (b) is equal to the set

$$\{\gamma t \ : \ t \text{ is a Gaussian integer}\}.$$

34.5. Find a greatest common divisor for each of the following pairs of Gaussian integers.
(a) $\alpha = 8 + 38i$ and $\beta = 9 + 59i$
(b) $\alpha = -9 + 19i$ and $\beta = -19 + 4i$
(c) $\alpha = 40 + 60i$ and $\beta = 117 - 26i$
(d) $\alpha = 16 - 120i$ and $\beta = 52 + 68i$

34.6. Let R be the following set of complex numbers:

$$R = \{a + bi\sqrt{5} \ : \ a \text{ and } b \text{ are ordinary integers}\}.$$

(a) Verify that R is a ring. That is, verify that the sum, difference, and product of elements of R are again in R.
(b) Show that the only solutions to $\alpha\beta = 1$ in R are $\alpha = \beta = 1$ and $\alpha = \beta = -1$. Conclude that 1 and -1 are the only units in the ring R.
(c) Let α and β be elements of R. We say that β *divides* α if there is an element γ in R satisfying $\alpha = \beta\gamma$. Show that $3 + 2i\sqrt{5}$ divides $85 - 11i\sqrt{5}$.

(d) We call an element α of R a *prime*[3] if its only divisors in R are ± 1 and $\pm\alpha$. Prove that the number 2 is a prime in R.

(e) We define the norm of an element $\alpha = a + bi\sqrt{5}$ in R to be $\mathrm{N}(\alpha) = a^2 + 5b^2$. Let $\alpha = 11 + 2i\sqrt{5}$ and $\beta = 1 + i\sqrt{5}$. Show that it is not possible to find elements γ and ρ in R satisfying

$$\alpha = \beta\gamma + \rho \qquad \text{and} \qquad \mathrm{N}(\rho) < \mathrm{N}(\beta).$$

Thus R does not have the Division with Remainder property. (*Hint*. Draw a picture illustrating the points in R and the complex number α/β.)

(f) The prime element 2 clearly divides the product

$$(1 + i\sqrt{5})(1 - i\sqrt{5}) = 6.$$

Show that 2 does not divide either of the factors $1 + i\sqrt{5}$ or $1 - i\sqrt{5}$.

(g) Show that the number 6 has two truly different factorizations into primes in R by verifying that the numbers in the factorizations

$$6 = 2 \cdot 3 = (1 + i\sqrt{5})(1 - i\sqrt{5})$$

are all primes.

(h) Find some other numbers α in R that have two truly different factorizations $\alpha = \pi_1\pi_2 = \pi_3\pi_4$, where $\pi_1, \pi_2, \pi_3, \pi_4$ are distinct primes in R.

(i) Can you find distinct primes $\pi_1, \pi_2, \pi_3, \pi_4, \pi_5, \pi_6$ in R with the property that $\pi_1\pi_2 = \pi_3\pi_4 = \pi_5\pi_6$?

34.7. During the proof of Legendre's Sum of Two Squares Theorem, we needed to know that different choices of the unit u and the exponents $x_1, \ldots, x_r$ in the formula

$$A + Bi = u(1+i)^t\big((a_1 + b_1 i)^{x_1}(a_1 - b_1 i)^{e_1 - x_1}\big) \cdots \big((a_r + b_r i)^{x_r}(a_r - b_r i)^{e_r - x_r}\big) q_1^{f_1/2} q_2^{f_2/2} \cdots q_s^{f_s/2}$$

yield different values of A and B. Prove that this is indeed the case.

34.8. (a) Make a list of all the divisors of the number $N = 2925$.

(b) Use (a) to compute D_1 and D_3, the number of divisors of 2925 congruent to 1 and 3 modulo 4, respectively.

(c) Use Legendre's Sum of Two Squares Theorem to compute $R(2925)$.

(d) Make a list of all the ways of writing 2925 as a sum of two squares and check that it agrees with your answer in (c).

---

[3]More properly, an element α whose only divisors are u and $u\alpha$ with u a unit is called an *irreducible* element. The name *prime* is reserved for an element α with the property that if it divides a product then it always divides at least one of the factors. For ordinary integers and for the Gaussian integers, we proved that every irreducible element is prime, but this is not true for the ring R in this exercise.

34.9. For each of the following values of N, compute the values of D_1 and D_3, check your answer by comparing the difference $D_1 - D_3$ to the formula given in the $D_1 - D_3$ Theorem, and use Legendre's Sum of Two Squares Theorem to compute $R(N)$. If $R(N) \neq 0$, find at least four distinct ways of writing $N = A^2 + B^2$ with $A > B > 0$.

(a) $N = 327026700$

(b) $N = 484438500$

Chapter 35

Irrational Numbers and Transcendental Numbers

In the historical development of numbers and mathematics, fractions (also called rational numbers since they are ratios) appeared quite early, having been used in ancient Egypt as early as 1700 BCE. Rational numbers come up very naturally as soon as a civilization needs to subdivide land or cloth or gold or whatever into pieces. Fractions also appear when two quantities are compared. To take a concrete example, the distance from Cairo to Luxor is more than twice as far as the distance from Cairo to Alexandria, but less than three times as far. Such a statement is helpful, but not particularly precise. On the other hand, for most practical purposes it suffices to say that the former distance is $17/6$ times the latter distance. This means that six times the distance from Cairo to Luxor is equal to seventeen times the distance from Cairo to Alexandria. We say that two quantities are *commensurable* if a nonzero integer multiple of the first is equal to a nonzero integer multiple of the second; or, equivalently, if their ratio is a rational number. Notice that this is how we measure distances today. When we say that it is 3.7 miles to the center of town, what we really mean is that 10 times the distance to the center of town is the same as 37 times the length of an idealized distance called a "mile."

For a very long time, people who gave the matter any thought seem to have assumed that every number is a rational number. In geometric terms, they assumed that any two distances were commensurable. The first indication that this might not be true appeared in Greece about 2500 years ago. Ironically, nonrational numbers made their debut in the Pythagorean Theorem, that gem of classical mathematics about which we have already waxed poetic in Chapter 2. Although the Pythagorean Theorem was known long before the time of Pythagoras, it was in ancient Greece that someone (possibly Pythagoras himself) first observed that the hypotenuse of an isosceles right triangle (see Figure 35.1) is not commensurable with the sides.

For example, the Pythagorean Theorem tells us that an isosceles right triangle whose sides have length 1 has a hypotenuse of length $\sqrt{2}$. It is not known exactly how the Pythagoreans deduced that the sides and the hypotenuse of such a triangle are incommensurable, but the following elegant proof of the irrationality of $\sqrt{2}$ is adapted from the tenth book of Euclid's *Elements*.

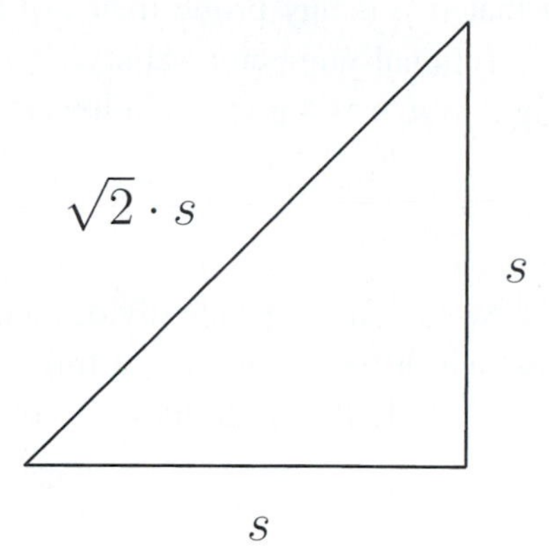

Figure 35.1: An Incommensurable Hypotenuse

Theorem 35.1 (Irrationality of $\sqrt{2}$ Theorem). *The square root of 2 is irrational. That is, there is no rational number r satisfying $r^2 = 2$.*

Proof. We assume that there does exist a rational number r satisfying $r^2 = 2$, and we use the supposed existence of r to end up with a contradictory statement, that is, with a statement that is clearly false. This contradiction shows that such an r does not exist. As noted in Chapter 34, this method of proof by contradiction (*reductio ad absurdum* in Latin) is a powerful tool in the mathematical arsenal.

Now for the details. We assume that r is a rational number satisfying $r^2 = 2$. Since r is a rational number, we can write r as a fraction $r = a/b$, and since we can always cancel factors that are common to the numerator and denominator, we can assume that a and b are relatively prime. In other words, we write r as a fraction in lowest terms.

The assumption that $r^2 = 2$ means that

$$a^2 = 2b^2.$$

In particular, a^2 is even, so a must be even, say $a = 2A$. If we substitute this in and cancel 2 from each side, we get

$$2A^2 = b^2,$$

so b must also be even. But a and b are relatively prime, so they can't both be even, which gives us the desired contradiction. Since the existence of r leads to a contradiction, we are forced to conclude that r cannot exist. Therefore, there is no rational number whose square equals 2. □

This proof of the irrationality of $\sqrt{2}$ can be generalized in many ways. For example, let's prove that if p is any prime then $\sqrt{p}$ is irrational. As before, we assume that there is a rational number r satisfying $r^2 = p$ and try to deduce a contradiction. Writing $r = a/b$ as a fraction in lowest terms, we obtain

$$a^2 = p \cdot b^2,$$

so p divides a^2.

In Chapter 7 we showed that if a prime divides a product of two numbers then it must divide at least one of the numbers. In this case, the prime p divides the product $a \cdot a$, so we conclude that p divides a, say $a = pA$. Substituting and canceling p gives

$$p \cdot A^2 = b^2,$$

so by the same reasoning we deduce that p divides b. Thus p divides both a and b, which contradicts the fact that a and b are relatively prime. Therefore, r cannot exist, which completes the proof that $\sqrt{p}$ is irrational.

Philosophical Interlude. The method of proof by contradiction (*reductio ad absurdum*) is based on the principle that if a statement leads to a false conclusion then the original statement is itself false. Although common sense says that this principle is valid, it actually depends on the underlying assumption that the original statement must be either true or false. This assumption that every statement is either true or false is called the *Law of the Excluded Middle*, and despite its grand-sounding name, the Law of the Excluded Middle is really an assumption (in mathematical terms, an axiom) that is used in the formal construction of mathematical systems.[1] Some mathematicians and logicians do not accept the Law of the Excluded Middle and have constructed mathematical theories without using proofs by contradiction.

[1] Actually, life is even more complicated than indicated in our brief philosophical digression. Kurt Gödel proved in the 1930s that any "interesting" mathematical system (for example, the theory of numbers) contains statements that are undecidable, which means that they are neither provably true nor provably false within the given mathematical system. A mind-twisting challenge for you is to try to imagine how one proves that certain statements cannot be proved! A further philosophical conundrum: Can a statement be true even if it is not possible to prove that it is true? What does "true" mean? If you believe that absolute mathematical knowledge already exists and is merely discovered, rather then created, by mathematicians (see exercise 33.2 and the footnote on page 240), then in some sense, isn't every statement either true or false?

When we say that $\sqrt{2}$ is irrational, we are really asserting that the polynomial

$$X^2 - 2$$

has no rational roots, and similarly the irrationality of $\sqrt{p}$ for primes p is the same as saying that $X^2 - p$ has no roots. In general, a polynomial

$$c_0X^d + c_1X^{d-1} + c_2X^{d-2} + \cdots + c_{d-1}X + c_d$$

with integer coefficients is likely to have many irrational roots, although it is frequently difficult to figure out what the roots look like. For example, one of the roots of the polynomial

$$X^{12} - 66X^{10} - 8X^9 + 1815X^8 - 26610X^6 + 5808X^5 + 218097X^4 \\ - 85160X^3 - 971388X^2 + 352176X + 1742288$$

is the horrible-looking number [2]

$$\sqrt{11} + \sqrt[3]{2 + \sqrt{7}}.$$

There are obviously many polynomials with integer coefficients, and most of them have irrational roots. We say that a number is *algebraic* if it is the root of a polynomial with integer coefficients. For example, the numbers

$$\frac{2}{7}, \quad \sqrt{2}, \quad \sqrt[3]{7}, \quad \sin(\pi/6), \quad \text{and even} \quad \sqrt{11} + \sqrt[3]{2 + \sqrt{7}}$$

are all algebraic numbers. Note that every rational number a/b is an algebraic number, since it is a root of the polynomial $bX - a$; but, as we have seen, many algebraic numbers are not rational numbers.

Given the seeming abundance of algebraic numbers, we might hope that every irrational number is algebraic; that is, we might hope that every irrational number is the root of a polynomial having integer coefficients. To take a specific example, do you think that the familiar number $\pi = 3.1415926\ldots$ is an algebraic number? In the mid-eighteenth century, Leonhard Euler suggested that it is not.[3] A number that is not algebraic is called a *transcendental number*, because it transcends the numbers that are roots of polynomials with integer coefficients.

---

[2]How do you think I found this complicated root of such a huge polynomial?

[3]Euler wrote (in 1755) that "it appears to be fairly certain that the periphery of a circle constitutes such a peculiar kind of transcendental quantity that it can in no way be compared with other quantities, either roots or other transcendentals." Legendre proved in 1794 that π^2 is irrational and noted that "it is probable that the number π is not even contained among the algebraic irrationalities ... but it seems to be very difficult to prove this strictly."

Euler and his contemporaries were not able to prove that π is transcendental, and indeed, it took more that 100 years before F. Lindemann proved the transcendence of π in 1882. Unfortunately, even with subsequent simplifications, the proof that π is transcendental is too complicated for us to give here. Actually, it's not easy to show that transcendental numbers exist at all! The first person to exhibit a transcendental number was Joseph Liouville in 1840. We follow Liouville's path by taking a particular number and proving that it is transcendental. *Liouville's number* is given by the nonrepeating decimal

$$\begin{array}{ll} \text{digit:} & \quad 1\,2 \quad 6 \qquad\qquad\qquad\qquad 24 \qquad\quad 120 \qquad\quad 720 \\ & \quad \downarrow\downarrow \quad \downarrow \qquad\qquad\qquad\qquad \downarrow \qquad\quad \downarrow \qquad\quad \downarrow \\ \beta & = 0.110001000000000000000000100\cdots 00100\cdots 00100\cdots. \end{array}$$

More precisely, the nth "one" in the decimal expansion of β appears as the $n!^{\text{th}}$ (that's n factorial) decimal digit, and all the rest of the decimal digits are zeros. Another way to write β is

$$\beta = \frac{1}{10} + \frac{1}{10^2} + \frac{1}{10^6} + \frac{1}{10^{24}} + \frac{1}{10^{120}} + \frac{1}{10^{720}} + \cdots,$$

or, using summation notation for infinite series,

$$\beta = \sum_{n=1}^{\infty} \frac{1}{10^{n!}}.$$

To show that β is transcendental, we need to show that β is not the root of any polynomial having integer coefficients. Just as in the proof of the irrationality of $\sqrt{2}$, we give a proof by contradiction, so we suppose that

$$f(X) = c_0X^d + c_1X^{d-1} + c_2X^{d-2} + \cdots + c_{d-1}X + c_d$$

is a polynomial with integer coefficients such that $f(\beta) = 0$. Liouville's brilliant idea is that if a number is the root of a polynomial then it cannot be too close to a rational number. So, before studying Liouville's number, we take a brief detour to discuss the question of approximating irrational numbers by rational numbers.

You may recall Dirichlet's Diophantine Approximation Theorem (see Chapter 31), which says that for any irrational number α there are infinitely many rational numbers a/b so that

$$\left|\frac{a}{b} - \alpha\right| < \frac{1}{b^2}.$$

In other words, we can find lots of rational numbers that are fairly close to α. We might ask whether we can get even closer. For example, are there infinitely many rational numbers a/b so that

$$\left|\frac{a}{b} - \alpha\right| < \frac{1}{b^3}?$$

The answer depends to some extent on the number α.

For example, suppose we take $\alpha = \sqrt{2}$. This means that α is a root of $f(X) = X^2 - 2$, so if a/b is close to α, then $f(a/b)$ should be fairly small. How can we quantify this observation? We can measure the smallness of $f(a/b)$ by factoring

$$f\left(\frac{a}{b}\right) = \left(\frac{a}{b}\right)^2 - 2 = \left(\frac{a}{b} + \sqrt{2}\right)\left(\frac{a}{b} - \sqrt{2}\right).$$

If a/b is close to $\sqrt{2}$, then the first factor $a/b + \sqrt{2}$ is close to $2\sqrt{2}$, so certainly it will be smaller than (say) 4. This allows us to estimate

$$\left|f\left(\frac{a}{b}\right)\right| \le 4\left|\frac{a}{b} - \sqrt{2}\right|.$$

On the other hand, we can write

$$f\left(\frac{a}{b}\right) = \left(\frac{a}{b}\right)^2 - 2 = \frac{a^2 - 2b^2}{b^2}.$$

Notice that the numerator $a^2 - 2b^2$ is a nonzero integer. (Why is it nonzero? Answer: Because $\sqrt{2}$ is irrational, and so cannot equal a/b.) Of course, we don't know the exact value of $a^2 - 2b^2$, but we do know that the absolute value of a nonzero integer must be at least 1.[4] Hence

$$\left|f\left(\frac{a}{b}\right)\right| = \left|\frac{a^2 - 2b^2}{b^2}\right| \ge \frac{1}{b^2}.$$

We now have an upper bound and a lower bound for $|f(a/b)|$, and if we put them together, we obtain the interesting inequality

$$\frac{1}{4b^2} \le \left|\frac{a}{b} - \sqrt{2}\right| \tag{1}$$

which is valid for *every* rational number a/b. Notice how this inequality complements Dirichlet's inequality

$$\left|\frac{a}{b} - \sqrt{2}\right| < \frac{1}{b^2}.$$

In particular, we can use (1) to show that a stronger inequality such as

$$\left|\frac{a}{b} - \sqrt{2}\right| < \frac{1}{b^3} \tag{2}$$

[4]The fact we are using here is the seemingly trivial observation that there are no whole numbers lying strictly between 0 and 1. Although it seems trivial, this fact lies at the heart of all proofs of transcendence. It is equivalent to the fancier-sounding *well-ordering property of the nonnegative integers*, which asserts that any set of nonnegative integers has a smallest element.

can have only finitely many solutions. To do this, we combine the inequalities (1) and (2) to obtain

$$\frac{1}{4b^2} < \frac{1}{b^3}, \qquad \text{and hence} \qquad b < 4.$$

This means that the only possibilities for b are $b = 1, 2, 3$, and then for each value of b, the inequality (2) allows at most a finite number of possible values for a. In fact, we find that (2) has exactly three solutions: $\frac{a}{b} = \frac{1}{1}$, $\frac{a}{b} = \frac{2}{1}$, and $\frac{a}{b} = \frac{3}{2}$

Let's review what we've done. We've used the fact that $\sqrt{2}$ is a root of the polynomial $X^2 - 2$ to deduce an inequality (1) that says that a rational number a/b cannot be too close to $\sqrt{2}$. Liouville's proof that the number β given above is transcendental rests on the following two legs (which might make an unsteady table, but is perfectly acceptable for a proof):

(i) If α is an algebraic number, that is, if α is a root of a polynomial with integer coefficients, then a rational number a/b cannot be too close to α.

(ii) For the number β given above, there are lots of rational numbers a/b that are extremely close to β.

Our aim is to take these two qualitative statements and make them precise. We start with statement (i), whose quantification takes the following form.

Theorem 35.2 (Liouville's Inequality). *Let α be an algebraic number; say α is a root of the polynomial*

$$f(X) = c_0X^d + c_1X^{d-1} + c_2X^{d-2} + \cdots + c_{d-1}X + c_d$$

having integer coefficients. Let D be any number with $D > d$ (i.e., D is larger than the degree of the polynomial f). Then there are only finitely many rational numbers a/b that satisfy the inequality

$$\left| \frac{a}{b} - \alpha \right| \le \frac{1}{b^D}. \qquad (*)$$

Proof. The fact that $X = \alpha$ is a root of $f(X)$ means that when we divide $f(X)$ by $X - \alpha$, we get a remainder of 0. In other words, $f(X)$ factors as

$$f(X) = (X - \alpha)g(X)$$

for some polynomial

$$g(X) = e_1X^{d-1} + e_2X^{d-2} + \cdots + e_{d-1}X + e_d.$$

For example, the algebraic number $\sqrt[3]{7}$ is a root of the polynomial $X^3 - 7$, and when we divide $X^3 - 7$ by $X - \sqrt[3]{7}$, we obtain the factorization

$$X^3 - 7 = (X - \sqrt[3]{7})(X^2 + \sqrt[3]{7}\,X + \sqrt[3]{49}).$$

Notice that the coefficients $e_1, \ldots, e_d$ won't be integers, but this won't cause any problems for us.

Suppose now that a/b is a solution to the inequality

$$\left|\frac{a}{b} - \alpha\right| \le \frac{1}{b^D}.$$

If we substitute $X = a/b$ into the factorization $f(X) = (X - \alpha)g(X)$ and take absolute values, we obtain the fundamental formula

$$\left|f\left(\frac{a}{b}\right)\right| = \left|\frac{a}{b} - \alpha\right| \cdot \left|g\left(\frac{a}{b}\right)\right|.$$

The importance of this formula is that the right-hand side is small if a/b is close to α, while the left-hand side is a rational number. The next two things we need to do are find an upper bound for $|g(a/b)|$ and a lower bound for $|f(a/b)|$.

We start with the latter. If we write out $f(a/b)$ and put it over a common denominator, we obtain

$$\begin{aligned} f\left(\frac{a}{b}\right) &= c_0\left(\frac{a}{b}\right)^d + c_1\left(\frac{a}{b}\right)^{d-1} + c_2\left(\frac{a}{b}\right)^{d-2} + \cdots + c_{d-1}\frac{a}{b} + c_d \\ &= \frac{c_0a^d + c_1a^{d-1}b + c_2a^{d-2}b^2 + \cdots + c_{d-1}ab^{d-1} + c_db^d}{b^d}. \end{aligned}$$

Note that the numerator of this fraction is an integer, and so as long as it isn't zero, we see that

$$\left|f\left(\frac{a}{b}\right)\right| \ge \frac{1}{b^d}.$$

[We deal later with the case that $f(a/b) = 0$.] We can illustrate this using our example $f(X) = X^3 - 7$ from before,

$$\left|f\left(\frac{a}{b}\right)\right| = \left|\left(\frac{a}{b}\right)^3 - 7\right| = \left|\frac{a^3 - 7b^3}{b^3}\right| \ge \frac{1}{b^3}.$$

Next we want an upper bound for $g(a/b)$. The fact that a/b is a solution to the inequality $(*)$ certainly implies that

$$|a/b| \le |\alpha| + 1,$$

so we can estimate

$$\begin{aligned}\left|g\left(\frac{a}{b}\right)\right| &\le e_1\left|\frac{a}{b}\right|^{d-1} + e_2\left|\frac{a}{b}\right|^{d-2} + e_3\left|\frac{a}{b}\right|^{d-3} + \cdots + e_{d-1}\left|\frac{a}{b}\right| + e_d \\ &\le e_1\bigl(|\alpha|+1\bigr)^{d-1} + e_2\bigl(|\alpha|+1\bigr)^{d-2} + e_3\bigl(|\alpha|+1\bigr)^{d-3} + \cdots \\ &\qquad + e_{d-1}\bigl(|\alpha|+1\bigr) + e_d\,.\end{aligned}$$

This last quantity is rather messy, but whatever it equals, it has one tremendously important property: *It doesn't depend on the rational number a/b.* In other words, we have shown that there is a positive number K so that, if a/b is *any* solution to the inequality $(*)$, then

$$\bigl|g(a/b)\bigr| \le K.$$

Again we illustrate this estimate with our example $f(X) = X^3 - 7$, where we use the bound $|a/b| \le \sqrt[3]{7} + 1$. Thus

$$\begin{aligned}\left|g\left(\frac{a}{b}\right)\right| &\le \left|\frac{a}{b}\right|^2 + \sqrt[3]{7}\left|\frac{a}{b}\right| + \sqrt[3]{49} \\ &\le (\sqrt[3]{7}+1)^2 + \sqrt[3]{7}(\sqrt[3]{7}+1) + \sqrt[3]{49} \\ &\le 17.717,\end{aligned}$$

so for this example we could take $K = 17.717$.

We now have

The Inequality $(*)$:	$\left\lvert\frac{a}{b} - \alpha\right\rvert \le \frac{1}{b^D}$
A Factorization Formula:	$\left\lvert f\left(\frac{a}{b}\right)\right\rvert = \left\lvert\frac{a}{b} - \alpha\right\rvert \cdot \left\lvert g\left(\frac{a}{b}\right)\right\rvert$
A Lower Bound:	$\left\lvert f\left(\frac{a}{b}\right)\right\rvert \ge \frac{1}{b^d}$
An Upper Bound:	$\left\lvert g\left(\frac{a}{b}\right)\right\rvert \le K.$

Putting them together yields

$$\frac{1}{b^d} \le \left|f\left(\frac{a}{b}\right)\right| = \left|\frac{a}{b} - \alpha\right| \cdot \left|g\left(\frac{a}{b}\right)\right| \le \frac{K}{b^D}.$$

Since we are told that $D > d$, we can isolate b on the left-hand side to obtain the upper bound

$$b \le K^{1/(D-d)}.$$

To illustrate this using our example $\alpha = \sqrt[3]{7}$ and $f(X) = X^3 - 7$, we have $d = 3$, and we found that we can take $K = 17.717$, so if we take (say) $D = 3.5$, then we obtain the bound

$$b \le 17.717^{1/(3.5-3)} \approx 313.89.$$

Now you can see why it was so important that the upper bound K not depend on the number a/b, since it is this fact that allows us to conclude that there are only finitely many allowable values for b. (Note that b is necessarily a positive integer, since it is the denominator of the fraction a/b written in lowest terms.) Furthermore, for each fixed choice of b, there are only finitely many values of a for which the inequality $(*)$ holds. (In fact, if $b^{D-1} > 2$, then there's at most one allowable a for a given b.) Returning to our example one last time, we see that the allowable b's are the integers $1 \le b \le 313$, and then for each particular b, the corresponding allowable a's [i.e., those that are solutions to the inequality $(*)$] are those satisfying

$$b\sqrt[3]{7} - \frac{1}{b^{2.5}} \le a \le b\sqrt[3]{7} + \frac{1}{b^{2.5}}.$$

This shows that there are only finitely many solutions, and a quick computation (on a computer) reveals that for this example there are only the two solutions $a/b = 1/1$ and $2/1$.

We have almost completed our proof that the inequality $(*)$ has only finitely many solutions a/b. If you review what we've done so far, you'll see that what we have actually proved is that $(*)$ has only finitely many solutions satisfying $f(a/b) \ne 0$. Thus we still need to deal with the roots of $f(X)$. But a polynomial of degree d has at most d roots of any sort, rational or irrational, so the finitely many rational roots of $f(X)$ don't change our conclusion that $(*)$ has only finitely many solutions. □

Liouville's Inequality says that an algebraic number α cannot be too closely approximated by rational numbers. The next Lemma, which is the second leg in our proof, says that Liouville's number β can be very closely approximated by lots of rational numbers.

Lemma 35.3 (Lemma on Good Approximations to β). *Let β be Liouville's number*

$$\beta = \sum_{n=1}^{\infty} \frac{1}{10^{n!}}$$

as described above. Then for every number $D \ge 1$ we can find infinitely many different rational numbers a/b that satisfy the inequality

$$\left| \frac{a}{b} - \beta \right| \le \frac{1}{b^D}.$$

Proof. Intuitively, this Lemma says that we can find rational numbers that are very, very close to β. How might we find such good approximations? The definition of

$$\beta = \frac{1}{10^{1!}} + \frac{1}{10^{2!}} + \frac{1}{10^{3!}} + \frac{1}{10^{4!}} + \frac{1}{10^{5!}} + \frac{1}{10^{6!}} + \cdots$$

provides the clue. The terms in this series are decreasing very rapidly, so if we just take the first few terms, we should get a pretty good approximation to β. For example, if we take the first four terms, then we get the rational number

$$r_4 = \frac{1}{10^{1!}} + \frac{1}{10^{2!}} + \frac{1}{10^{3!}} + \frac{1}{10^{4!}} = 0.110001000000000000000001.$$

Then $|r_4 - \beta|$ has a decimal expansion whose first 119 decimal digits are all zero, so $|r_4 - \beta| < 2 \cdot 10^{-120}$, which is certainly very small. On the other hand, if we write r_4 as a fraction a_4/b_4, we find that

$$r_4 = \frac{a_4}{b_4} = \frac{110001000000000000000001}{1000000000000000000000000},$$

so its denominator b_4 is "only" 10^{24}. This may seem large, but notice that $|r_4 - \beta| < 2 \cdot 10^{-120} < 1/b_4^5$, so r_4 is a rather good approximation to β.

More generally, suppose we take the first N terms in the series and add them to form the rational number

$$r_N = \frac{a_N}{b_N} = \frac{1}{10^{1!}} + \frac{1}{10^{2!}} + \frac{1}{10^{3!}} + \cdots + \frac{1}{10^{N!}}.$$

We need to estimate the size of b_N and also how close r_N is to β.

The denominators of the fractions we're adding to form r_N are all powers of 10, so the least common denominator is the last one,

$$b_N = 10^{N!}.$$

On the other hand, the difference $\beta - r_N$ looks like

$$\beta - r_N = \frac{1}{10^{(N+1)!}} + \frac{1}{10^{(N+2)!}} + \frac{1}{10^{(N+3)!}} + \cdots.$$

Thus the first nonzero digit in the decimal expansion of $\beta - r_N$ occurs at the $(N+1)!^{\text{th}}$ digit, and this digit is a 1. This shows that the difference $\beta - r_N$ is certainly smaller than the number that has a 2 as its $(N+1)!^{\text{th}}$ digit. In other words,

$$0 < \beta - r_N < \frac{2}{10^{(N+1)!}}.$$

To relate this to the value of b_N, we observe that

$$10^{(N+1)!} = (10^{N!})^{N+1} = b_N^{N+1},$$

so we find that

$$0 < \beta - r_N < \frac{2}{b_N^{N+1}}.$$

To recapitulate, for every $N \geq 1$ we have found a rational number a_N/b_N so that

$$0 < \beta - \frac{a_N}{b_N} < \frac{2}{b_N^{N+1}} = \frac{2}{b_N} \cdot \frac{1}{b_N^N} < \frac{1}{b_N^N}.$$

Furthermore, these rational numbers are all different, since their denominators $b_N = 10^{N!}$ are different. Hence the rational numbers a_N/b_N with $N \geq D$ provide infinitely many solutions to the inequality

$$\left|\frac{a}{b} - \beta\right| \leq \frac{1}{b^D},$$

which completes the proof of the Lemma. □

We now have the two ingredients needed to prove that β is transcendental.

Theorem 35.4 (Transcendence of β Theorem). *Liouville's number*

$$\beta = \sum_{n=1}^{\infty} \frac{1}{10^{n!}}$$

is transcendental.

Proof. We give a proof by contradiction, so we start by assuming that β is actually algebraic and try to derive a false statement. The assumption that β is algebraic means that it is a root of a polynomial

$$f(X) = c_0 X^d + c_1 X^{d-1} + c_2 X^{d-2} + \cdots + c_{d-1} X + c_d$$

having integer coefficients. Let $D = d + 1$. Then Liouville's Inequality tells us that there are only finitely many rational numbers a/b that satisfy the inequality

$$\left|\frac{a}{b} - \beta\right| \leq \frac{1}{b^D}.$$

On the other hand, the Lemma tells us that there are infinitely many rational numbers satisfying this inequality. This contradiction shows that β cannot be an algebraic number, which completes the proof that β must be a transcendental number. □

The proof that Liouville's number is transcendental is not easy, and you are to be congratulated at having reached the end of our transcendental expedition. But be aware that we have surveyed only a tiny sliver of the vast continent of transcendental numbers.

One of the most beautiful theorems in transcendence theory was proved independently by A.O. Gelfond and T. Schneider in 1934. They showed that, if a is any algebraic number other than 0 or 1 and if b is any irrational algebraic number, then the number a^b is transcendental. For example, the number $2^{\sqrt{2}}$ is transcendental. Amazingly, the Gelfond–Schneider theorem is true even if a and b are complex numbers. Thus the number e^π is transcendental,[5] since e^π is equal to $(-1)^{-i}$.

Transcendence theory is today an active field of mathematical research with many innocuous-sounding open problems. For example, it is not known if the number $\pi + e$ is transcendental; indeed, it is not even known if $\pi + e$ is irrational!

Exercises

35.1. (a) Suppose that N is a positive integer that is not a perfect square. Prove that $\sqrt{N}$ is irrational. (Be careful not to prove too much. For example, check to make sure that your proof won't show that $\sqrt{4}$ is irrational.)

(b) Let $n \ge 2$ be an integer and let p be a prime. Prove that $\sqrt[n]{p}$ is irrational.

(c) Let $n \ge 2$ and $N \ge 2$ be integers. Describe when $\sqrt[n]{N}$ is irrational and prove that your description is correct.

35.2. Let A, B, C be integers with $A \ne 0$. Let r_1 and r_2 be the roots of the polynomial $Ax^2 + Bx + C$. Explain under what conditions r_1 and r_2 are rational. In particular, explain why they are either both rational or both irrational.

35.3. Give an example of a polynomial of degree 3 with integer coefficients having:

(a) three rational roots.

(b) one rational root and two irrational roots.

(c) no rational roots.

(d) Can a polynomial of degree 3 have two rational roots and one irrational root? Either give an example of such a polynomial or prove that none exists.

35.4. (a) Find a polynomial with integer coefficients that has the number $\sqrt{2} + \sqrt[3]{3}$ as one of its roots.

(b) Find a polynomial with integer coefficients that has the number $\sqrt{5} + i$ as one of its roots, where $i = \sqrt{-1}$.

35.5. Suppose that $f(X) = X^n + c_1 X^{d-1} + c_2 X^{d-2} + \cdots + c_{d-1} X + c_d$ is a polynomial of degree d whose coefficients $c_1, c_2, \ldots, c_d$ are all integers. Suppose that r is a rational number that is a root of $f(X)$.

(a) Prove that r must in fact be an integer.

(b) Prove that r must divide c_d.

---

[5]Here $e = 2.7182818\ldots$ is the base of the natural logarithms. Hermite proved that e is transcendental in 1873. The equality $(-1)^{-i} = e^\pi$ follows from Euler's identity $e^{i\theta} = \cos(\theta) + i\sin(\theta)$. Putting $\theta = \pi$ gives $e^{i\pi} = -1$, and raising both sides to the $-i$ power gives the desired formula.

35.6. Use the previous exercise to solve the following problems.

(a) Find all the rational roots of $X^5 - X^4 - 3X^3 - 2X^2 - 19X - 6$.

(b) Find all the rational roots of $X^5 + 63X^4 + 135X^3 + 785X^2 - 556X - 4148$. (*Hint.* You can cut down on the amount of work if, as soon as you find a root r, you divide the polynomial by $X - r$ to get rid of that root.)

(c) For what integer value(s) of c does the following polynomial have a rational root: $X^5 + 2X^4 - cX^3 + 3cX^2 + 3$?

35.7. (a) Suppose that $f(X) = c_0X^n + c_1X^{d-1} + c_2X^{d-2} + \cdots + c_{d-1}X + c_d$ is a polynomial of degree d whose coefficients $c_0, c_1, c_2, \ldots, c_d$ are all integers. Suppose that $r = a/b$ is a rational number that is a root of $f(X)$. Prove that a must divide c_d and that b must divide c_0.

(b) Use (a) to find all rational roots of the polynomial
$$8x^7 - 10x^6 - 3x^5 + 24x^4 - 30x^3 - 33x^2 + 30x + 9.$$

(c) Let p be a prime number. Prove that the polynomial $pX^5 - X - 1$ has no rational roots.

35.8. Let α be an algebraic number.

(a) Prove that $\alpha + 2$ and 2α are algebraic numbers.

(b) Prove that $\alpha + \frac{2}{3}$ and $\frac{2}{3}\alpha$ are algebraic numbers.

(c) More generally, let r be any rational number and prove that $\alpha + r$ and $r\alpha$ are algebraic numbers.

(d) Prove that $\alpha + \sqrt{2}$ and $\sqrt{2} \cdot \alpha$ are algebraic numbers.

(e) More generally, let A be an integer and prove that $\alpha + \sqrt{A}$ and $\sqrt{A} \cdot \alpha$ are algebraic numbers.

(f) Try to generalize this exercise as much as you can.

35.9. The number $\alpha = \sqrt{2} + \sqrt{3}$ is a root of the polynomial $f(X) = X^4 - 10X^2 + 1$.

(a) Find a polynomial $g(X)$ so that $f(X)$ factors as $f(X) = (X - \alpha)g(X)$.

(b) Find a number K so that if a/b is any rational number with $|a/b - \alpha| \le 1$ and $f(a/b) \ne 0$, then $|g(a/b)| \le K$.

(c) Find all rational numbers a/b satisfying the inequality
$$\left|\frac{a}{b} - (\sqrt{2} + \sqrt{3})\right| \le \frac{1}{b^5}.$$

(d) If you know how to program , redo (c) with $1/b^5$ replaced by $1/b^{4.5}$.

35.10. Let β_1 and β_2 be the numbers
$$\beta_1 = \sum_{n=1}^{\infty} \frac{1}{k^{n!}} \quad \text{and} \quad \beta_2 = \sum_{n=1}^{\infty} \frac{1}{10^{n^n}}.$$

Here k is some fixed integer with $k \ge 2$.

(a) Prove that β_1 is transcendental. (If you find it confusing to work with a general value for k, first try to do $k = 2$. Note that we already did the case $k = 10$.)

(b) Prove that β_2 is transcendental.

35.11. Let β_3 and β_4 be the numbers

$$\beta_3 = \sum_{n=0}^{\infty} \frac{1}{n!} \quad \text{and} \quad \beta_4 = \sum_{n=1}^{\infty} \frac{1}{10^{10^n}}.$$

(a) Try to use the methods of this chapter to prove that β_3 is transcendental. At what point does the proof break down?

(b) Prove that β_3 is irrational. (*Hint.* Assume that β_3 is rational, say $\beta_3 = a/b$, and look at the highest power of 2 that must divide b.) You may have recognized the famous number $\beta_3 = e = 2.7182818\ldots$. It turns out that e is indeed transcendental, but it wasn't until 33 years after Liouville's result that Hermite proved the transcendence of e.

(c) Try to use the methods of this chapter to prove that β_4 is transcendental. At what point does the proof break down?

(d) Prove that β_4 is not the root of a polynomial with integer coefficients of degree 9 or smaller.

35.12. Let $\alpha = r/s$ be a rational number written in lowest terms.

(a) Show that there is exactly one rational number a/b satisfying the inequality

$$|a/b - \alpha| < 1/sb.$$

(b) Show that the equality $|a/b - \alpha| = 1/sb$ is true for infinitely many different rational numbers a/b.

35.13. (a) Prove that $1/8b^2 < |a/b - \sqrt{10}|$ holds for every rational number a/b.

(b) Use (a) to find all rational numbers a/b satisfying $|a/b - \sqrt{10}| \le 1/b^3$.

35.14. (a) If N is not a perfect square, find a specific value for K so that the inequality $K/b^2 < |a/b - \sqrt{N}|$ holds for every rational number a/b. (The value of K will depend on N, but not on a or b.)

(b) Use (a) to find all rational numbers a/b satisfying each of the following inequalities:
  (i) $|a/b - \sqrt{7}| \le 1/b^3$
  (ii) $|a/b - \sqrt{5}| \le 1/b^{8/3}$

(c) Write a computer program that takes as input three numbers (N, C, e) and prints as output all rational numbers a/b satisfying $|a/b - \sqrt{N}| \le C/b^e$. Your program should check that N is a positive integer and that $C > 0$ and $e > 2$. (If $e < 2$, your program should tell the user that she won't get to see all the solutions, since there are infinitely many!) Use your program to find all solutions in rational numbers a/b to the following inequalities:
  (i) $|a/b - \sqrt{573}| \le 1/b^3$
  (ii) $|a/b - \sqrt{19}| \le 1/b^{2.5}$
  (iii) $|a/b - \sqrt{6}| \le 8/b^{2.3}$

[You'll need a moderately fast computer for (iii) if you try to do it directly.]

35.15. Determine which of the following numbers are algebraic and which are transcendental. Be sure to explain your reasoning. You may use the fact that π is transcendental, and you may use the Gelfond–Schneider theorem, which says that if a is any algebraic number other than 0 or 1 and if b is any irrational algebraic number then the number a^b is transcendental. (*Hint.* To keep you on your toes, I've thrown one number into the list for which the answer isn't known!)

(a) $\sqrt{2}^{\cos(\pi)}$ (b) $\sqrt{2}^{\sqrt{3}}$ (c) $(\tan \pi/4)^{\sqrt{2}}$ (d) π^{17}

(e) $\sqrt{\pi}$ (f) π^{π} (g) $\cos(\pi/5)$ (h) $2^{\sin(\pi/4)}$

35.16. A set S of (real) numbers is said to have the *well-ordering property* if every subset of S has a smallest element. (A subset T of S has a smallest element if there is an element $a \in T$ so that $a \le b$ for every other $b \in T$.)

(a) Using the fact that there are no integers lying strictly between 0 and 1, prove that the set of nonnegative integers has the well-ordering property.

(b) Show that the set of nonnegative rational numbers does not have the well-ordering property by writing down a specific subset that does not have a smallest element.

Chapter 36

Binomial Coefficients and Pascal's Triangle

We begin this chapter with a short list of powers of $A+B$.

$$\begin{aligned}
(A+B)^0 &= 1\\
(A+B)^1 &= A+B\\
(A+B)^2 &= A^2+2AB+B^2\\
(A+B)^3 &= A^3+3A^2B+3AB^2+B^3\\
(A+B)^4 &= A^4+4A^3B+6A^2B^2+4AB^3+B^4
\end{aligned}$$

There are many beautiful patterns lurking in this list, some fairly obvious, others extremely subtle. Before reading further, you should spend a few minutes looking for patterns on your own.

In this chapter we investigate what happens when the quantity $(A+B)^n$ is multiplied out. It is clear from the above examples that we get an expression that looks like

$$\begin{aligned}
(A+B)^n = \square A^n + \square A^{n-1}B + \square A^{n-2}B^2 + \square A^{n-3}B^3 + \cdots\\
+ \square A^2B^{n-2} + \square AB^{n-1} + \square B^n,
\end{aligned}$$

where the empty boxes need to be filled in with some integers.

Clearly, the first and last boxes are filled in with the number 1, and from the examples it appears that the second and next-to-last boxes should contain the number n. Unfortunately, it's not at all clear what should go into the other boxes, but our lack of knowledge doesn't prevent us from giving these numbers a name. The integers that appear in the expansion of $(A+B)^n$ are called *binomial coefficients*, because $A + B$ is a binomial (i.e., a quantity consisting of two terms), and the numbers we're studying appear as coefficients when the binomial $A + B$ is raised to a power. There are a variety of different symbols commonly used for binomial coefficients.[1] We use the symbol

$$\binom{n}{k} = \text{Coefficient of } A^{n-k}B^k \text{ in } (A+B)^n.$$

So using binomial coefficient symbols, the expansion of $(A+B)^n$ looks like

$$\binom{n}{0}A^n + \binom{n}{1}A^{n-1}B + \binom{n}{2}A^{n-2}B^2 + \cdots + \binom{n}{k}A^{n-k}B^k + \cdots + \binom{n}{n}B^n.$$

To study binomial coefficients, it is convenient to arrange them in the form of a triangle, where the nth row of the triangle contains the binomial coefficients appearing in the expansion of $(A+B)^n$. This arrangement is called *Pascal's Triangle* after the seventeenth-century French mathematician and natural philosopher Blaise Pascal.

$$\begin{array}{lccccccccc}
(A+B)^0: & & & & & \binom{0}{0} & & & & \\
(A+B)^1: & & & & \binom{1}{0} & & \binom{1}{1} & & & \\
(A+B)^2: & & & \binom{2}{0} & & \binom{2}{1} & & \binom{2}{2} & & \\
(A+B)^3: & & \binom{3}{0} & & \binom{3}{1} & & \binom{3}{2} & & \binom{3}{3} & \\
(A+B)^4: & \binom{4}{0} & & \binom{4}{1} & & \binom{4}{2} & & \binom{4}{3} & & \binom{4}{4}
\end{array}$$

The First Five Rows of Pascal's Triangle

We can use the list appearing at the beginning of this chapter to fill in the values.

$$\begin{array}{lccccccccc}
(A+B)^0: & & & & & 1 & & & & \\
(A+B)^1: & & & & 1 & & 1 & & & \\
(A+B)^2: & & & 1 & & 2 & & 1 & & \\
(A+B)^3: & & 1 & & 3 & & 3 & & 1 & \\
(A+B)^4: & 1 & & 4 & & 6 & & 4 & & 1
\end{array}$$

How might we form the next row of Pascal's Triangle? One method is simply to multiply out the quantity $(A+B)^5$ and record the coefficients. A simpler method

[1]The binomial coefficient $\binom{n}{k}$ is also called a *combinatorial number* and assigned the symbol ${}_nC_k$.

is to take the already known expansion of $(A+B)^4$ and multiply it by $A+B$ to get $(A+B)^5$. Thus

$$
\begin{array}{rr}
(A+B)^4 & A^4 + 4A^3B + 6A^2B^2 + 4AB^3 + B^4 \\
\times\ A+B & \times \qquad\qquad\qquad\qquad\qquad A + B \\
\hline
(A+B)^5 & A^4B + 4A^3B^2 + 6A^2B^3 + 4AB^4 + B^5 \\
 & A^5 + 4A^4B + 6A^3B^2 + 4A^2B^3 + AB^4 \\
\hline
 & A^5 + 5A^4B + 10A^3B^2 + 10A^2B^3 + 5AB^4 + B^5
\end{array}
$$

So the next row of Pascal's Triangle is

$$1 \quad 5 \quad 10 \quad 10 \quad 5 \quad 1$$

We can use this simple idea,

$$(A+B)^{n+1} = (A+B)\cdot(A+B)^n,$$

to derive a fundamental relationship for the binomial coefficients. If we multiply $(A+B)^n$ by $A+B$ as before and equate the result with $(A+B)^{n+1}$, we find that

$$
\begin{array}{l}
\qquad\binom{n}{0}A^n + \binom{n}{1}A^{n-1}B + \cdots + \binom{n}{n-1}AB^{n-1} + \binom{n}{n}B^n \\
\times \qquad\qquad\qquad\qquad\qquad\qquad\qquad A + B \\
\hline
\qquad\binom{n}{0}A^nB + \binom{n}{1}A^{n-1}B^2 + \cdots + \binom{n}{n-1}AB^n + \binom{n}{n}B^{n+1} \\
\binom{n}{0}A^{n+1} + \binom{n}{1}A^nB + \binom{n}{2}A^{n-1}B^2 + \cdots + \binom{n}{n}AB^n \\
\hline
\binom{n+1}{0}A^{n+1} + \binom{n+1}{1}A^nB + \binom{n+1}{2}A^{n-1}B^2 + \cdots + \binom{n+1}{n}AB^n + \binom{n+1}{n+1}B^{n+1}
\end{array}
$$

Thus, for example,

$$\binom{n}{0} + \binom{n}{1} = \binom{n+1}{1} \quad \text{and} \quad \binom{n}{1} + \binom{n}{2} = \binom{n+1}{2}, \quad \text{and so on.}$$

In general, we get the following fundamental formula:

Theorem 36.1 (Addition Formula for Binomial Coefficients). *Let $n \geq k \geq 0$ be integers. Then*

$$\binom{n}{k-1} + \binom{n}{k} = \binom{n+1}{k}.$$

The addition formula describes a wonderful property of Pascal's Triangle: each entry in the triangle is equal to the sum of the two entries above it. For example, we found earlier that the $n = 5$ row of Pascal's triangle is

1	5	10	10	5	1

,

so the $n = 6$ row can be computed by adding adjacent pairs in the $n = 5$ row, as illustrated here:

$$\begin{array}{lccccccccccccc} [n=5\text{ Row}] & & 1 & & 5 & & 10 & & 10 & & 5 & & 1 & \\ & \swarrow & & \searrow^{+}\swarrow & & \searrow^{+}\swarrow & & \searrow^{+}\swarrow & & \searrow^{+}\swarrow & & \searrow^{+}\swarrow & & \searrow \\ [n=6\text{ Row}] & 1 & & 6 & & 15 & & 20 & & 15 & & 6 & & 1 \end{array}$$

This shows that

$$(A+B)^6 = A^6 + 6A^5B + 15A^4B^2 + 20A^3B^3 + 15A^2B^4 + 6AB^5 + B^6$$

without doing any algebra at all! Here is a picture of Pascal's Triangle illustrating the binomial coefficient addition formula.

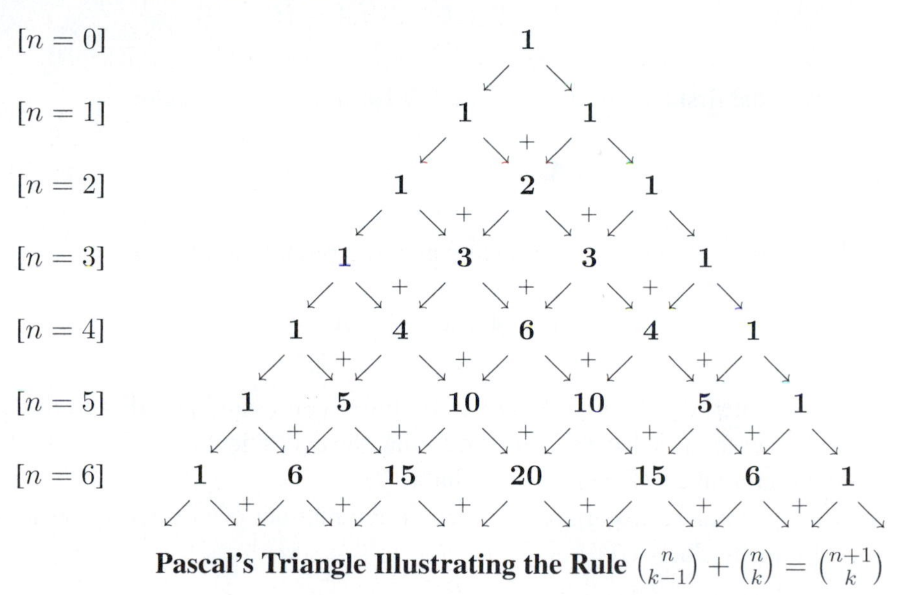

Pascal's Triangle Illustrating the Rule $\binom{n}{k-1} + \binom{n}{k} = \binom{n+1}{k}$

Our next task is to derive a completely different sort of formula for the binomial coefficients. This illustrates a simple but amazingly powerful method that appears time and again in the development of modern mathematics. Anytime you can compute a quantity in two different ways, comparison of the resulting formulas yields interesting and useful information.

To find a new formula for the binomial coefficient $\binom{n}{k}$, we consider what happens when we multiply out the quantity

$$(A+B)^n = (A+B)(A+B)(A+B)\cdots(A+B)(A+B).$$

Let's start with the particular case of

$$(A+B)^3 = (A+B)(A+B)(A+B).$$

The product consists of a bunch of terms formed by choosing either A or B from the first factor, then choosing either A or B from the second factor, and finally choosing either A or B from the third factor. This gives a total of 8 terms. How many of those terms are equal to A^3? The only way to get A^3 is to choose A from every factor, so there is only one way to get A^3.

Next consider the number of ways to get A^2B. We can get A^2B in the following ways:

- A from the first and second factors and B from the third factor,

$$(\underset{\uparrow}{\boldsymbol{A}}+B)(\underset{\uparrow}{\boldsymbol{A}}+B)(A+\underset{\uparrow}{\boldsymbol{B}});$$

- A from the first and third factors and B from the second factor,

$$(\underset{\uparrow}{\boldsymbol{A}}+B)(A+\underset{\uparrow}{\boldsymbol{B}})(\underset{\uparrow}{\boldsymbol{A}}+B);$$

- A from the second and third factors and B from the first factor,

$$(A+\underset{\uparrow}{\boldsymbol{B}})(\underset{\uparrow}{\boldsymbol{A}}+B)(\underset{\uparrow}{\boldsymbol{A}}+B).$$

We have illustrated the three possibilities by highlighting and pointing to the A's and B's being used in each case. This shows that the coefficient of A^2B in $(A+B)^3$ is 3, so the binomial coefficient $\binom{3}{1}$ is equal to 3.

Now we generalize this argument to count the number of different ways to get A^kB^{n-k} in the product

$$\overbrace{(A+B)(A+B)(A+B)\cdots(A+B)(A+B)}^{n\text{ factors}}.$$

We can get A^kB^{n-k} by choosing A from any k of the factors and then choosing B from the remaining $n-k$ factors. So we need to count the number of ways to select k of the factors. Let's make one selection at a time.

We have n choices for the first factor. Once we've made that choice, then there are $n-1$ factors left from which to make our second choice. After we've made these two choices, there are $n-2$ factors left from which to make our third choice, and so on. There are thus

$$n(n-1)(n-2)\cdots(n-k+1)$$

ways to choose k of the factors from among the collection of n factors.

Unfortunately, we've overcounted the number of ways to get A^kB^{n-k}, because we've made our choices in a particular order. To illustrate the problem, we return to our $n = 3$ example. In this case one way to get A^2B is to take A from the first and the third factors, but we've counted this choice twice, because we counted it once as

"first choose the first factor, next choose the third factor"

and we counted it a second time as

"first choose the third factor, next choose the first factor."

Thus the actual number of A^kB^{n-k} terms in $(A+B)^n$ is

$$n(n-1)(n-2)\cdots(n-k+1)$$

divided by the number of different orders in which we can make our choices. Remember we're making k choices, so the number of different orders for these choices is $k!$, since we can put any of the k choices first, then any of the remaining $k-1$ choices second, and so on. Therefore, the number of ways to get the term A^kB^{n-k} in the product $(A+B)^n$ is equal to

$$\frac{n(n-1)(n-2)\cdots(n-k+1)}{k!}.$$

Of course, this is precisely the binomial coefficient $\binom{n}{k}$, so we have proved the celebrated Binomial Theorem.

Theorem 36.2 (Binomial Theorem). *The binomial coefficients in the expansion*

$$(A+B)^n = \binom{n}{0}A^n + \binom{n}{1}A^{n-1}B + \binom{n}{2}A^{n-2}B^2 + \cdots + \binom{n}{n}B^n$$

are given by the formula

$$\binom{n}{k} = \frac{n(n-1)(n-2)\cdots(n-k+1)}{k!} = \frac{n!}{k!(n-k)!}.$$

Proof. We have already done the hard work of proving the first equality. To get the second formula, we simply multiply the numerator and denominator of the first fraction by $(n-k)!$ to get

$$\frac{n(n-1)(n-2)\cdots(n-k+1)}{k!}\cdot\frac{(n-k)!}{(n-k)!} = \frac{n!}{k!(n-k)!}.$$ □

For example, the coefficient of A^4B^3 in $(A+B)^7$ is equal to

$$\binom{7}{3} = \frac{7\cdot 6\cdot 5}{3!} = \frac{210}{6} = 35.$$

As another example, the coefficient of A^8B^{11} in $(A+B)^{19}$ is

$$\begin{aligned}\binom{19}{11} &= \frac{19\cdot 18\cdot 17\cdot 16\cdot 15\cdot 14\cdot 13\cdot 12\cdot 11\cdot 10\cdot 9}{11!}\\ &= \frac{3016991577600}{39916800}\\ &= 75582.\end{aligned}$$

Of course, if k is larger than $n/2$, as in this last example, then it is easier to first use the following *Binomial Coefficient Symmetry Formula*:

$$\binom{n}{k} = \binom{n}{n-k}.$$

The symmetry formula simply says that when $(A+B)^n$ is multiplied out, the two terms A^kB^{n-k} and $A^{n-k}B^k$ have the same coefficient. This is clearly true since there is nothing to distinguish A and B from one another. Using the symmetry formula, we can compute

$$\begin{aligned}\binom{19}{11} = \binom{19}{8} &= \frac{19\cdot 18\cdot 17\cdot 16\cdot 15\cdot 14\cdot 13\cdot 12}{8!}\\ &= \frac{3047466240}{40320} = 75582.\end{aligned}$$

Binomial Coefficients Modulo p

What happens if we reduce a binomial coefficient $\binom{n}{k}$ modulo p, where p is a prime number? Here is what the first few lines of Pascal's triangle look like modulo 5 and modulo 7.

```
              1                                    1
            1   1                                1   1
          1   2   1                            1   2   1
        1   3   3   1                        1   3   3   1
      1   4   1   4   1                    1   4   6   4   1
    1   0   0   0   0   1                1   5   3   3   5   1
  1   1   0   0   0   1   1            1   6   1   6   1   6   1
1   2   1   0   0   1   2   1        1   0   0   0   0   0   0   1
```

Pascal's Triangle Modulo 5 | **Pascal's Triangle Modulo 7**

Notice that the $n = 5$ line of the modulo 5 Pascal triangle is 1 0 0 0 0 1, and similarly the $n = 7$ line of the modulo 7 Pascal triangle is 1 0 0 0 0 0 0 1. This suggests that $\binom{p}{k}$ should equal 0 modulo p if $1 \le k \le p-1$. Having made this observation, it is easy to prove, and it gives a wonderfully simple version of the binomial theorem modulo p.

Theorem 36.3 (Binomial Theorem Modulo p). *Let p be a prime number.*
(a) *The binomial coefficient $\binom{p}{k}$ is congruent to*

$$\binom{p}{k} \equiv \begin{cases} 0 \pmod{p} & \text{if } 1 \le k \le p-1, \\ 1 \pmod{p} & \text{if } k = 0 \text{ or } k = p. \end{cases}$$

(b) *For any numbers A and B, we have*

$$(A+B)^p \equiv A^p + B^p \pmod{p}.$$

Proof. (a) If $k = 0$ or $k = p$, then we know that $\binom{p}{k} = 1$. So the interesting problem is to find out what happens when k is between 1 and $p-1$. Let's take a particular example, say $\binom{7}{5}$, and try to understand what's going on. Our formula for this binomial coefficient is

$$\binom{7}{5} = \frac{7 \cdot 6 \cdot 5 \cdot 4 \cdot 3}{5 \cdot 4 \cdot 3 \cdot 2 \cdot 1}.$$

Notice that the number 7 appears in the numerator and that there are no 7's in the denominator to cancel the 7 in the numerator. Thus $\binom{7}{5}$ is divisible by 7, which is the same as saying that it is congruent to 0 modulo 7.

This idea works in complete generality. The binomial coefficient $\binom{p}{k}$ is equal to

$$\binom{p}{k} = \frac{p \cdot (p-1) \cdot (p-2) \cdots (p-k+1)}{k \cdot (k-1) \cdot (k-2) \cdots 2 \cdot 1}.$$

Thus $\binom{p}{k}$ has a p in the numerator (provided $k \ge 1$), and there are no p's in the denominator to cancel it (provided $k \le p-1$). Hence $\binom{p}{k}$ is divisible by p, so it is congruent to 0 modulo p.

Do you see where we are using the fact that p is a prime? If it weren't a prime, then it might happen that some of the smaller numbers in the denominator could combine to cancel part or all of p. Thus our proof does not work for composite numbers. In other words, we have not proved that $\binom{n}{k} \equiv 0 \pmod{n}$ for composite numbers n. Do you think that this more general statement is true?

(b) Using the Binomial Theorem and part (a), it is easy to compute

$$\begin{aligned}(A+B)^p &= \binom{p}{0}A^p + \binom{p}{1}A^{p-1}B + \binom{p}{2}A^{p-2}B^2 \\ &\quad + \cdots + \binom{p}{p-2}A^2B^{p-2} + \binom{p}{p-1}AB^{p-1} + \binom{p}{p}B^p \\ &\equiv 1\cdot A^p + 0\cdot A^{p-1}B + 0\cdot A^{p-2}B^2 \\ &\qquad + \cdots + 0\cdot A^2B^{p-2} + 0\cdot AB^{p-1} + 1\cdot B^p \pmod{p} \\ &\equiv A^p + B^p \pmod{p}.\end{aligned}$$

□

The formula

$$(A+B)^p \equiv A^p + B^p \pmod{p}$$

is one of the most important formulas in all of number theory. It says that the pth power of a sum is congruent to the sum of the pth powers. We conclude by using this formula to give a new proof of Fermat's Little Theorem. You should compare this proof with the one that we gave in Chapter 9. Each proof reveals different aspects of the underlying formula. Which proof do you like best?

Theorem 36.4 (Fermat's Little Theorem). *Let p be a prime number, and let a be any number with $a \not\equiv 0 \pmod{p}$. Then*

$$a^{p-1} \equiv 1 \pmod{p}.$$

Proof by Induction. We start by using induction to prove that the formula

$$a^p \equiv a \pmod{p}$$

is true for all numbers a. This formula is clearly true for $a = 0$, which gets our induction started. Next suppose that we know it is true for some particular value of a. Then

$$\begin{aligned}(a+1)^p &\equiv a^p + 1^p \pmod{p} && \text{using the Binomial Theorem Modulo } p \\ & && \quad\text{with } A = a \text{ and } B = 1, \\ &\equiv a + 1 \pmod{p} && \text{since } a^p \equiv a \pmod{p} \\ & && \quad\text{by the induction hypothesis.}\end{aligned}$$

This completes the proof by induction of the formula

$$a^p \equiv a \pmod{p}.$$

This means that p divides $a^p - a$, so

$$p \quad \text{divides} \quad a(a^{p-1} - 1).$$

Since p does not divide a by assumption, we conclude that

$$a^{p-1} \equiv 1 \pmod{p},$$

which completes the proof of Fermat's Little Theorem. □

Exercises

36.1. Compute each of the following binomial coefficients.

$$\text{(a)} \binom{10}{5} \qquad \text{(b)} \binom{20}{10} \qquad \text{(c)} \binom{15}{11} \qquad \text{(d)} \binom{300}{297}$$

36.2. Use the formula $\binom{n}{k} = \frac{n!}{k!(n-k)!}$ to prove the addition formula

$$\binom{n}{k-1} + \binom{n}{k} = \binom{n+1}{k}.$$

36.3. What is the value obtained if we add up a row

$$\binom{n}{0} + \binom{n}{1} + \binom{n}{2} + \binom{n}{3} + \cdots + \binom{n}{n-1} + \binom{n}{n}$$

of Pascal's Triangle? Compute some values, formulate a conjecture, and prove that your conjecture is correct.

36.4. If we use the formula

$$\binom{n}{k} = \frac{n(n-1)(n-2)\cdots(n-k+1)}{k!}$$

to define the binomial coefficient $\binom{n}{k}$, then the binomial coefficient makes sense for any value of n as long as k is a nonnegative integer.

(a) Find a simple formula for $\binom{-1}{k}$ and prove that your formula is correct.

(b) Find a formula for $\binom{-1/2}{k}$ and prove that your formula is correct.

36.5. This exercise presupposes some knowledge of calculus. If n is a positive integer, then putting $A = 1$ and $B = x$ in the formula for $(A + B)^n$ gives

$$(1+x)^n = \binom{n}{0} + \binom{n}{1}x + \binom{n}{2}x^2 + \binom{n}{3}x^3 + \cdots + \binom{n}{n-1}x^{n-1} + \binom{n}{n}x^n.$$

In the previous exercise we noted that the binomial coefficient $\binom{n}{k}$ makes sense even if n is not a positive integer. Assuming that n is not a positive integer, prove that the infinite series

$$\binom{n}{0} + \binom{n}{1}x + \binom{n}{2}x^2 + \binom{n}{3}x^3 + \cdots$$

converges to the value $(1+x)^n$ provided that x satisfies $|x| < 1$.

36.6. We proved that if p is a prime number and if $1 \le k \le p-1$, then the binomial coefficient $\binom{p}{k}$ is divisible by p.

(a) Find an example of integers n and k with $1 \le k \le n-1$ and $\binom{n}{k}$ not divisible by n.

(b) For each composite number $n = 4$, 6, 8, 10, 12, and 14, compute $\binom{n}{k}$ modulo n for each $1 \le k \le n-1$ and pick out the ones that are 0 modulo n.

(c) Use your data from (b) to make a conjecture as to when the binomial coefficient $\binom{n}{k}$ is divisible by n.

(d) Prove that your conjecture in (c) is correct.

36.7. (a) Compute the value of the quantity

$$\binom{p-1}{k} \pmod{p}$$

for a selection of prime numbers p and integers $0 \le k \le p-1$, and make a conjecture as to its value. Prove that your conjecture is correct.

(b) Find a similar formula for the value of

$$\binom{p-2}{k} \pmod{p}.$$

36.8. We proved that $(A+B)^p \equiv A^p + B^p \pmod{p}$.

(a) Generalize this result to a sum of n numbers. That is, prove that

$$(A_1 + A_2 + A_3 + \cdots + A_n)^p \equiv A_1^p + A_2^p + A_3^p + \cdots + A_n^p \pmod{p}.$$

(b) Is the corresponding multiplication formula true,

$$(A_1 \cdot A_2 \cdot A_3 \cdots A_n)^p \equiv A_1^p \cdot A_2^p \cdot A_3^p \cdots A_n^p \pmod{p}?$$

Either prove that it is true, or give a counterexample.

Chapter 37

Fibonacci's Rabbits and Linear Recurrence Sequences

In 1202 Leonardo of Pisa (also known as Leonardo Fibonacci) published his *Liber Abbaci*, a highly influential book of practical mathematics. In this book Leonardo introduced the elegant Hindu/Arabic numerical system (the digits $1, 2, \ldots, 9$ and a symbol/placeholder for 0) to Europeans who were still laboring under the handicap of doing calculations with Roman numerals. Leonardo's book also contains the following curious Rabbit Problem.

> In the first month, start with a pair of baby rabbits. One month later they have grown up. The following month the pair of grown rabbits produce a pair of babies, so now we have one pair of grown rabbits and one pair of baby rabbits. Each month thereafter, each pair of grown rabbits produces a new pair of babies, and every pair of baby rabbits grows up. How many pairs of rabbits will there be at the end of one year?

The first few months of rabbit procreation are illustrated in Figure 37.1, where each bunny image in Figure 37.1 represents a pair of rabbits. If we let

$$F_n = \text{Number of pairs of rabbits after } n \text{ months},$$

and if we remember that each month the baby pairs grow up and that each month the grown pairs produce new baby pairs, we can compute the number of pairs of rabbits (baby and grown) in each subsequent month. Thus $F_1 = 1$ (one baby pair) and $F_2 = 1$ (one grown pair) and $F_3 = 2$ (one grown pair plus a new baby pair) and $F_4 = 3$ (two grown pairs plus a new baby pair). Continuing with this computation,

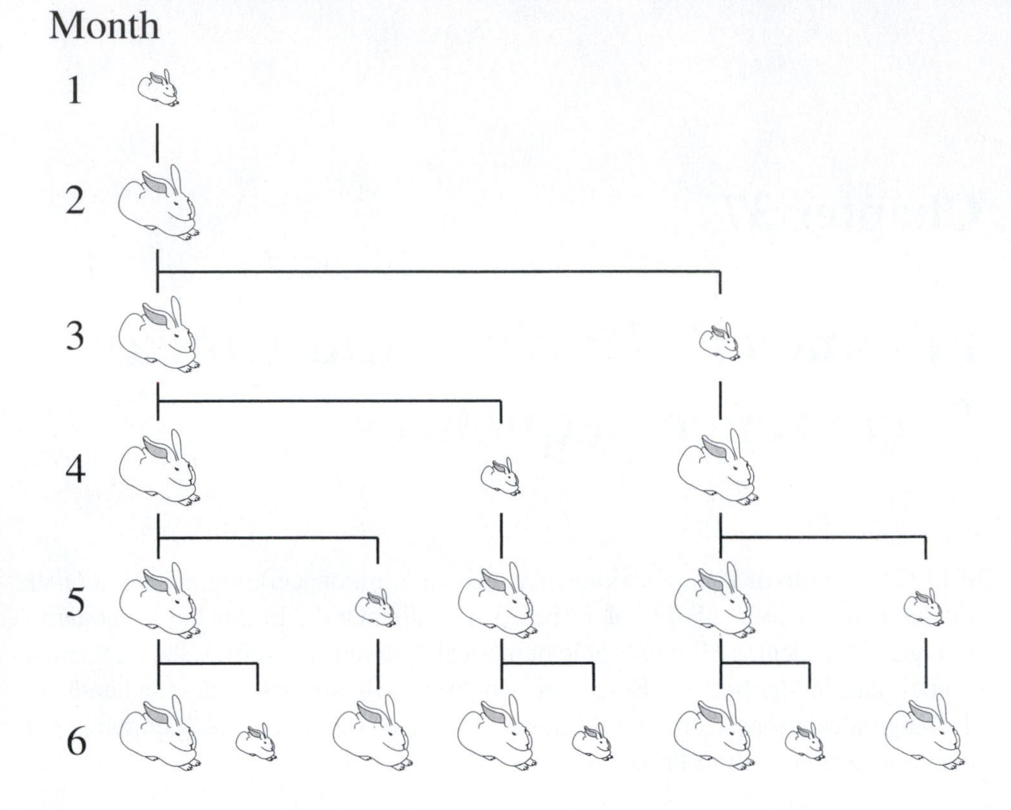

Figure 37.1: Fibbonaci's Rabbits (each rabbit image represents a pair)

we find that

$$\begin{aligned}
F_1 &= 0 \text{ Grown Pairs} &+ 1 \text{ Baby Pair} &= 1 \text{ Pair}\\
F_2 &= 1 \text{ Grown Pair} &+ 0 \text{ Baby Pairs} &= 1 \text{ Pair}\\
F_3 &= 1 \text{ Grown Pair} &+ 1 \text{ Baby Pair} &= 2 \text{ Pairs}\\
F_4 &= 2 \text{ Grown Pairs} &+ 1 \text{ Baby Pair} &= 3 \text{ Pairs}\\
F_5 &= 3 \text{ Grown Pairs} &+ 2 \text{ Baby Pairs} &= 5 \text{ Pairs}\\
F_6 &= 5 \text{ Grown Pairs} &+ 3 \text{ Baby Pairs} &= 8 \text{ Pairs}\\
F_7 &= 8 \text{ Grown Pairs} &+ 5 \text{ Baby Pairs} &= 13 \text{ Pairs}\\
F_8 &= 13 \text{ Grown Pairs} &+ 8 \text{ Baby Pairs} &= 21 \text{ Pairs}\\
F_9 &= 21 \text{ Grown Pairs} &+ 13 \text{ Baby Pairs} &= 34 \text{ Pairs}\\
F_{10} &= 34 \text{ Grown Pairs} &+ 21 \text{ Baby Pairs} &= 55 \text{ Pairs}\\
F_{11} &= 55 \text{ Grown Pairs} &+ 34 \text{ Baby Pairs} &= 89 \text{ Pairs}\\
F_{12} &= 89 \text{ Grown Pairs} &+ 55 \text{ Baby Pairs} &= 144 \text{ Pairs}\\
F_{13} &= 144 \text{ Grown Pairs} &+ 89 \text{ Baby Pairs} &= 233 \text{ Pairs.}
\end{aligned}$$

This answers Fibonacci's question. At the end of the year (after the 12th month is completed) there are 233 pairs of rabbits. The *Fibonacci sequence* of numbers

$$1, 1, 2, 3, 5, 8, 13, 21, \ldots$$

arising from Fibonacci's Rabbit Problem has intrigued people from the thirteenth century up to the present day.[1]

Suppose that we want to extend our list of *Fibonacci numbers* F_n beyond the 12th month. Looking at our list, we see that each Fibonacci number is simply the sum of the previous two Fibonacci numbers. In symbols, this becomes the formula

$$F_n = F_{n-1} + F_{n-2}.$$

Notice that this isn't really a *formula* for F_n, because it doesn't directly give the value of F_n. Instead it gives a rule telling us how to compute the nth Fibonacci number from the previous numbers. The fancy mathematical word for this sort of rule is a *recursion* or a *recursive formula.*

We can use the recursive formula for F_n to create a table of values.

n	F_n
1	1
2	1
3	2
4	3
5	5
6	8
7	13
8	21
9	34
10	55

n	F_n
11	89
12	144
13	233
14	377
15	610
16	987
17	1,597
18	2,584
19	4,181
20	6,765

n	F_n
21	10,946
22	17,711
23	28,657
24	46,368
25	75,025
26	121,393
27	196,418
28	317,811
29	514,229
30	832,040

The Fibonacci Numbers F_n

The Fibonacci numbers appear to grow very rapidly. Indeed, the 31st Fibonacci number is already larger than 1 million,

$$F_{31} = 1{,}346{,}269;$$

[1] There is even a journal called the *Fibonacci Quarterly* that was started in 1962 and is devoted to Fibonacci's sequence and its generalizations.

and in 45 months (less than 4 years),

$$F_{45} = 1{,}134{,}903{,}170,$$

and we have more than one billion pairs of rabbits! Now look how large the numbers become before we reach even the 200th Fibonacci number:

$$\begin{aligned}
F_{60} &= 1{,}548{,}008{,}755{,}920\\
F_{74} &= 1{,}304{,}969{,}544{,}928{,}657\\
F_{88} &= 1{,}100{,}087{,}778{,}366{,}101{,}931\\
F_{103} &= 1{,}500{,}520{,}536{,}206{,}896{,}083{,}277\\
F_{117} &= 1{,}264{,}937{,}032{,}042{,}997{,}393{,}488{,}322\\
F_{131} &= 1{,}066{,}340{,}417{,}491{,}710{,}595{,}814{,}572{,}169\\
F_{146} &= 1{,}454{,}489{,}111{,}232{,}772{,}683{,}678{,}306{,}641{,}953\\
F_{160} &= 1{,}226{,}132{,}595{,}394{,}188{,}293{,}000{,}174{,}702{,}095{,}995\\
F_{174} &= 1{,}033{,}628{,}323{,}428{,}189{,}498{,}226{,}463{,}595{,}560{,}281{,}832\\
F_{189} &= 1{,}409{,}869{,}790{,}947{,}669{,}143{,}312{,}035{,}591{,}975{,}596{,}518{,}914.
\end{aligned}$$

Number theory is all about patterns, but how can we possibly find a pattern in numbers that grow so rapidly? One thing we can do is try to discover just how fast the Fibonacci numbers are growing. For example, how much larger than its predecessor is each successive Fibonacci number? This is measured by the ratio F_n/F_{n-1}, so we compute the first few values.

$$\begin{aligned}
F_3/F_2 &= 2.00000 & F_{11}/F_{10} &= 1.61818\\
F_4/F_3 &= 1.50000 & F_{12}/F_{11} &= 1.61797\\
F_5/F_4 &= 1.66666 & F_{13}/F_{12} &= 1.61805\\
F_6/F_5 &= 1.60000 & F_{14}/F_{13} &= 1.61802\\
F_7/F_6 &= 1.62500 & F_{15}/F_{14} &= 1.61803\\
F_8/F_7 &= 1.61538 & F_{16}/F_{15} &= 1.61803\\
F_9/F_8 &= 1.61904 & F_{17}/F_{16} &= 1.61803\\
F_{10}/F_9 &= 1.61764 & F_{18}/F_{17} &= 1.61803
\end{aligned}$$

It looks like the ratio F_n/F_{n-1} is getting closer and closer to some number around 1.61803. It's hard to guess exactly what number this is, so let's see how we might figure it out.

The last table suggests that F_n is approximately equal to αF_{n-1} for some fixed number α whose value we don't know. So we write

$$F_n \approx \alpha F_{n-1},$$

where the squiggly equals sign means "approximately equal to." The same reasoning tells us that

$$F_{n-1} \approx \alpha F_{n-2},$$

and if we substitute this into $F_n \approx \alpha F_{n-1}$, we get

$$F_n \approx \alpha F_{n-1} \approx \alpha^2 F_{n-2}.$$

So we suspect that $F_n \approx \alpha^2 F_{n-2}$ and $F_{n-1} \approx \alpha F_{n-2}$. We also know the Fibonacci recursive equation $F_n = F_{n-1} + F_{n-2}$, so we find that

$$\alpha^2 F_{n-2} \approx \alpha F_{n-2} + F_{n-2}.$$

Dividing by F_{n-2} and moving everything to one side yields the equation

$$\alpha^2 - \alpha - 1 \approx 0.$$

We know how to solve an equation like this: use the quadratic formula.

$$\alpha = \frac{1+\sqrt{5}}{2} \quad \text{or} \quad \frac{1-\sqrt{5}}{2}$$

We were looking for the value of α, but we seem to have hit the jackpot and found two values! Both of these values satisfy the equation $\alpha^2 = \alpha + 1$, so for any number n, they both satisfy the equation

$$\alpha^n = \alpha^{n-1} + \alpha^{n-2}.$$

This looks a lot like the Fibonacci recursive equation $F_n = F_{n-1} + F_{n-2}$. In other words, if we let $G_n = \alpha^n$ for either of the values of α listed above, then $G_n = G_{n-1} + G_{n-2}$.

In fact, we can do even better by using both of the values

$$\alpha_1 = \frac{1+\sqrt{5}}{2} \qquad \text{and} \qquad \alpha_2 = \frac{1-\sqrt{5}}{2}$$

and letting

$$H_n = c_1\alpha_1^n + c_2\alpha_2^n.$$

Then

$$\begin{aligned} H_{n-1} + H_{n-2} &= (c_1\alpha_1^{n-1} + c_2\alpha_2^{n-1}) + (c_1\alpha_1^{n-2} + c_2\alpha_2^{n-2}) \\ &= c_1(\alpha_1^{n-1} + \alpha_1^{n-2}) + c_2(\alpha_2^{n-1} + \alpha_2^{n-2}) \\ &= c_1\alpha_1^n + c_2\alpha_2^n \\ &= H_n, \end{aligned}$$

so H_n satisfies the same recursive formula as the Fibonacci sequence, and we are free to choose the numbers c_1 and c_2 to have any values that we want.

The idea now is to choose c_1 and c_2 so that the H_n sequence and the Fibonacci sequence start with the same two values. In other words, we want to choose c_1 and c_2 so that

$$H_1 = F_1 = 1 \qquad \text{and} \qquad H_2 = F_2 = 1.$$

This means we need to solve

$$c_1\alpha_1 + c_2\alpha_2 = 1 \qquad \text{and} \qquad c_1\alpha_1^2 + c_2\alpha_2^2 = 1.$$

(Remember that α_1 and α_2 are specific numbers.) These two equations are easy to solve. For example, subtracting α_2 times the first equation from the second equation gives

$$c_1\alpha_1^2 - c_1\alpha_1\alpha_2 = 1 - \alpha_2, \qquad \text{so} \qquad c_1 = \frac{1-\alpha_2}{\alpha_1^2 - \alpha_1\alpha_2}.$$

This last expression can be evaluated by substituting in the values $\alpha_1 = \frac{1+\sqrt{5}}{2}$ and $\alpha_2 = \frac{1-\sqrt{5}}{2}$, but it is better to be clever and use the formulas $\alpha_1 + \alpha_2 = 1$ and $\alpha_1 - \alpha_2 = \sqrt{5}$. These formulas allow us to simplify

$$1 - \alpha_2 = \alpha_1 \qquad \text{and} \qquad \alpha_1^2 - \alpha_1\alpha_2 = \alpha_1(\alpha_1 - \alpha_2) = \alpha_1\sqrt{5},$$

and thus

$$c_1 = \frac{1-\alpha_2}{\alpha_1^2 - \alpha_1\alpha_2} = \frac{\alpha_1}{\alpha_1\sqrt{5}} = \frac{1}{\sqrt{5}}.$$

A similar calculation yields the value of c_2,

$$c_2 = \frac{1-\alpha_1}{\alpha_2^2 - \alpha_2\alpha_1} = \frac{\alpha_2}{\alpha_2(\alpha_2 - \alpha_1)} = -\frac{1}{\sqrt{5}}.$$

The culmination of our calculations is the following beautiful formula for the nth term of the Fibonacci sequence. It is named after Binet, who published it in 1843, although the formula was known to Euler and Daniel Bernoulli at least 100 years earlier.

Theorem 37.1 (Binet's Formula). *The Fibonacci sequence F_n is described by the recursion*

$$F_1 = F_2 = 1 \qquad \textit{and} \qquad F_n = F_{n-1} + F_{n-2} \quad \textit{for } n = 3, 4, 5, \ldots.$$

Then the nth term of the Fibonacci sequence is given by the formula

$$F_n = \frac{1}{\sqrt{5}}\left\{\left(\frac{1+\sqrt{5}}{2}\right)^n - \left(\frac{1-\sqrt{5}}{2}\right)^n\right\}.$$

Proof. For each number $n = 1, 2, 3, \ldots$, let H_n be the number

$$H_n = \frac{1}{\sqrt{5}} \left\{ \left(\frac{1+\sqrt{5}}{2} \right)^n - \left(\frac{1-\sqrt{5}}{2} \right)^n \right\}.$$

We will prove by induction on n that $H_n = F_n$ for every number n.

First we check that

$$H_1 = \frac{1}{\sqrt{5}} \left\{ \left(\frac{1+\sqrt{5}}{2} \right) - \left(\frac{1-\sqrt{5}}{2} \right) \right\} = \frac{1}{\sqrt{5}} \cdot \sqrt{5} = 1$$

and

$$H_2 = \frac{1}{\sqrt{5}} \left\{ \left(\frac{1+\sqrt{5}}{2} \right)^2 - \left(\frac{1-\sqrt{5}}{2} \right)^2 \right\}$$
$$= \frac{1}{\sqrt{5}} \left\{ \frac{6+2\sqrt{5}}{4} - \frac{6-2\sqrt{5}}{4} \right\} = \frac{1}{\sqrt{5}} \cdot \frac{4\sqrt{5}}{4} = 1.$$

This shows that $H_1 = F_1$ and $H_2 = F_2$.

Now suppose that $n \geq 3$ and that $H_i = F_i$ for every value of i between 1 and $n-1$. In particular, $H_{n-1} = F_{n-1}$ and $H_{n-2} = F_{n-2}$. We need to prove that $H_n = F_n$. But we have already checked that

$$H_n = H_{n-1} + H_{n-2},$$

and we know from the definition of the Fibonacci sequence that

$$F_n = F_{n-1} + F_{n-2},$$

so we see that $H_n = F_n$. This completes our induction proof that $H_n = F_n$ for every value of n. □

Historical Interlude. The number

$$\frac{1+\sqrt{5}}{2} = 1.61803\ldots$$

was called the *Golden Ratio* (or the Divine Proportion) by the ancient Greeks, who attributed aesthetic merit to artistic compositions in the divine proportion. For example, it has been suggested that the Parthenon (Figure 37.2) was designed so that its exterior dimensions are in the golden ratio. Here is a small rectangle ▭ whose sides are in the golden ratio, and here is a larger divinely proportioned rectangle ▭. Do you find the proportions of these rectangles to be especially pleasing to the eye?

Figure 37.2: The Parthenon

The Fibonacci sequence is an example of a *Linear Recurrence Sequence*. The word *linear* in this context means that the nth term of the sequence is a linear combination of some of the previous terms. Here are examples of some other linear recurrence sequences:

$$A_n = 3A_{n-1} + 10A_{n-2} \qquad A_1 = 1 \qquad A_2 = 3$$
$$B_n = 2B_{n-1} - 4B_{n-2} \qquad B_1 = 0 \qquad B_2 = -2$$
$$C_n = 4C_{n-1} - C_{n-2} - 6C_{n-3} \qquad C_1 = 0 \qquad C_2 = 0 \qquad C_3 = 1$$

The method that we used to derive Binet's Formula for the nth Fibonacci number can be used, *mutatis mutandis*,[2] to find a formula for the nth term of any linear recurrence sequence. Of course, not all recurrence sequences are linear. Here are some examples of recurrence sequences that are not linear:

$$D_n = D_{n-1} + D_{n-2}^2 \qquad D_1 = 1 \qquad D_2 = 1$$
$$E_n = E_{n-1}E_{n-2} + E_{n-3} \qquad E_1 = 1 \qquad E_2 = 2 \qquad E_3 = 1$$

In general, there is no simple expression for the nth term of a nonlinear recurrence sequence. This does not mean that nonlinear sequences are uninteresting (quite the contrary), but it does mean that they are much harder to analyze than linear recurrence sequences.

The Fibonacci Sequence Modulo m

What happens to the numbers in the Fibonacci sequence if we reduce them modulo m? There are only finitely many different numbers modulo m, so the values do not get larger and larger. As always, we start by computing some examples.

[2] A useful Latin phrase meaning "the necessary changes having been made." The implication, of course, is that the necessary changes are relatively minor.

Here's what the Fibonacci sequence modulo m looks like for the first few values of m.

$F_n \pmod 2$ **1,1,0,**1,1,0,1,1,0,1,1,0...

$F_n \pmod 3$ **1,1,2,0,2,2,1,0,**1,1,2,0,2,2,1...

$F_n \pmod 4$ **1,1,2,3,1,0,**1,1,2,3,1,0,1,1,2...

$F_n \pmod 5$ **1,1,2,3,0,3,3,1,4,0,4,4,3,2,0,2,2,4,1,0,**1,1,2...

$F_n \pmod 6$ **1,1,2,3,5,2,1,3,4,1,5,0,5,5,4,3,1,4,5,3,2,5,1,0,**1,1,2,3...

Notice in each case that the Fibonacci sequence eventually starts to repeat. In other words, when we compute the Fibonacci sequence modulo m, we eventually find two consecutive 1's appearing, and as soon this happens, the sequence repeats. (We leave as an exercise for you to prove that this always happens.) Thus there is an integer $N \geq 1$ so that

$$F_{n+N} \equiv F_n \pmod m \qquad \text{for all } n = 1, 2, \ldots.$$

The smallest such integer N is called the *period of the Fibonacci sequence modulo* m. We denote it by $N(m)$. The preceding examples give us the following short table:

m	2	3	4	5	6
$N(m)$	3	8	6	20	24

The period of the Fibonacci sequence modulo m exhibits many interesting patterns, but our brief table is much too short to use in making conjectures. The periods $N(m)$ for all $m \leq 100$ are listed in Table 37.1 (page 308). Exercises 37.11–37.14 ask you to use this list to search for patterns, make conjectures, and prove that (some of) your conjectures are correct. Happy Hunting!

Exercises

37.1. (a) Look at a table of Fibonacci numbers and compare the values of F_m and F_{mn} for various choices of m and n. Try to find a pattern. (*Hint.* Look for a divisibility pattern.)

(b) Prove that the pattern you found in (a) is true.

(c) If $\gcd(m, n) = 1$, try to find a stronger pattern involving the values of F_m, F_n and F_{mn}.

(d) Is the pattern that you found in (c) still true if $\gcd(m, n) \neq 1$?

(e) Prove that the pattern you found in (c) is true.

37.2. (a) Find as many square Fibonacci numbers as you can. Do you think that there are finitely many or infinitely many square Fibonacci numbers?

(b) Find as many triangular Fibonacci numbers as you can. Do you think there are finitely many or infinitely many triangular Fibonacci numbers?

37.3. (a) Make a list of Fibonacci numbers F_n that are prime.

(b) Using your data, fill in the box to make an interesting conjecture:

$$\text{If } F_n \text{ is prime, then } n \text{ is } \boxed{}.$$

(*Hint.* Actually, your conjecture should be that the statement is true with one exception.)

(c) Does your conjecture in (b) work in the other direction? In other words, is the following statement true, where the box is the same as in (b)?

$$\text{If } n \text{ is } \boxed{}, \text{ then } F_n \text{ is prime.}$$

(d) Prove that your conjecture in (b) is correct.

37.4. The *Lucas sequence* is the sequence of numbers L_n given by the rules $L_1 = 1$, $L_2 = 3$, and $L_n = L_{n-1} + L_{n-2}$.

(a) Write down the first 10 terms of the Lucas sequence.

(b) Find a simple formula for L_n, similar to Binet's Formula for the Fibonacci number F_n.

(c) Compute the value of $L_n^2 - 5F_n^2$ for each $1 \le n \le 10$. Make a conjecture about this value. Prove that your conjecture is correct.

(d) Show that L_{3n} and F_{3n} are even for all values of n. Combining this fact with the formula you discovered in (c), find an interesting equation satisfied by the pair of numbers $\left(\frac{1}{2}L_{3n}, \frac{1}{2}F_{3n}\right)$. Relate your answer to the material in Chapters 30 and 32.

37.5. Write down the first few terms for each of the following linear recursion sequences, and then find a formula for the nth term similar to Binet's formula for the nth Fibonacci number. Be sure to check that your formula is correct for the first few values.

(a) $A_n = 3A_{n-1} + 10A_{n-2}$ $\quad A_1 = 1 \quad A_2 = 3$

(b) $B_n = 2B_{n-1} - 4B_{n-2}$ $\quad B_1 = 0 \quad B_2 = -2$

(c) $C_n = 4C_{n-1} - C_{n-2} - 6C_{n-3}$ $\quad C_1 = 0 \quad C_2 = 0 \quad C_3 = 1$

[*Hint.* For (b), you'll need to use complex numbers. For (c), the cubic polynomial has some small integer roots.]

37.6. Let P_n be the linear recursion sequence defined by

$$P_n = P_{n-1} + 4P_{n-2} - 4P_{n-3}, \qquad P_1 = 1, \quad P_2 = 9, \quad P_3 = 1.$$

(a) Write down the first 10 terms of P_n.

(b) Does the sequence behave in a strange manner?

(c) Find a formula for P_n that is similar to Binet's formula. Does your formula for P_n explain the strange behavior that you noted in (b)?

37.7. (This question requires some elementary calculus.)

(a) Compute the value of the limit

$$\lim_{n\to\infty} \frac{\log(F_n)}{n}.$$

Here F_n is the nth Fibonacci number.

(b) Compute $\lim_{n\to\infty}(\log(A_n))/n$, where A_n is the sequence in Exercise 37.5(a).

(c) Compute $\lim_{n\to\infty}(\log(|B_n|))/n$, where B_n is the sequence in Exercise 37.5(b).

(d) Compute $\lim_{n\to\infty}(\log(C_n))/n$, where C_n is the sequence in Exercise 37.5(c).

37.8. Write down the first few terms for each of the following nonlinear recursion sequences. Can you find a simple formula for the nth term? Can you find any patterns in the list of terms?

(a) $D_n = D_{n-1} + D_{n-2}^2$ $\quad D_1 = 1 \quad D_2 = 1$

(b) $E_n = E_{n-1}E_{n-2} + E_{n-3}$ $\quad E_1 = 1 \quad E_2 = 2 \quad E_3 = 1$

37.9. We have said that the Fibonacci sequence modulo m always eventually repeats itself with two consecutive 1's, but we did not prove that this is true. Do you think that this is obvious? If so, why? If not, either prove that it is true or find a counterexample.

37.10. Let $N = N(m)$ be the period of Fibonacci sequence modulo m.

(a) What is the value of F_N modulo m? What is the value of F_{N-1} modulo m?

(b) Write out the Fibonacci sequence modulo m in reverse direction:

$$F_{N-1},\quad F_{N-2},\quad F_{N-3},\quad \ldots\quad F_3,\quad F_2,\quad F_1 \pmod{m}.$$

Do this for several values of m, and try to find a pattern. (*Hint*. The pattern will be more evident if you take some of the values modulo m to lie between $-m$ and -1, instead of between 1 and m.)

(c) Prove that the pattern you found in (b) is correct.

37.11. The material in Table 37.1 suggests that if $m \geq 3$ then the period $N(m)$ of the Fibonacci sequence modulo m is always an even number. Prove that this is true, or find a counterexample.

37.12. Let $N(m)$ be the period of the Fibonacci sequence modulo m.

(a) Use Table 37.1 to compare the values of $N(m_1)$, $N(m_2)$, and $N(m_1m_2)$ for various values of m_1 and m_2, especially for $\gcd(m_1, m_2) = 1$.

(b) Make a conjecture relating $N(m_1)$, $N(m_2)$, and $N(m_1m_2)$ when m_1 and m_2 satisfy $\gcd(m_1, m_2) = 1$.

(c) Use your conjecture from (b) to guess the values of $N(5184)$ and $N(6887)$. (*Hint*. $6887 = 71 \cdot 97$.)

(d) Prove that your conjecture in (b) is correct.

37.13. Let $N(m)$ be the period of the Fibonacci sequence modulo m.

m	$N(m)$	m	$N(m)$	m	$N(m)$	m	$N(m)$	m	$N(m)$
1	—	21	16	41	40	61	60	81	216
2	3	22	30	42	48	62	30	82	120
3	8	23	48	43	88	63	48	83	168
4	6	24	24	44	30	64	96	84	48
5	20	25	100	45	120	65	140	85	180
6	24	26	84	46	48	66	120	86	264
7	16	27	72	47	32	67	136	87	56
8	12	28	48	48	24	68	36	88	60
9	24	29	14	49	112	69	48	89	44
10	60	30	120	50	300	70	240	90	120
11	10	31	30	51	72	71	70	91	112
12	24	32	48	52	84	72	24	92	48
13	28	33	40	53	108	73	148	93	120
14	48	34	36	54	72	74	228	94	96
15	40	35	80	55	20	75	200	95	180
16	24	36	24	56	48	76	18	96	48
17	36	37	76	57	72	77	80	97	196
18	24	38	18	58	42	78	168	98	336
19	18	39	56	59	58	79	78	99	120
20	60	40	60	60	120	80	120	100	300

Table 37.1: The Period $N(m)$ of the Fibonacci Sequence Modulo m

(a) Use Table 37.1 to compare the values of $N(p)$ and $N(p^2)$ for various primes p.

(b) Make a conjecture relating the value of $N(p)$ and $N(p^2)$ when p is a prime.

(c) More generally, make a conjecture relating the value of $N(p)$ to the values of all the higher powers $N(p^2)$, $N(p^3)$, $N(p^4)$, ...

(d) Use your conjectures from (b) and (c) to guess the values of $N(2209)$, $N(1024)$, and $N(729)$. (*Hint.* $2209 = 47^2$. You can factor 1024 and 729 yourself!)

(e) Try to prove your conjectures in (b) and/or (c).

37.14. Let $N(m)$ be the period of the Fibonacci sequence modulo m.

(a) Use Table 37.1 to make a list of the periods $N(p)$ of the Fibonacci sequence modulo p when p is a prime number.

(b) There are two primes for which $N(p)$ is somewhat exceptional. One of these primes is (as usual) $p = 2$. Pick out the other exceptional prime, and explain why you think it is exceptional.

(c) No exact formula is known for the value of $N(p)$, but there are a number of divisibility patterns. For example, from the five values

$$N(11) = 10, \quad N(31) = 30, \quad N(41) = 40, \quad N(61) = 60, \quad N(71) = 70$$

in the table, we might guess that $N(p) = p - 1$ whenever $p \equiv 1 \pmod{10}$ (i.e., whenever the last digit of p is a 1). Unfortunately, this guess is incorrect, since if we extend the table to larger primes, we find that

$$\begin{array}{lllll} N(101) = 50, & N(131) = 130, & N(151) = 50, & N(181) = 90, & N(191) = 190 \\ N(211) = 42, & N(241) = 240, & N(251) = 250, & N(271) = 270, & N(281) = 56 \\ N(311) = 310, & N(331) = 110, & N(401) = 200, & N(421) = 84, & N(431) = 430. \end{array}$$

But we do notice that $N(p)$ always seems to divide $p - 1$. Try to find similar divisibility patterns for other sorts of primes (i.e., for primes whose last digit isn't 1). It might be helpful to extend your table in (a), especially if you can program a computer to help you.

(d) Prove the pattern noted in (c): if $p \equiv 1 \pmod{10}$ then $N(p)$ divides $p - 1$. (*Hint.* This is a difficult problem. One way to solve it is to use Binet's formula for the nth Fibonacci number, but first you'll need to find a number modulo p to play the role of $\sqrt{5}$. You will also need to use Fermat's Little Theorem.)

37.15. The Fibonacci numbers satisfy many amazing identities.

(a) Compute the quantity $F_{n+1}^2 - F_{n-1}^2$ for the first few integers $n = 2, 3, 4, \ldots$ and try to guess its value. (*Hint.* It is equal to a Fibonacci number.) Prove that your guess is correct.

(b) Same question (and same hint!) for the quantity $F_{n+1}^3 + F_n^3 - F_{n-1}^3$.

(c) Same question (but not the same hint!) for the quantity $F_{n-1}F_{n+1} - F_n^2$.

(d) Same question for $4F_nF_{n-1} + F_{n-2}^2$. (*Hint.* Compare the value with the square of a Fibonacci number.)

(e) Same question for the quantity $F_{n+4}^4 - 4F_{n+3}^4 - 19F_{n+2}^4 - 4F_{n+1}^4 + F_n^4$.

Chapter 38

Oh, What a Beautiful Function

A long time ago[1] we found a formula for the sum of the first n integers:

$$1+2+3+\cdots+(n-1)+n=\frac{n(n+1)}{2}=\frac{1}{2}n^2+\frac{1}{2}n.$$

This is a very beautiful and completely accurate formula, but there might be situations where we'd prefer a formula that is less complicated, even at the cost of losing some accuracy. Thus we can say that $1+2+\cdots+n$ is approximately equal to $\frac{1}{2}n^2$, since when n is large, the $\frac{1}{2}n^2$ term is much larger than the $\frac{1}{2}n$ term.

Similarly, there is an exact formula for the sum of the first n squares,[2]

$$1^2+2^2+\cdots+(n-1)^2+n^2=\frac{n(n+1)(2n+1)}{6}.$$

If we multiply out the right-hand side, this becomes

$$1^2+2^2+\cdots+(n-1)^2+n^2=\frac{1}{3}n^3+\frac{1}{2}n^2+\frac{1}{6}n.$$

The $\frac{1}{3}n^3$ term is much larger than the other terms when n is large, so we can say that $1^2+2^2+\cdots+n^2$ is approximately equal to $\frac{1}{3}n^3$, and if we want to be more precise, we can say that the difference between $1^2+2^2+\cdots+n^2$ and $\frac{1}{3}n^3$ is more or less a multiple of n^2.

Approximate formulas of this sort appear quite frequently in number theory, as well as in other areas of mathematics and computer science. They take the form

$$\left(\begin{array}{c}\text{Complicated}\\ \text{function of } n\end{array}\right)=\left(\begin{array}{c}\text{Simple}\\ \text{function of } n\end{array}\right)+\left(\begin{array}{c}\text{A bound for the}\\ \text{size of the error}\\ \text{in terms of } n\end{array}\right)$$

[1] A long time ago in a chapter far, far away...

[2] You already proved this by induction if you did Exercise 7.3. If not, you'll find a proof using a different method in Chapter 42.

For example,

$$\underbrace{1^2 + 2^2 + \cdots + (n-1)^2 + n^2}_{\text{complicated function of n}} = \underbrace{\frac{1}{3}n^3}_{\substack{\text{simple} \\ \text{function of } n}} + \left(\begin{array}{c}\text{Error that is} \\ \text{not much} \\ \text{larger than } n^2\end{array}\right)$$

The mathematical way to write this approximate formula is with "big-Oh" notation. Using big-Oh notation, the previous formula is written

$$1^2 + 2^2 + \cdots + (n-1)^2 + n^2 = \frac{1}{3}n^3 + O(n^2).$$

Informally, this means that the difference between $1^2 + 2^2 + \cdots + n^2$ and $\frac{1}{3}n^3$ is smaller than some fixed multiple of n^2.

The formal definition of big-Oh notation is somewhat abstract and can be confusing at first. But if you keep the $1^2 + 2^2 + \cdots + n^2$ example in mind, you will find that big-Oh notation is not that complicated and, after some practice, it becomes very natural.

Definition. Suppose that $f(n)$, $g(n)$, and $h(n)$ are functions. The formula

$$f(n) = g(n) + O(h(n))$$

means that there is a constant C and a starting value n_0 so that

$$\left|f(n) - g(n)\right| \le C\left|h(n)\right| \qquad \text{for all } n \ge n_0.$$

In words, the difference between $f(n)$ and $g(n)$ is no larger than a constant multiple of $h(n)$. When reading the formula $f(n) = g(n) + O(h(n))$ aloud, we say that

"$f(n)$ equals $g(n)$ plus big-Oh of $h(n)$."

Sometimes the function $g(n)$ is absent, which is the same as saying that $g(n) = 0$ for all n, so the formula

$$f(n) = O(h(n))$$

means that there is a constant C and a starting value n_0 so that

$$|f(n)| \le C|h(n)| \qquad \text{for all } n \ge n_0.$$

For example,

$$n^3 = O(2^n),$$

since[3]

$$n^3 \le 2^n \qquad \text{for all } n \ge 10.$$

It is also very common to have formulas with $h(n)$ equal to the constant function $h(n) = 1$. A common mistake is to believe that the formula

$$f(n) = O(1)$$

means that $f(n)$ is itself constant. Nothing could be further from the truth. The formula $f(n) = O(1)$ means that $|f(n)|$ is smaller than a constant C. For example, the function $f(n) = \dfrac{2n+1}{n+2}$ is certainly not constant, but it is true that

$$\frac{2n+3}{n+2} = O(1),$$

since

$$\left|\frac{2n+3}{n+2}\right| \le 2 \qquad \text{for all } n \ge 1.$$

The Fibonacci sequence

$$1, 1, 2, 3, 5, 8, 13, \ldots$$

provides another opportunity to use big-Oh notation. In Chapter 37 we proved Binet's beautiful formula for the nth Fibonacci number,

$$F_n = \frac{1}{\sqrt{5}}\left\{\left(\frac{1+\sqrt{5}}{2}\right)^n - \left(\frac{1-\sqrt{5}}{2}\right)^n\right\}.$$

The two quantities appearing in this formula have the values

$$\frac{1+\sqrt{5}}{2} = 1.618039\ldots \qquad \text{and} \qquad \frac{1-\sqrt{5}}{2} = -0.618039\ldots.$$

When we take $\frac{1-\sqrt{5}}{2}$ and raise it to a large power, we get something that is very small, while $\frac{1+\sqrt{5}}{2}$ raised to a large power is very large. So an approximate, but still useful, version of Binet's formula says that

$$F_n = \frac{1}{\sqrt{5}}\left(\frac{1+\sqrt{5}}{2}\right)^n + O(0.61304^n).$$

[3]Note that there are lots of possible choices for C and n_0. For example, we could say that $n^3 = O(2^n)$ since $n^3 \le 10 \cdot 2^n$ for all $n \ge 1$. But we cannot say that $n^3 = O(n^2)$, since there is no choice of C that makes n^3 smaller than Cn^2 when n is large.

In this case, the error term $O(0.613046^n)$ actually approaches zero extremely rapidly as n gets larger and larger. This contrasts with the big-Oh formula for the sum $1^2 + 2^2 + \cdots + n^2$, where the error $O(n^2)$ got larger as n increased, albeit at a slower rate than the main term $\frac{1}{3}n^3$.

There are many methods for discovering and for proving big-Oh formulas. One of the most powerful uses geometry and a little bit of calculus. We demonstrate this geometric method by finding a big-Oh formula for the sum

$$1^k + 2^k + \cdots + n^k.$$

We already know that

$$1 + 2 + \cdots + n = \frac{1}{2}n^2 + O(n) \qquad \text{and} \qquad 1^2 + 2^2 + \cdots + n^2 = \frac{1}{3}n^3 + O(n^2),$$

so we might guess that

$$1^k + 2^k + \cdots + n^k \stackrel{?}{=} \frac{1}{k+1}n^{k+1} + O(n^k).$$

This is indeed the case.

Theorem 38.1. *Fix a power $k \geq 1$. Then*

$$1^k + 2^k + 3^k + \cdots + n^k = \frac{1}{k+1}n^{k+1} + O(n^k).$$

Proof. Let $S(n) = 1^k + 2^k + \cdots + n^k$ denote the sum that we are trying to estimate. We draw a bunch of rectangles. The first rectangle has base 1 and height 1^k, the second rectangle has base 1 and height 2^k, the third rectangle has base 1 and height 3^k, and so on. Placing these rectangles side-by-side, we get the picture illustrated in Figure 38.1.[4] Notice that if we add up the area of all the rectangles we get precisely the quantity $S(n)$.

Rather than exactly computing the area inside the rectangles, we approximate the total rectangle area by instead computing the area of a simpler region. If we draw the curve $y = x^k$, then, as you can see in Figure 38.2, the rectangles fit fairly snugly under the curve. And since the rectangles in Figure 38.2 lie *underneath* the curve $y = x^k$, we know that the area inside the rectangles is smaller than the area under the curve.

[4]Note that to fit the diagram on the page, we have not drawn the rectangles in Figure 38.1 to be the correct size, so you should use the picture only as an aid in gaining understanding of the general idea. Feel free to draw your own pictures to the proper scale, say with $k = 2$, but be prepared to use a large piece of paper!

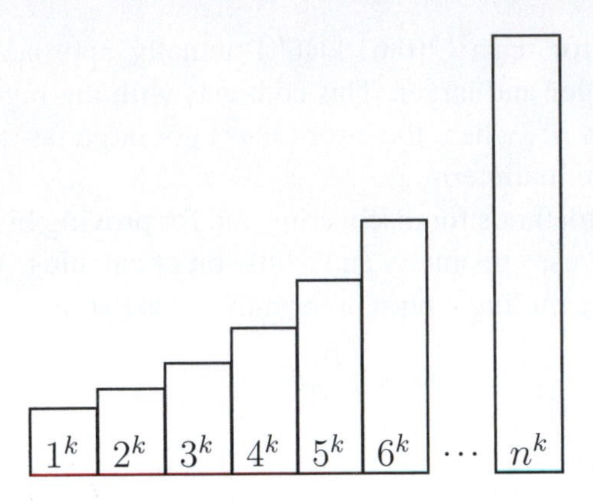

Figure 38.1: Rectangles[4] Whose Total Area Is $1^k + 2^k + \cdots + n^k$

In other words, from the picture we have deduced that

$$1^k + 2^k + \cdots + n^k = \left(\begin{array}{c} \text{Area of} \\ \text{rectangles} \end{array} \right) < \left(\begin{array}{c} \text{Area under the curve } y = x^k \\ \text{with } 1 \le x \le n+1 \end{array} \right).$$

We can use basic calculus to compute the area under the curve.

$$\begin{aligned} \left(\begin{array}{c} \text{Area under the curve } y = x^k \\ \text{with } 1 \le x \le n+1 \end{array} \right) &= \int_1^{n+1} x^k \, dx \\ &= \left. \frac{x^{k+1}}{k+1} \right|_1^{n+1} = \frac{1}{k+1}\left((n+1)^{k+1} - 1\right). \end{aligned}$$

This gives us an upper bound

$$1^k + 2^k + \cdots + n^k < \frac{1}{k+1}(n+1)^{k+1}.$$

(We have dropped the -1 on the right-hand side, since leaving it in gives a slightly stronger estimate.)

Similarly, if we slide the rectangles over to the left one unit, then, as illustrated in Figure 38.3, the rectangles completely cover the area under the curve $y = x^k$ between $0 \le x \le n$. This means that the area of the rectangles is larger than the

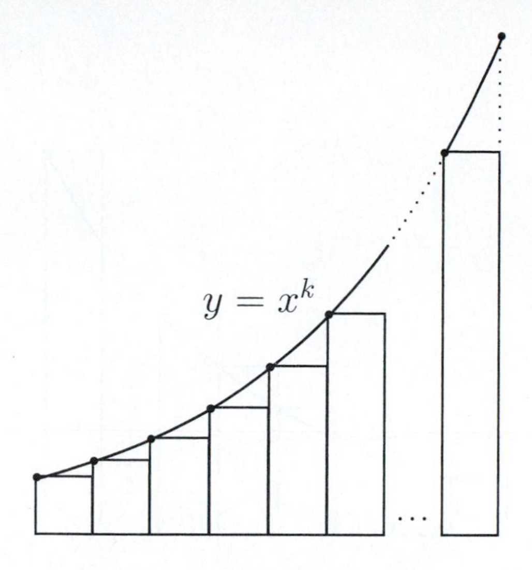

Figure 38.2: Area Under Curve Is Larger Than Area of Rectangles

area under this part of the curve, so we get a corresponding lower bound

$$\begin{aligned} 1^k + 2^k + \cdots + n^k &= \begin{pmatrix} \text{Area of} \\ \text{rectangles} \end{pmatrix} \\ &> \begin{pmatrix} \text{Area under the curve } y = x^k \\ \text{with } 0 \le x \le n \end{pmatrix} \\ &= \int_0^n x^k \, dx \\ &= \left. \frac{x^{k+1}}{k+1} \right|_0^n \\ &= \frac{1}{k+1} n^{k+1}. \end{aligned}$$

Putting together our upper and lower bounds, we have proved that

$$\frac{1}{k+1} n^{k+1} < 1^k + 2^k + \cdots + n^k < \frac{1}{k+1}(n+1)^{k+1}.$$

We subtract $\frac{1}{k+1} n^{k+1}$ to get

$$0 < (1^k + 2^k + \cdots + n^k) - \frac{1}{k+1} n^{k+1} < \frac{1}{k+1}\big((n+1)^{k+1} - n^{k+1}\big). \qquad (*)$$

We need to show that the upper bound is not too large. To do this, we use the

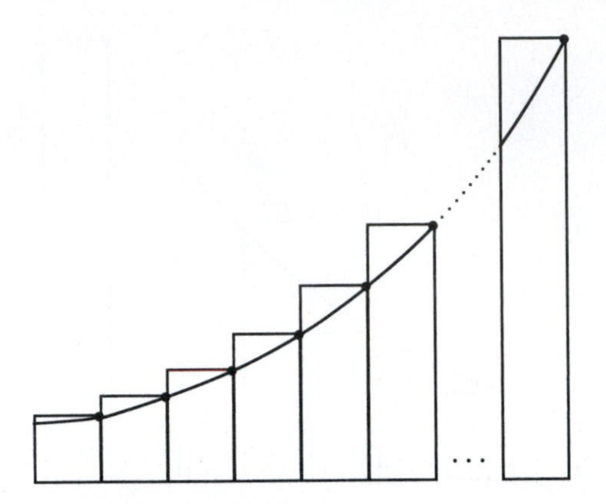

Figure 38.3: Area Under Curve Is Smaller Than Area of Rectangles

binomial expansion (Chapter 36) for $(n+1)^{k+1}$,

$$(n+1)^{k+1} = n^{k+1} + \binom{k+1}{1}n^k + \binom{k+1}{2}n^{k-1} + \cdots + \binom{k+1}{k}n + \binom{k+1}{k+1}.$$

The crucial point is that the largest term is n^{k+1} and all the other terms involve smaller powers of n. Hence

$$\begin{aligned}
(n+1)^{k+1} - n^{k+1} \\
&= \binom{k+1}{1}n^k + \binom{k+1}{2}n^{k-1} + \cdots + \binom{k+1}{k}n + \binom{k+1}{k+1} \\
&\le \binom{k+1}{1}n^k + \binom{k+1}{2}n^k + \cdots + \binom{k+1}{k}n^k + \binom{k+1}{k+1}n^k \\
&= (\text{some mess involving } k) \cdot n^k.
\end{aligned}$$

We combine this estimate with our earlier inequality $(*)$ to obtain

$$0 < (1^k + 2^k + \cdots + n^k) - \frac{1}{k+1}n^{k+1} < (\text{some mess involving } k) \cdot n^k.$$

Of course, the "new mess" is $\frac{1}{k+1}$ times the "old mess," but in any case, it only involves k and does not depend on n. This proves that

$$(1^k + 2^k + \cdots + n^k) = \frac{1}{k+1}n^{k+1} + O(n^k),$$

and indeed it proves something a bit stronger, since it shows that the sum of the kth powers $1^k + 2^k + \cdots + n^k$ is always strictly larger than $\frac{1}{k+1} n^{k+1}$. This completes the proof of the theorem. □

Describing How Long a Computation Will Take

Big-Oh notation is frequently used to describe how long it takes to do a certain computation using a particular method. For example, suppose that we fix a number a and a modulus m and that we want to compute the value of $a^n \pmod{m}$ for some large value of the exponent n. How long does it take us?

One way to do the computation is to compute

$$a_1 \equiv a \pmod{m}, \quad \text{and then}$$
$$a_2 \equiv a \cdot a_1 \pmod{m}, \quad \text{and then}$$
$$a_3 \equiv a \cdot a_2 \pmod{m}, \quad \text{and then} \ldots$$

Eventually we get to a_n, which is equal to $a^n \pmod{m}$. We end up having to do n steps, where each step consists of one multiplication and one reduction modulo m. Assuming that each step takes a more or less fixed amount of time, the total time is a constant multiple of n. So the running time of this method is $O(n)$.

Of course, no one who has read Chapter 16 would ever use this absurdly inefficient method to compute $a^n \pmod{m}$. The method of successive squaring allows us to compute $a^n \pmod{m}$ much more rapidly. As described in Chapter 16, the method of successive squares has three pieces:

1. Write n as a sum of powers of 2 (the binary expansion)

$$n = u_0 + u_1 \cdot 2 + u_2 \cdot 4 + u_3 \cdot 8 + \cdots + u_r \cdot 2^r,$$

where $u_r = 1$ and every u_i is either 0 or 1.

2. Create a table of values

$$A_0 = a, \quad A_1 = A_0^2 \pmod{m}, \quad A_2 = A_1^2 \pmod{m},$$
$$A_3 = A_2^2 \pmod{m}, \quad \ldots, \quad A_r = A_{r-1}^2 \pmod{m}.$$

3. Compute the product

$$A_0^{u_0} \cdot A_1^{u_1} \cdot A_2^{u_2} \cdots A_r^{u_r} \pmod{m}. \tag{38.1}$$

Each of the three pieces takes approximately r steps, so the total time is a constant multiple of r. Do you see how the number r is related to the exponent n? From the binary expansion (38.1), the number n is at least 2^r, so if we take logarithms, we see that[5]

$$r \leq \log_2(n).$$

Therefore, the method of successive squares allows us to compute $a^n \pmod{m}$ in time $O(\log_2(n))$. This is immensely faster than time $O(n)$, since $\log_2(n)$ is much smaller than n when n is large.

The number $\log_2(n)$ is the number of binary digits in n; that is, it's the number of digits when we write n using binary notation. Similarly, $\log_{10}(n)$ is the number of decimal digits in n. So, roughly speaking, $\log(n)$ tells us each of the following pieces of information:

- how much time it takes to write down the number n;
- how long it takes us to describe the number n to another person;
- how long it takes to input the number n into a computer or to output the number n from a computer

We can summarize this by saying that it takes time $O(\log(n))$ to describe the number n.

It is thus interesting and somewhat surprising that we can compute the quantity $a^n \pmod{m}$ in time $O(\log(n))$, since it already takes time $O(\log(n))$ simply to input the number n. The successive squaring method is said to take *linear time* because the total running time is at most a constant multiple of the time it takes to input the initial information.

Let's look at another problem, that of multiplying two polynomials of degree d,

$$F(X) = a_0 + a_1X + \cdots + a_dX^d \quad \text{and} \quad G(X) = b_0 + b_1X + \cdots + b_dX^d.$$

There are $2d + 2$ coefficients, so it takes time $O(d)$ to describe the polynomials.[6]

[5]The function $\log_2(x)$ is the logarithm to the base 2. By definition, the value of $\log_2(x)$ is the number y that is needed to make $2^y = x$.

[6]Notice that the degree of a polynomial plays a role similar to that played by the logarithm of a number. Another property shared by the degree and the logarithm is illustrated by

$$\deg(F(X)G(X)) = \deg(F(X)) + \deg(G(X)) \qquad \text{and} \qquad \log(MN) = \log(M) + \log(N).$$

Thus both the degree and the logarithm convert multiplication into addition.

The product of $F(X)$ and $G(X)$ is given by the formula

$$H(X) = F(X)G(X) = c_0 + c_1X + c_2X^2 + \cdots + c_{2d}X^{2d}$$

$$\text{with} \quad c_j = \begin{cases} a_0b_j + a_1b_{j-1} + \cdots + a_jb_0 & \text{if } 0 \le j \le d, \\ a_{j-d}b_d + a_{j-d+1}b_{d-1} + \cdots + a_db_{j-d} & \text{if } d < j \le 2d. \end{cases}$$

If $0 \le j \le d$, then computing c_j requires j additions and $j+1$ multiplications, so it takes time $O(j)$ to compute c_j. And if $d < j \le 2d$, then it takes time $O(2d - j + 1)$ to compute c_j. Hence the total time to compute the product $H(X)$ is

$$\begin{aligned} \sum_{j=0}^{d} O(j) + \sum_{j=d+1}^{2d} O(2d-j) &= O\left(\sum_{j=0}^{d} j + \sum_{j=d+1}^{2d} (2d-j+1)\right) \\ &= O\left(\frac{d^2+d}{2} + \frac{d^2+d}{2}\right) = O(d^2). \end{aligned}$$

Thus the time to compute $F(X)G(X)$ is big-Oh of the *square* of the amount of time it takes to input the initial data, so we say that it takes *quadratic time* to compute the product of two polyomials.[7]

Exercises

38.1. (a) Suppose that

$$f_1(n) = g_1(n) + O(h(n)) \qquad \text{and} \qquad f_2(n) = g_2(n) + O(h(n)).$$

Prove that

$$f_1(n) + f_2(n) = g_1(n) + g_2(n) + O(h(n)).$$

(b) More generally, if a and b are any constants, prove that

$$af_1(n) + bf_2(n) = ag_1(n) + bg_2(n) + O(h(n)).$$

(Note that the constant C appearing in the definition of big-Oh notation is allowed to depend on the constants a and b. The only requirement is that there be one fixed value of C that works for all sufficiently large values of n.)

---

[7]Of course, we really mean that this particular method of computing $F(X)G(X)$ takes quadratic time. There are other methods, such as Karatsuba multiplication and Fast Fourier Transforms, that are much faster. These fancier methods are able to multiply two polyomials in time $O(d \log d)$, so just slightly slower than linear time. The advantage of linear time over quadratic time is not too important if d is small, say $d = 10$ or $d = 15$, but if $d = 10000$, there is a considerable difference.

(c) The formula that you proved in part (b) shows that big-Oh formulas (with the same h) can be added, subtracted, and multiplied by constants. Is it also okay to multiply them by quantities that are not constant? In other words, if $f(n) = g(n) + O(h(n))$ and if $k(n)$ is another function of n, is it true that

$$k(n)f(n) = k(n)g(n) + O(h(n))?$$

If not, how about

$$k(n)f(n) = k(n)g(n) + O(k(n)h(n))?$$

38.2. Suppose that

$$f_1(n) = g_1(n) + O(h_1(n)) \qquad \text{and} \qquad f_2(n) = g_2(n) + O(h_2(n)).$$

Prove that

$$f_1(n) + f_2(n) = g_1(n) + g_2(n) + O(\max\{h_1(n), h_2(n)\}).$$

38.3. Which of the following functions are $O(1)$? Why?

(a) $f(n) = \dfrac{3n+17}{2n-1}$ (b) $f(n) = \dfrac{3n^2+17}{2n-1}$ (c) $f(n) = \dfrac{3n+17}{2n^2-1}$

(d) $f(n) = \cos(n)$ (e) $f(n) = \dfrac{1}{\sin(1/n)}$ (f) $f(n) = \dfrac{1}{n \cdot \sin(1/n)}$

38.4. Find a big-Oh estimate for the sum of square roots, that is, fill in the boxes in the following formula:

$$\sqrt{1} + \sqrt{2} + \sqrt{3} + \cdots + \sqrt{n} = \square\, n^{\square} + O\left(n^{\square}\right).$$

38.5. (a) Prove the following big-Oh estimate for the sum of the reciprocals of the integers:

$$1 + \frac{1}{2} + \frac{1}{3} + \frac{1}{4} + \frac{1}{5} + \cdots + \frac{1}{n} = \ln(n) + O(1).$$

[Here $\ln(x)$ is the natural logarithm of x.]

(b) Prove the stronger statement that there is a constant γ so that

$$1 + \frac{1}{2} + \frac{1}{3} + \frac{1}{4} + \frac{1}{5} + \cdots + \frac{1}{n} = \ln(n) + \gamma + O\left(\frac{1}{n}\right).$$

The number γ, which is equal to $0.577215664\ldots$, is called *Euler's constant*. Very little is known about Euler's constant. For example, it is not known whether or not γ is a rational number.

38.6. Bob and Alice play the following guessing game. Alice picks a number between 1 and n. Bob starts guessing numbers and, after each guess, Alice tells him whether he is right or wrong. Let $G(n)$ be the most guesses it can take Bob to guess Alice's number, assuming that he uses the best possible strategy.

(a) Prove that $G(n) = O(n)$.
(b) Prove that $G(n)$ is not $O(\sqrt{n})$.
(c) More generally, if $G(n) = O(h(n))$, what can you say about the function $h(n)$?
(d) Suppose that we change the rules of the game so that, after Bob guesses a number, Alice tells him whether his guess is too high, too low, or exactly right. Describe a strategy for Bob so that his number of guesses before winning satisfies $G(n) = O(\log_2(n))$. (*Hint*. Eliminate half the remaining numbers with each guess.)

38.7. Bob knows that the number n is composite and he wants to find a nontrivial factor. He employs the following strategy: Check if 2 divides n, then check if 3 divides n, then check if 4 divides n, etc. Let $F(n)$ be the number of steps it takes until he finds a factor of n.
(a) Prove that $F(n) = O(\sqrt{n})$.
(b) Suppose that, instead of checking every number $2, 3, 4, 5, 6, \ldots$, Bob only checks if n is divisible by primes $2, 3, 5, 7, 11, \ldots$. Explain why this strategy still works and show that the number of steps $F(n)$ now satisfies $F(n) = O\left(\frac{\sqrt{n}}{\ln(n)}\right)$. (*Hint*. You'll need to use the Prime Number Theorem 13.1.) Do you think that this new strategy is actually practical?
(c) Faster methods are known for solving this problem, such as the *Quadratic Sieve* and the *Elliptic Curve Method*. The number of steps $L(n)$ that these methods require satisfies
$$L(n) = O\left(e^{c\sqrt{\ln(n)\cdot\ln\ln(n)}}\right),$$
where c is a small constant. Prove that this is faster than the method in (a) by showing that
$$\lim_{n\to\infty} \frac{e^{c\sqrt{\ln(n)\cdot\ln\ln(n)}}}{\sqrt{n}} = 0.$$
More generally, show that the limit is 0 even if the $\sqrt{n}$ in the denominator is replaced by n^ϵ for some (small) $\epsilon > 0$.
(d) The fastest known method to solve this problem for large numbers n is called the *Number Field Sieve* (NFS). The number of steps $M(n)$ required by the NFS is
$$M(n) = O\left(e^{c'\sqrt[3]{(\ln n)(\ln\ln n)^2}}\right),$$
where again c' is a small constant. Prove that for large values of n the function $M(n)$ is much smaller than the big-O estimate for $L(n)$ in (c).

Big-Oh notation is so useful that mathematicians and computer scientists have devised similar notation to describe some other typical situations. In the next few exercises, we introduce some of these concepts and ask you to work out some examples.

38.8. Small-oh Notation. Intuitively, the notation $o(h(n))$ indicates a quantity that is much smaller than $h(n)$. The precise definition is that
$$f(n) = g(n) + o(h(n)) \qquad \text{means that} \qquad \lim_{n\to\infty} \frac{f(n) - g(n)}{h(n)} = 0.$$

(a) Prove that $n^{10} = o(2^n)$.
(b) Prove that $2^n = o(n!)$.
(c) Prove that $n! = o(2^{n^2})$.
(d) What does the formula $f(n) = o(1)$ mean? Which of the following functions are $o(1)$?

$$\text{(i)}\quad f(n) = \frac{1}{\sqrt{n}} \qquad \text{(ii)}\quad f(n) = \frac{1}{\sin(n)} \qquad \text{(iii)}\quad f(n) = 2^{n-n^2}$$

38.9. Big-Omega Notation. Big-Omega notation is very similar to Big-Oh notation, except that the inequality is reversed.[8] In other words,

$$f(n) = g(n) + \Omega(h(n))$$

means that there is a positive constant C and a starting value n_0 so that

$$\bigl|f(n) - g(n)\bigr| \ge C\bigl|h(n)\bigr| \qquad \text{for all } n \ge n_0.$$

Frequently g is zero, in which case $f(n) = \Omega(h(n))$ means that $|f(n)| \ge C|h(n)|$ for all sufficiently large values of n.

(a) Prove that each of the following formulas is true.

$$\text{(i)}\quad n^2 - n = \Omega(n) \qquad \text{(ii)}\quad n! = \Omega(2^n) \qquad \text{(iii)}\quad \frac{5^n - 3^n}{2^n} = \Omega(2^n)$$

(b) If $f(n) = \Omega(h(n))$ and $h(n) = \Omega(k(n))$, prove that $f(n) = \Omega(k(n))$.
(c) If $f(n) = \Omega(h(n))$, is it then always true that $h(n) = O(f(n))$?
(d) Let $f(n) = n^3 - 3n^2 + 7$. For what values of d is it true that $f(n) = \Omega(n^d)$?
(e) For what values of d is it true that $\sqrt{n} = \Omega((\log_2 n)^d)$?
(f) Prove that the function $f(n) = n\cdot\sin(n)$ does not satisfy $f(n) = \Omega(\sqrt{n})$. [*Hint*. Use Dirichlet's Diophantine Approximation Theorem 31.2 to find fractions p/q satisfying $|p - 2\pi q| < 1/q$, let $n = p$, and use the fact that $\sin(x) \approx x$ when x is small.]

38.10. Big-Theta Notation. Big-Theta notation combines both Big-Oh and Big-Omega. One way to define Big-Theta is to use the earlier definitions and say that

$$f(n) = g(n) + \Theta(h(n))$$

if both

$$f(n) = g(n) + O(h(n)) \quad \text{and} \quad f(n) = g(n) + \Omega(h(n)).$$

Or we can write everything out explicitly and define

$$f(n) = g(n) + \Theta(h(n))$$

---

[8] Warning: Exercise 38.9 describes what Ω means to computer scientists. Mathematicians typically assign a different meaning to Ω. They take it to mean that there is a positive constant C and infinitely many values of n so that $|f(n) - g(n)| \ge C|h(n)|$. Notice the important distinction between a statement being true for all (large) values of n and merely being true for infinitely many values of n.

to mean that there are postive constants C_1 and C_2 and a starting value n_0 so that

$$C_1|h(n)| \leq |f(n) - g(n)| \leq C_2|h(n)| \qquad \text{for all } n \geq n_0.$$

(a) Prove that

$$\ln\left(1 + \frac{1}{n}\right) = \Theta\left(\frac{1}{n}\right).$$

[*Hint.* Use the Taylor series expansion of $\ln(1+t)$ to estimate its value when t is small.]

(b) Use (a) to prove that

$$\ln|n^3 - n^2 + 3| = 3\ln(n) + \Theta\left(\frac{1}{n}\right).$$

(c) Generalize (b) and prove that if $f(x)$ is a polynomial of degree d then

$$\log|f(n)| = d\ln(n) + \Theta\left(\frac{1}{n}\right).$$

(d) If $f_1(n) = g_1(n) + \Theta(h(n))$ and $f_2(n) = g_2(n) + \Theta(h(n))$, prove that

$$f_1(n) = f_2(n) = g_1(n) + g_2(n) + \Theta(h(n))$$

(e) If $f(n) = \Theta(h(n))$, is it then necessarily true that $h(n) = \Theta(f(n))$?

Chapter 39

The Topsy-Turvy World of Continued Fractions

'You are old, Father William,' the young man said,
'And your hair has become very white;
And yet you incessantly stand on your head–
Do you think, at your age, it is right?'
'In my youth,' Father William replied to his son,
'I feared it might injure the brain;
But, now that I'm perfectly sure I have none,
Why, I do it again and again.'

Lewis Caroll, *Alice in Wonderland*, Illustration by Sir John Tenniel

The famous number π has a never-ending, never-repeating decimal expansion

$$\pi = 3.141592653589793238462643 3\ldots.$$

If we are willing to sacrifice accuracy for brevity, we might say that

$$\pi = 3 + \text{"a little bit more."}$$

The "little bit more" is a number between 0 and 1. We are going to follow the example of Old Father William and stand the "little bit more" on its head. When the small number $0.14159\ldots$ is turned on its head, it becomes the reciprocal of a number that is larger than 1. Thus

$$\begin{aligned}\pi &= 3 + 0.14159265358979323846264 33\ldots \\ &= 3 + \cfrac{1}{\cfrac{1}{0.14159265358979323846264 33\ldots}} \\ &= 3 + \frac{1}{7.0625133059310457697930051\ldots} \\ &= 3 + \frac{1}{7 + 0.0625133059310457697930051\ldots} \\ &= 3 + \frac{1}{7 + \text{“a little bit more”}}.\end{aligned}$$

Notice that if we ignore the “little bit more” in this last equation, we find that π is approximately equal to $3 + \frac{1}{7}$, that is, to $\frac{22}{7}$. You may have learned in high school that $\frac{22}{7}$ is a fairly good approximation to π.

Let's repeat the process. We take the “little bit more” in the last equation and turn it on its head,

$$\begin{aligned}0.0625133059310457697930051\ldots &= \cfrac{1}{\cfrac{1}{0.0625133059310457697930051\ldots}} \\ &= 15.9965944066857198889230 60\ldots.\end{aligned}$$

Now we substitute this into the earlier formula, which gives a double-decker fraction,

$$\pi = 3 + \cfrac{1}{7 + \cfrac{1}{15 + 0.9965944066857198889230 60\ldots}}$$

The bottom level of this fraction is $15.99659\ldots$, which is very close to 16. If we replace it with 16, we get a rational number that is quite close to π,

$$3 + \cfrac{1}{7 + \cfrac{1}{16}} = \frac{355}{113} = 3.14159292035398230088495 57\ldots.$$

The fraction $\frac{355}{113}$ agrees with π to six decimal places.

Continuing on our merry way, we compute

$$0.9965944066857198889230 60\ldots = \frac{1}{1.0034172310133726034641468\ldots}$$

to get the triple-decker fraction

$$\pi = 3 + \cfrac{1}{7 + \cfrac{1}{15 + \cfrac{1}{1 + 0.0034172310133726034641468\ldots}}},$$

and then

$$0.0034172310133726034641468\ldots = \frac{1}{292.63459101439547237857177\ldots}$$

to add yet another layer to our fraction,

$$\pi = 3 + \cfrac{1}{7 + \cfrac{1}{15 + \cfrac{1}{1 + \cfrac{1}{292 + 0.63459101439547237857177\ldots}}}}$$

Now let's see what happens if we round that last denominator to 293. We get the rational number

$$3 + \cfrac{1}{7 + \cfrac{1}{15 + \cfrac{1}{1 + \cfrac{1}{293}}}} = \frac{104348}{33215} = 3.14159265392142104470871 59\cdots.$$

So the fraction $\frac{104348}{33215}$ agrees with the value of π to nine decimal places.

Just how accurate is nine decimal places? Suppose that we are given that the distance from Earth to the sun is approximately 145,000,000 kilometers and that we want to calculate the length of Earth's orbit using the formula[1]

$$\text{Circumference} = 2 \times \pi \times \text{Radius}.$$

Then the error in the circumference if you use $\frac{104348}{33215}$ instead of π will amount to a little under a tenth of a kilometer. So, unless you've managed to measure the distance to the sun to within a fraction of a kilometer, it's fine to use the approximation $\pi \approx \frac{104348}{33215}$.

These multistory, topsy-turvy fractions have a name. They are called

---

[1]All right, all right, you caught me, Earth's orbit is an ellipse, not a circle. So we're really calculating the circumference of an invisible circle whose radius is approximately 145,000,000 kilometers.

Continued Fractions

We can form the continued fraction for any number by repeatedly flipping and separating off the whole integer part. The first few steps in the computation of the continued fraction for the cube root of 2 are given in full in Figure 39.1. In a similar fashion, we compute the continued fraction of $\sqrt{2}$,

$$\sqrt{2} = 1 + \cfrac{1}{2 + \cfrac{1}{2 + \cfrac{1}{2 + \cfrac{1}{2 + \cfrac{1}{\ddots}}}}}$$

and the continued fraction of $e = 2.7182818\ldots$ (the base of the natural logarithms),

$$e = 2 + \cfrac{1}{1 + \cfrac{1}{2 + \cfrac{1}{1 + \cfrac{1}{1 + \cfrac{1}{4 + \cfrac{1}{1 + \cfrac{1}{1 + \cfrac{1}{6 + \cfrac{1}{1 + \cfrac{1}{1 + \cfrac{1}{8 + \cfrac{1}{\ddots}}}}}}}}}}}}$$

Continued fractions are visually striking as they slide down to the right, but writing them as fractions takes a lot of ink and a lot of space. There must be a more convenient way to describe a continued fraction. All the numerators are 1's, so all we need to do is list the denominators. We write

$$[a_0, a_1, a_2, a_3, a_4, \ldots]$$

$$\sqrt[3]{2} = 1.259921\ldots$$
$$= 1 + \cfrac{1}{3.847322\ldots}$$
$$= 1 + \cfrac{1}{3 + \cfrac{1}{1.180189\ldots}}$$
$$= 1 + \cfrac{1}{3 + \cfrac{1}{1 + \cfrac{1}{5.549736\ldots}}}$$
$$= 1 + \cfrac{1}{3 + \cfrac{1}{1 + \cfrac{1}{5 + \cfrac{1}{1.819053\ldots}}}}$$
$$= 1 + \cfrac{1}{3 + \cfrac{1}{1 + \cfrac{1}{5 + \cfrac{1}{1 + \cfrac{1}{1.220922\ldots}}}}}$$
$$= 1 + \cfrac{1}{3 + \cfrac{1}{1 + \cfrac{1}{5 + \cfrac{1}{1 + \cfrac{1}{1 + \cfrac{1}{4.526491\ldots}}}}}}$$

Figure 39.1: The Continued Fraction Expansion of $\sqrt[3]{2}$

as shorthand for the continued fraction

$$a_0 + \cfrac{1}{a_1 + \cfrac{1}{a_2 + \cfrac{1}{a_3 + \cfrac{1}{a_4 + \cfrac{1}{a_5 + \cfrac{1}{a_6 + \ddots}}}}}}.$$

Using this new notation, our earlier continued fractions expansions (extended a bit further) can be written succintly as

$$\begin{aligned}
\pi &= [3, 7, 15, 1, 292, 1, 1, 1, 2, 1, 3, 1, 14, 2, 1, 1, 2, \ldots], \\
\sqrt[3]{2} &= [1, 3, 1, 5, 1, 1, 4, 1, 1, 8, 1, 14, 1, 10, 2, 1, 4, 12, 2, 3, 2, \ldots], \\
\sqrt{2} &= [1, 2, 2, 2, 2, 2, 2, 2, 2, 2, 2, 2, 2, 2, 2, 2, 2, 2, 2, 2, 2, \ldots], \\
e &= [2, 1, 2, 1, 1, 4, 1, 1, 6, 1, 1, 8, 1, 1, 10, 1, 1, 12, 1, 1, 14, \ldots].
\end{aligned}$$

Now that we've looked at several examples of continued fractions, it's time to work out some of the general theory. If a number α has a continued fraction expansion

$$\alpha = [a_0, a_1, a_2, a_3, \ldots],$$

then we have seen that cutting off after a few terms gives a rational number that is quite close to α. The nth *convergent* to α is the rational number

$$\frac{p_n}{q_n} = [a_0, a_1, \ldots, a_n] = a_0 + \cfrac{1}{a_1 + \cfrac{1}{a_2 + \cfrac{1}{\ddots + \cfrac{1}{a_n}}}}$$

obtained by using the terms up to a_n. For example, the first few convergents to

$\sqrt{2} = [1, 2, 2, 2, 2, \ldots]$ are

$$\frac{p_0}{q_0} = 1,$$

$$\frac{p_1}{q_1} = 1 + \frac{1}{2} = \frac{3}{2},$$

$$\frac{p_2}{q_2} = 1 + \cfrac{1}{2 + \cfrac{1}{2}} = 1 + \frac{2}{5} = \frac{7}{5},$$

$$\frac{p_3}{q_3} = 1 + \cfrac{1}{2 + \cfrac{1}{2 + \cfrac{1}{2}}} = 1 + \cfrac{1}{2 + \cfrac{2}{5}} = 1 + \frac{5}{12} = \frac{17}{12}.$$

A longer list of convergents to $\sqrt{2}$ is given in Table 39.1.

n	0	1	2	3	4	5	6	7	8	9	10
$\frac{p_n}{q_n}$	$\frac{1}{1}$	$\frac{3}{2}$	$\frac{7}{5}$	$\frac{17}{12}$	$\frac{41}{29}$	$\frac{99}{70}$	$\frac{239}{169}$	$\frac{577}{408}$	$\frac{1393}{985}$	$\frac{3363}{2378}$	$\frac{8119}{5741}$

Table 39.1: Convergents to $\sqrt{2}$

Staring at the list of convergents to $\sqrt{2}$ is not particularly enlightening, but it would certainly be useful to figure out how successive convergents are generated from the earlier ones. It is easier to spot the pattern if we look at $[a_0, a_1, a_2, a_3, \ldots]$ using symbols, rather than looking at any particular example.

$$\frac{p_0}{q_0} = \frac{a_0}{1},$$

$$\frac{p_1}{q_1} = \frac{a_1a_0 + 1}{a_1},$$

$$\frac{p_2}{q_2} = \frac{a_2a_1a_0 + a_2 + a_0}{a_2a_1 + 1},$$

$$\frac{p_3}{q_3} = \frac{a_3a_2a_1a_0 + a_3a_2 + a_3a_0 + a_1a_0 + 1}{a_3a_2a_1 + a_3 + a_1}.$$

Let's concentrate for the moment on the numerators $p_0, p_1, p_2, \ldots$. Table 39.2 gives the values of $p_0, p_1, p_2, p_3, p_4, p_5$.

At first glance, the formulas in Table 39.2 look horrible, but you might notice that p_0 appears at the tail end of p_2, that p_1 appears at the tail end of p_3, that p_2

n	p_n
p_0	a_0
p_1	$a_1a_0 + 1$
p_2	$a_2a_1a_0 + a_2 + a_0$
p_3	$a_3a_2a_1a_0 + a_3a_2 + a_3a_0 + a_1a_0 + 1$
p_4	$a_4a_3a_2a_1a_0 + a_4a_3a_2 + a_4a_3a_0 + a_4a_1a_0 + a_2a_1a_0 + a_4 + a_2 + a_0$
p_5	$a_5a_4a_3a_2a_1a_0 + a_5a_4a_3a_2 + a_5a_4a_3a_0 + a_5a_4a_1a_0 + a_5a_2a_1a_0$ $+a_3a_2a_1a_0 + a_5a_4 + a_5a_2 + a_3a_2 + a_5a_0 + a_3a_0 + a_1a_0 + 1$

Table 39.2: Numerator of the Continued Fraction $[a_0, a_1, \ldots, a_n]$

appears at the tail end of p_4, and that p_3 appears in the tail end of p_5. In other words, it seems that p_n is equal to p_{n-2} plus some other stuff. Here's a list of the "other stuff" for the first few values of n:

$$\begin{aligned}
p_2 - p_0 &= a_2a_1a_0 + a_2 \\
&= a_2(a_1a_0 + 1) \\
p_3 - p_1 &= a_3a_2a_1a_0 + a_3a_2 + a_3a_0 \\
&= a_3(a_2a_1a_0 + a_2 + a_0) \\
p_4 - p_2 &= a_4a_3a_2a_1a_0 + a_4a_3a_2 + a_4a_3a_0 + a_4a_1a_0 + a_4 \\
&= a_4(a_3a_2a_1a_0 + a_3a_2 + a_3a_0 + a_1a_0 + 1).
\end{aligned}$$

Looking back at Table 39.2, it seems that the "other stuff" for $p_n - p_{n-2}$ is simply a_n multiplied by the quantity p_{n-1}. We can describe this observation by the formula

$$p_n = a_np_{n-1} + p_{n-2}.$$

This is an example of a *recursion formula*, because it gives the successive values of $p_0, p_1, p_2, \ldots$ recursively in terms of the previous values. It is very similar to the recursion formula for the Fibonacci numbers that we investigated in Chapter 37.[2] Of course, this recursion formula needs two initial values to get started,

$$p_0 = a_0 \qquad \text{and} \qquad p_1 = a_1a_0 + 1.$$

A similar investigation of the denominators $q_0, q_1, q_2, \ldots$ reveals an analogous recursion. In fact, if you make a table for $q_0, q_1, q_2, \ldots$ similar to Table 39.2, you

[2]Indeed, if all the a_n's are equal to 1, then the sequence of p_n's is precisely the Fibonacci sequence. You can study the connection between continued fractions and Fibonacci numbers by doing Exercise 39.7.

will find that $q_0, q_1, q_2, \ldots$ seem to obey exactly the same recursion formula as the one obeyed by $p_0, p_1, p_2, \ldots$, but $q_0, q_1, q_2, \ldots$ use different starting values for q_0 and q_1. We summarize our investigations in the following important theorem.

Theorem 39.1 (Continued Fraction Recursion Formula). *Let*

$$\frac{p_n}{q_n} = [a_0, a_1, \ldots, a_n] = a_0 + \cfrac{1}{a_1 + \cfrac{1}{a_2 + \cfrac{1}{\ddots + \cfrac{1}{a_n}}}},$$

where we treat $a_0, a_1, a_2, \ldots$ as variables, rather than as specific numbers. Then the numerators $p_0, p_1, p_2, \ldots$ are given by the recursion formula

$$p_0 = a_0, \qquad p_1 = a_1 a_0 + 1, \qquad \textit{and} \qquad p_n = a_n p_{n-1} + p_{n-2} \quad \textit{for } n \geq 2,$$

and the denominators $q_0, q_1, q_2, \ldots$ are given by the recursion formula

$$q_0 = 1, \qquad q_1 = a_1, \qquad \textit{and} \qquad q_n = a_n q_{n-1} + q_{n-2} \quad \textit{for } n \geq 2,$$

Proof. When a sequence is defined by a recursive formula, it is often easiest to use induction to prove facts about the sequence. To get our induction started, we need to check that

$$\frac{p_0}{q_0} = [a_0] \qquad \text{and} \qquad \frac{p_1}{q_1} = [a_0, a_1].$$

We are given that $p_0 = a_0$ and $q_0 = 1$, so $p_0/q_0 = a_0$, which verifies the first equation. Similarly, we are given that $p_1 = a_1 a_0 + 1$ and $q_1 = a_1$, so

$$[a_0, a_1] = a_0 + \frac{1}{a_1} = \frac{a_1 a_0 + 1}{a_1} = \frac{p_1}{q_1},$$

which verifies the second equation.

Now we assume that the theorem is true when $n = N$ and use that assumption to prove that it is also true when $n = N+1$. A key observation is that the continued fraction

$$[a_0, a_1, a_2, \ldots, a_N, a_{N+1}]$$

can be written as a continued fraction with one less term by combining the last two terms,[3]

$$[a_0, a_1, a_2, \ldots, a_N, a_{N+1}] = \left[a_0, a_1, a_2, \ldots, a_N + \frac{1}{a_{N+1}}\right].$$

[3]Don't let the complicated last term confuse you. If you think about writing everything out as a fraction, you'll see immediately that both sides are equal. Try it for $N = 2$ and for $N = 3$ if it's still not clear.

To simplify the notation, let's use different letters for the terms in the continued fraction on the right, say

$$[b_0, b_1, \ldots, b_N]$$

$$\text{with} \quad b_0 = a_0, \quad b_1 = a_1, \ldots, b_{N-1} = a_{N-1}, \quad \text{and} \quad b_N = a_N + \frac{1}{a_{N+1}}.$$

Notice that $[b_0, b_1, \ldots, b_N]$ is a continued fraction with one fewer term than the continued fraction $[a_0, a_1, \ldots, a_{N+1}]$, so our induction hypothesis says that the theorem is true for $[b_0, b_1, \ldots, b_N]$. To avoid confusion, we use capital letters P_n/Q_n for the convergents of $[b_0, b_1, \ldots, b_N]$. Then the induction hypothesis tells us that the P_n's and the Q_n's satisfy the recursion formulas

$$P_n = b_n P_{n-1} + P_{n-2} \qquad \text{and} \qquad Q_n = b_n Q_{n-1} + Q_{n-2} \qquad \text{for all } 2 \le n \le N.$$

Therefore

$$[b_0, b_1, \ldots, b_N] = \frac{P_N}{Q_N} = \frac{b_N P_{N-1} + P_{N-2}}{b_N Q_{N-1} + Q_{N-2}}. \tag{$*$}$$

How are the convergents for $[a_0, a_1, \ldots, a_{N+1}]$ and $[b_0, b_1, \ldots, b_N]$ related? We know that $b_n = a_n$ for all $0 \le n \le N-1$, so the nth convergents are the same for all $0 \le n \le N-1$. This means that we can make the following substitutions into the formula $(*)$:

$$P_{N-1} = p_{N-1}, \quad P_{N-2} = p_{N-2}, \quad Q_{N-1} = q_{N-1}, \quad Q_{N-2} = q_{N-2}.$$

Since we also know that

$$[b_0, b_1, \ldots, b_N] = [a_0, a_1, \ldots, a_{N+1}] \qquad \text{and} \qquad b_N = a_N + \frac{1}{a_{N+1}},$$

we find that

$$\begin{aligned} [a_0, a_1, \ldots, a_{N+1}] &= \frac{b_N P_{N-1} + P_{N-2}}{b_N Q_{N-1} + Q_{N-2}} \\ &= \frac{\left(a_N + \dfrac{1}{a_{N+1}}\right) p_{N-1} + p_{N-2}}{\left(a_N + \dfrac{1}{a_{N+1}}\right) q_{N-1} + q_{N-2}} \\ &= \frac{a_{N+1}(a_N p_{N-1} + p_{N-2}) + p_{N-1}}{a_{N+1}(a_N q_{N-1} + q_{N-2}) + q_{N-1}}. \end{aligned} \tag{$**$}$$

The induction hypothesis applied to the continued fraction $[a_0, a_1, \ldots, a_N]$ tells us that its convergents satisfy

$$p_N = a_N p_{N-1} + p_{N-2} \qquad \text{and} \qquad q_N = a_N q_{N-1} + q_{N-2},$$

which allows us to simplify the formula (**) to read

$$[a_0, a_1, \ldots, a_{N+1}] = \frac{a_{N+1} p_N + p_{N-1}}{a_{N+1} q_N + q_{N-1}}.$$

But by definition, the $(N+1)^{\text{st}}$ convergent is

$$[a_0, a_1, \ldots, a_{N+1}] = \frac{p_{N+1}}{q_{N+1}}.$$

Comparing these two expressions for $[a_0, a_1, \ldots, a_{N+1}]$, we see that

$$p_{N+1} = a_{N+1} p_N + p_{N-1} \qquad \text{and} \qquad q_{N+1} = a_{N+1} q_N + q_{N-1}.$$

We have now shown that if the recursion relations are true for $n = N$ they are also true for $n = N + 1$. This completes our induction proof that they are true for all values of n. □

We expect that the convergents to a number such as $\sqrt{2}$ should get closer and closer to $\sqrt{2}$, so it might be interesting to see how close the convergents are to one another:

$$\begin{aligned}
\frac{p_0}{q_0} - \frac{p_1}{q_1} &= \frac{1}{1} - \frac{3}{2} = -\frac{1}{2} \\
\frac{p_1}{q_1} - \frac{p_2}{q_2} &= \frac{3}{2} - \frac{7}{5} = \frac{1}{10} \\
\frac{p_2}{q_2} - \frac{p_3}{q_3} &= \frac{7}{5} - \frac{17}{12} = -\frac{1}{60} \\
\frac{p_3}{q_3} - \frac{p_4}{q_4} &= \frac{17}{12} - \frac{41}{29} = \frac{1}{348} \\
\frac{p_4}{q_4} - \frac{p_5}{q_5} &= \frac{41}{29} - \frac{99}{70} = -\frac{1}{2030} \\
\frac{p_5}{q_5} - \frac{p_6}{q_6} &= \frac{99}{70} - \frac{239}{169} = \frac{1}{11830}
\end{aligned}$$

The Difference Between Successive Convergents of $\sqrt{2}$

The difference between successive convergents does indeed seem to be getting smaller and smaller, but an even more interesting pattern has emerged. It seems that all the numerators are equal to 1 and that the values alternate between positive and negative.

Let's try another example, say the continued fraction expansion of π. We find that

$$\begin{aligned}
\frac{p_0}{q_0} - \frac{p_1}{q_1} &= \frac{3}{1} - \frac{22}{7} = -\frac{1}{7} \\
\frac{p_1}{q_1} - \frac{p_2}{q_2} &= \frac{22}{7} - \frac{333}{106} = \frac{1}{742} \\
\frac{p_2}{q_2} - \frac{p_3}{q_3} &= \frac{333}{106} - \frac{355}{113} = -\frac{1}{11978} \\
\frac{p_3}{q_3} - \frac{p_4}{q_4} &= \frac{355}{113} - \frac{103993}{33102} = \frac{1}{3740526} \\
\frac{p_4}{q_4} - \frac{p_5}{q_5} &= \frac{103993}{33102} - \frac{104348}{33215} = -\frac{1}{1099482930} \\
\frac{p_5}{q_5} - \frac{p_6}{q_6} &= \frac{104348}{33215} - \frac{208341}{66317} = \frac{1}{2202719155}
\end{aligned}$$

The Difference Between Successive Convergents of π

The exact same pattern has appeared. So let's buckle down and prove a theorem.

Theorem 39.2 (Difference of Successive Convergents Theorem). *As usual, let* $\frac{p_0}{q_0}, \frac{p_1}{q_1}, \frac{p_2}{q_2}, \ldots$ *be the convergents to the continued fraction* $[a_0, a_1, a_2, \ldots]$. *Then*

$$p_{n-1}q_n - p_nq_{n-1} = (-1)^n \qquad \textit{for all } n = 1, 2, 3, \ldots.$$

Equivalently, dividing both sides by $q_{n-1}q_n$,

$$\frac{p_{n-1}}{q_{n-1}} - \frac{p_n}{q_n} = \frac{(-1)^n}{q_{n-1}q_n} \qquad \textit{for all } n = 1, 2, 3, \ldots.$$

Proof. This theorem is quite easy to prove using induction and the Continued Fraction Recursion Formula (Theorem 39.1). First we check that it is true for $n = 1$:

$$p_0q_1 - p_1q_0 = a_0 \cdot a_1 - (a_1a_0 + 1) \cdot 1 = -1.$$

This gets our induction started.

Now we assume that the theorem is true for $n = N$, and we need to prove that it is true for $n = N + 1$. We compute

$$\begin{aligned} p_N q_{N+1} - p_{N+1} q_N &= p_N \left(a_N q_N + q_{N-1}\right) - \left(a_N p_N + p_{N-1}\right) q_N \\ &\quad \text{using the Continued Fraction Recursion Theorem,} \\ &= p_N q_{N-1} - p_{N-1} q_N \qquad \text{since the } a_N p_N q_N \text{ terms cancel,} \\ &= -\left(p_{N-1} q_N - p_N q_{N-1}\right) \\ &= -(-1)^N \qquad \text{from the induction hypothesis with } n = N, \\ &= (-1)^{N+1}. \end{aligned}$$

We have now shown that the desired formula is true for $n = 1$ and that, if it is true for $n = N$, then it is also true for $n = N + 1$. Therefore, by induction it is true for all values of $n \geq 1$, which completes the proof of the theorem. □

Exercises

39.1. (a) Compute the first 10 terms in the continued fractions of $\sqrt{3}$ and $\sqrt{5}$.
(b) Do the terms in the continued fraction of $\sqrt{3}$ appear to follow a repetitive pattern? If so, prove that they really do repeat.
(c) Do the terms in the continued fraction of $\sqrt{5}$ appear to follow a repetitive pattern? If so, prove that they really do repeat.

39.2. The continued fraction of π^2 is

$$[\_\_, \_\_, \_\_, 1, 2, 47, 1, 8, 1, 1, 2, 2, 1, 1, 8, 3, 1, 10, 5, 1, 3, 1, 2, 1, 1, 3, 15, 1, 1, 2, \ldots].$$

(a) Fill in the three initial missing entries.
(b) Do you see any sort of pattern in the continued fraction of π^2?
(c) Use the first 5 terms in the continued fraction to find a rational number that is close to π^2. How close do you come?
(d) Same question as (c), but use the first 6 terms.

39.3. The continued fraction of $\sqrt{2} + \sqrt{3}$ is

$$[\_\_, \_\_, \_\_, 5, 7, 1, 1, 4, 1, 38, 43, 1, 3, 2, 1, 1, 1, 1, 2, 4, 1, 4, 5, 1, 5, 1, 7, \ldots].$$

(a) Fill in the three initial missing entries.
(b) Do you see any sort of pattern in the continued fraction of $\sqrt{2} + \sqrt{3}$?
(c) For each $n = 1, 2, 3, \ldots, 7$, compute the nth convergent

$$\frac{p_n}{q_n} = [a_0, a_1, \ldots, a_n]$$

to $\sqrt{2} + \sqrt{3}$.

(d) The fractions that you computed in (b) should give more and more accurate approximations to $\sqrt{2}+\sqrt{3}$. Verify this by making a table of values

$$\left|\sqrt{2}+\sqrt{3}-\frac{p_n}{q_n}\right| = \frac{1}{10^{\text{some power}}}$$

for $n = 1, 2, 3, \ldots, 7$.

39.4. Let p_n/q_n be the nth convergent to α. For each of the following values of α, make a table listing the value of the quantity

$$q_n\,|p_n - q_n\alpha| \qquad \text{for } n = 1, 2, 3, \ldots, N.$$

(The continued fraction expansions of $\sqrt{2}$, $\sqrt[3]{2}$, and π are listed on page 329, so you can use that information to compute the associated convergents.)

(a) $\alpha = \sqrt{2}$ up to $N = 8$.
(b) $\alpha = \sqrt[3]{2}$ up to $N = 7$.
(c) $\alpha = \pi$ up to $N = 5$.
(d) Your data from (a) suggest that not only is $|p_n - q_n\sqrt{2}|$ bounded, but it actually approaches a limit as $n \to \infty$. Try to guess what that limit equals, and then prove that your guess is correct.
(e) Recall that Dirichlet's Diophantine Approximation Theorem 31.2 says that for any irrational number α, there are infinitely many pairs of positive integers x and y satisfying

$$|x - y\alpha| < 1/y. \tag{39.1}$$

Your data from (a), (b), and (c) suggest that if p_n/q_n is a convergent to α then (p_n, q_n) provides a solution to the inquality (39.1). Prove that this is true.

39.5. The Continued Fraction Recursion Formula (Theorem 39.1) gives a procedure for generating two lists of numbers $p_0, p_1, p_2, p_3, \ldots$ and $q_0, q_1, q_2, q_3, \ldots$ from two initial values a_0 and a_1. The fraction p_n/q_n is then the nth convergent to some number α. Prove that the fraction p_n/q_n is already in lowest terms, that is, prove that $\gcd(p_n, q_n) = 1$. (*Hint*. Use the Difference of Successive Convergents Theorem 39.2.)

39.6. We proved that successive convergents p_{n-1}/q_{n-1} and p_n/q_n satisfy

$$p_{n-1}q_n - p_nq_{n-1} = (-1)^n.$$

In this exercise you will figure out what happens if instead we take every other convergent.

(a) Compute the quantity

$$p_{n-2}q_n - p_nq_{n-2} \tag{$*$}$$

for the convergents of the partial fraction $\sqrt{2} = [1, 2, 2, 2, 2, \ldots]$. Do this for $n = 2, 3, \ldots, 6$.
(b) Compute the quantity $(*)$ for $n = 2, 3, \ldots, 6$ for the convergents of the partial fraction

$$\pi = [3, 7, 15, 1, 292, 1, 1, 1, 2, \ldots].$$

(c) Using your results from (a) and (b) (and any other data that you want to collect), make a conjecture for the value of the quantity (*) for a general continued fraction $[a_0, a_1, a_2, \ldots]$.

(d) Prove that your conjecture in (c) is correct. (*Hint.* The Continued Fraction Recursion Formula may be useful.)

39.7. The "simplest" continued fraction is the continued fraction $[1, 1, 1, \ldots]$ consisting entirely of 1's.

(a) Compute the first 10 convergents of $[1, 1, 1, \ldots]$.

(b) Do you recognize the numbers appearing in the numerators and denominators of the fractions that you computed in (a)? (If not, look back at Chapter 37.)

(c) What is the exact value of the limit

$$\lim_{n\to\infty} \frac{p_n}{q_n}$$

of the convergents for the continued fraction $[1, 1, 1, \ldots]$?

39.8. In Table 39.2 we listed the numerator p_n of the continued fraction $[a_0, a_1, \ldots, a_n]$ for the first few values of n.

(a) How are the numerator of $[a, b]$ and $[b, a]$ related to one another?

(b) How are the numerator of $[a, b, c]$ and $[c, b, a]$ related to one another?

(c) More generally, how do the numerator of

$$[a_0, a_1, a_2, \ldots, a_{n-1}, a_n] \qquad \text{and} \qquad [a_n, a_{n-1}, \ldots, a_2, a_1, a_0]$$

seem to be related to one another?

(d) Prove that your conjecture in (c) is correct.

39.9. Write a program that takes as input a decimal number A and an integer n and returns the following values:

(a) the first $n + 1$ terms $[a_0, a_1, \ldots, a_n]$ of the continued fraction of A;

(b) the nth convergent p_n/q_n of A, as a fraction;

(c) the difference between A and p_n/q_n, as a decimal.

39.10. Use your program from Exercise 39.9 to make a table of (at least) the first 10 terms of the continued fraction expansion of $\sqrt{D}$ for $2 \le D \le 30$. What sort of pattern(s) can you find? (You can check your output by comparing with Table 40.1 in the next chapter.)

39.11. Same question as Exercise 39.10, but with cube roots. In other words, make a table of (at least) the first 10 terms of the continued fraction expansion of $\sqrt[3]{D}$ for each value of D satisfing $2 \le D \le 20$. Do you see any patterns?

39.12. (Advanced Calculus Exercise) Let $a_0, a_1, a_2, a_3, \ldots$ be a sequence of real numbers satisfying $a_i \geq 1$. Then, for each $n = 0, 1, 2, 3, \ldots$, we can compute the real number

$$u_n = [a_0, a_1, a_2, \ldots, a_n] = a_0 + \cfrac{1}{a_1 + \cfrac{1}{a_2 + \cfrac{1}{\ddots + \cfrac{1}{a_n}}}}.$$

Prove that the limit $\lim_{n \to \infty} u_n$ exists. (*Hint.* Use Theorems 39.1 and 39.2 to prove that the sequence $u_1, u_2, u_3, \ldots$ is a Cauchy sequence.)

Chapter 40

Continued Fractions, Square Roots, and Pell's Equation

The continued fraction for $\sqrt{2}$,

$$[1, 2, 2, 2, 2, 2, 2, 2, 2, 2, 2, 2, 2, 2, 2, 2, 2, 2, 2, 2, 2, \ldots],$$

certainly appears to be quite repetitive. Let's see if we can prove that the continued fraction of $\sqrt{2}$ really does consist of the number 1 followed entirely by 2's. Since $\sqrt{2} = 1.414\ldots$, the first step in the continued fraction algorithm is to write

$$\sqrt{2} = 1 + (\sqrt{2} - 1) = 1 + \frac{1}{1/(\sqrt{2} - 1)}.$$

Next we simplify the denominator,

$$\frac{1}{\sqrt{2} - 1} = \frac{1}{\sqrt{2} - 1} \cdot \frac{\sqrt{2} + 1}{\sqrt{2} + 1} = \frac{\sqrt{2} + 1}{\sqrt{2}^2 - 1} = \sqrt{2} + 1.$$

Substituting this back in above yields

$$\sqrt{2} = 1 + \frac{1}{\sqrt{2} + 1}.$$

The number $\sqrt{2} + 1$ is between 2 and 3, so we write it as

$$\sqrt{2} + 1 = 2 + (\sqrt{2} - 1) = 2 + \frac{1}{1/(\sqrt{2} - 1)}.$$

But we already checked that $1/(\sqrt{2} - 1)$ is equal to $\sqrt{2} + 1$, so we find that

$$\sqrt{2} + 1 = 2 + \frac{1}{\sqrt{2} + 1}, \qquad (*)$$

and hence that

$$\sqrt{2} = 1 + \cfrac{1}{2 + \cfrac{1}{\sqrt{2}+1}}.$$

Now we can use the formula (*) again to obtain

$$\sqrt{2} = 1 + \cfrac{1}{2 + \cfrac{1}{2 + \cfrac{1}{\sqrt{2}+1}}},$$

and yet again to obtain

$$\sqrt{2} = 1 + \cfrac{1}{2 + \cfrac{1}{2 + \cfrac{1}{2 + \cfrac{1}{\sqrt{2}+1}}}}.$$

Continuing to employ the formula (*), we find that the continued fraction of $\sqrt{2}$ does indeed consist of a single 1 followed entirely by 2's.

Emboldened by our success with $\sqrt{2}$, can we find other numbers whose continued fractions are similarly repetitive (or, to employ proper mathematical terminology, whose continued fractions are *periodic*)? If you have done Exercise 39.10, which asks you to compute the continued fraction of $\sqrt{D}$ for $D = 2, 3, 4, \ldots, 20$, you found some examples. Collecting further data of this sort, Table 40.1 lists the continued fractions for $\sqrt{p}$ for each prime p less than 40.

Let's turn the question on its head and ask what we can deduce about a continued fraction that happens to be repetitive. We start with a simple example. Suppose that the number A has as its continued fraction

$$A = [a, b, b, b, b, b, b, b, \ldots].$$

Table 40.1 includes several numbers of this sort, including $\sqrt{2}$, $\sqrt{5}$, and $\sqrt{37}$. We can write A in the form

$$A = a + \frac{1}{[b, b, b, b, b, \ldots]},$$

so we really need to determine the value of the continued fraction

$$B = [b, b, b, b, b, b, b, b, \ldots].$$

D	Continued fraction of $\sqrt{D}$	Period
2	$[1, 2, 2, 2, 2, 2, 2, 2, 2, 2, 2, 2, 2, 2, 2, 2, 2, 2, 2, \cdots]$	1
3	$[1, 1, 2, 1, 2, 1, 2, 1, 2, 1, 2, 1, 2, 1, 2, 1, 2, 1, \cdots]$	2
5	$[2, 4, 4, 4, 4, 4, 4, 4, 4, 4, 4, 4, 4, 4, 4, 4, 4, 4, 4, \cdots]$	1
7	$[2, 1, 1, 1, 4, 1, 1, 1, 4, 1, 1, 1, 4, 1, 1, 1, 4, 1, \cdots]$	4
11	$[3, 3, 6, 3, 6, 3, 6, 3, 6, 3, 6, 3, 6, 3, 6, 3, 6, 3, 6, \cdots]$	2
13	$[3, 1, 1, 1, 1, 6, 1, 1, 1, 1, 6, 1, 1, 1, 1, 6, 1, 1, \cdots]$	5
17	$[4, 8, 8, 8, 8, 8, 8, 8, 8, 8, 8, 8, 8, 8, 8, \cdots]$	1
19	$[4, 2, 1, 3, 1, 2, 8, 2, 1, 3, 1, 2, 8, 2, 1, 3, 1, 2, 8, \cdots]$	6
23	$[4, 1, 3, 1, 8, 1, 3, 1, 8, 1, 3, 1, 8, 1, 3, 1, 8, 1, \cdots]$	4
29	$[5, 2, 1, 1, 2, 10, 2, 1, 1, 2, 10, 2, 1, 1, 2, 10, 2, 1, 1, \cdots]$	5
31	$[5, 1, 1, 3, 5, 3, 1, 1, 10, 1, 1, 3, 5, 3, 1, 1, 10, 1, 1, \cdots]$	8
37	$[6, 12, 12, 12, 12, 12, 12, 12, 12, 12, 12, 12, \cdots]$	1

Table 40.1: Continued Fractions of Square Roots

Just as we did for A, we can pull off the first entry of B and write B as

$$B = b + \frac{1}{[b, b, b, b, b, \ldots]}.$$

But "Lo and Behold," the denominator $[b, b, b, b, \ldots]$ is simply B itself, so we find that

$$B = b + \frac{1}{B}.$$

Now we can multiply through by B to get $B^2 = bB + 1$ and then use the quadratic formula to solve for B,

$$B = \frac{b + \sqrt{b^2+4}}{2}.$$

(Note that we use the plus sign, since we need B to be positive.) Finally, we compute the value of A,

$$\begin{aligned} A = a + \frac{1}{B} &= a + \frac{2}{b + \sqrt{b^2+4}} \\ &= a + \frac{2}{b + \sqrt{b^2+4}} \cdot \left(\frac{b - \sqrt{b^2+4}}{b - \sqrt{b^2+4}} \right) \\ &= a - \frac{b - \sqrt{b^2+4}}{2} \\ &= \frac{2a-b}{2} + \frac{\sqrt{b^2+4}}{2} \end{aligned}$$

We summarize our calculations, including two very interesting special cases.

Proposition 40.1. *For any positive integers a and b, we have the continued fraction formula*

$$\frac{2a-b}{2} + \frac{\sqrt{b^2+4}}{2} = [a, b, b, b, b, b, b, b, \ldots].$$

In particular, taking $a = b$ gives the formula

$$\frac{b+\sqrt{b^2+4}}{2} = [b, b, b, b, b, b, b, \ldots]$$

and taking $b = 2a$ gives the formula

$$\sqrt{a^2+1} = [a, 2a, 2a, 2a, 2a, 2a, 2a, 2a, \ldots].$$

What happens if we have a continued fraction that repeats in a more complicated fashion? Let's do an example to try to gain some insight. Suppose that A has the continued fraction

$$A = [1, 2, 3, 4, 5, 4, 5, 4, 5, 4, 5, \ldots],$$

where the subsequent terms continue to alternate 4 and 5. The first thing to do is to pull off the nonrepetitive part,

$$A = 1 + \cfrac{1}{2 + \cfrac{1}{3 + \cfrac{1}{[4, 5, 4, 5, 4, 5, 4, 5, \ldots]}}}.$$

So now we need to figure out the value of the *purely periodic* continued fraction

$$B = [4, 5, 4, 5, 4, 5, 4, 5, 4, 5, \ldots].$$

(A continued fraction is called *purely periodic* if it repeats from the very beginning.) We can write B as

$$B = 4 + \cfrac{1}{5 + \cfrac{1}{[4, 5, 4, 5, 4, 5, 4, 5, 4, 5, \ldots]}}.$$

As in our earlier example, we recognize that the bottommost denominator is equal to B, so we have shown that

$$B = 4 + \cfrac{1}{5 + \cfrac{1}{B}}.$$

Now we simplify this complicated fraction to get an equation for B,

$$B = 4 + \cfrac{1}{5 + \cfrac{1}{B}} = 4 + \cfrac{1}{\cfrac{5B+1}{B}} = 4 + \frac{B}{5B+1} = \frac{21B+4}{5B+1}.$$

Cross-multiplying by $5B+1$, moving everything to one side, and doing a little bit of algebra, we find the equation

$$5B^2 - 20B - 4 = 0,$$

and then the good old quadratic formula yields

$$B = \frac{20 + \sqrt{400+80}}{10} = \frac{10 + 2\sqrt{30}}{5}.$$

Next we find the value of A by substituting the value of B into our earlier formula and using elementary algebra to repeatedly flip, combine, and simplify.

$$A = 1 + \cfrac{1}{2 + \cfrac{1}{3 + \cfrac{1}{B}}} = 1 + \cfrac{1}{2 + \cfrac{1}{3 + \cfrac{1}{\cfrac{10+2\sqrt{30}}{5}}}}$$

$$= 1 + \cfrac{1}{2 + \cfrac{1}{3 + \cfrac{5}{10+2\sqrt{30}}}} = 1 + \cfrac{1}{2 + \cfrac{1}{\cfrac{35+6\sqrt{30}}{10+2\sqrt{30}}}}$$

$$= 1 + \cfrac{1}{2 + \cfrac{10+2\sqrt{30}}{35+6\sqrt{30}}} = 1 + \cfrac{1}{\cfrac{80+14\sqrt{30}}{35+6\sqrt{30}}}$$

$$= 1 + \frac{35+6\sqrt{30}}{80+14\sqrt{30}} = \frac{115+20\sqrt{30}}{80+14\sqrt{30}}.$$

Finally, we rationalize the denominator of A by multiplying the numerator and denominator by $80 - 14\sqrt{30}$,

$$A = \frac{115+20\sqrt{30}}{80+14\sqrt{30}} \cdot \left(\frac{80-14\sqrt{30}}{80-14\sqrt{30}}\right) = \frac{800-10\sqrt{30}}{520} = \frac{80-\sqrt{30}}{52}.$$

If you have done Exercise 39.9, try running your program with the input

$$\frac{80 - \sqrt{30}}{52} = 1.4331302774028526704 89\ldots$$

and check that you indeed get the continued fraction $[1, 2, 3, 4, 5, 4, 5, 4, 5, \ldots]$.

A continued fraction is called *periodic* if it looks like

$$[\underbrace{a_1, a_2, \ldots, a_\ell}_{\text{initial part}}, \underbrace{b_1, b_2, \ldots, b_m}_{\text{periodic part}}, \underbrace{b_1, b_2, \ldots, b_m}_{\text{periodic part}}, \underbrace{b_1, b_2, \ldots, b_m}_{\text{periodic part}}, \ldots].$$

In other words, it is periodic if, after some initial terms, it consists of a finite list of terms that are repeated over and over again. The number of repeated terms m is called the *period*. For example, $\sqrt{2} = [1, 2, 2, 2, \ldots]$ has period 1 and $\sqrt{23} = [4, 1, 3, 1, 8, 1, 3, 1, 8, \ldots]$ has period 4. Other examples are given in Table 40.1. A convenient notatation, which makes the periodicity more visible, is to place a bar over the repeating part to indicate that it repeats indefinitely. For example

$$\sqrt{2} = [1, \overline{2}], \qquad \sqrt{23} = [4, \overline{1, 3, 1, 8}], \qquad \frac{80 - \sqrt{30}}{52} = [1, 2, 3, \overline{4, 5}].$$

Similarly, a general periodic continued fraction is written as

$$[a_1, a_2, \ldots, a_\ell, \overline{b_1, b_2, \ldots, b_m}].$$

The examples that we have done suggest that the following theorem might be true. We prove the first part and leave the second part as a (challenging) exercise.

Theorem 40.2 (Periodic Continued Fraction Theorem).

(a) *Suppose that the number A has a periodic continued fraction*

$$A = [a_1, a_2, \ldots, a_\ell, \overline{b_1, b_2, \ldots, b_m}].$$

Then A is equal to a number of the form

$$A = \frac{r + s\sqrt{D}}{t} \qquad \text{with } r, s, t, D \text{ integers and } D > 0.$$

(b) *Let r, s, t, D be integers with $D > 0$. Then the number*

$$\frac{r + s\sqrt{D}}{t}$$

has a periodic continued fraction.

Proof. (a) Let's start with the purely periodic continued fraction

$$B = [\,\overline{b_1, b_2, \ldots, b_m}\,].$$

If we write out the first m steps, we find that

$$B = b_1 + \cfrac{1}{b_2 + \cfrac{1}{\ddots + \cfrac{1}{b_m + \cfrac{1}{[\,\overline{b_1, b_2, \ldots, b_m}\,]}}}} = b_1 + \cfrac{1}{b_2 + \cfrac{1}{\ddots + \cfrac{1}{b_m + \cfrac{1}{B}}}}.$$

We now simplify the right-hand side by repeatedly combining terms and flipping fractions, where we treat B as a variable and the quantities $b_1, \ldots, b_m$ as numbers. After much algebra, our equation eventually simplifies to

$$B = \frac{uB + v}{wB + z}, \qquad (*)$$

where u, v, w, z are certain integers that depend on $b_1, b_2, \ldots, b_m$. Furthermore, it is clear that u, v, w, z are all positive numbers, since $b_1, b_2, \ldots$ are positive.

To illustrate this procedure, we do the case $m = 2$.

$$b_1 + \cfrac{1}{b_2 + \cfrac{1}{B}} = b_1 + \cfrac{1}{\cfrac{b_2 B + 1}{B}} = b_1 + \frac{B}{b_2 B + 1} = \frac{(b_1 b_2 + 1)B + b_1}{b_2 B + 1}.$$

Returning to the general case, we cross-multiply equation $(*)$ and move everything to one side, which gives the equation

$$wB^2 + (z - u)B - v = 0.$$

Now the quadratic formula yields

$$B = \frac{-(z - u) + \sqrt{(z - u)^2 + 4vw}}{2w},$$

so B has the form

$$B = \frac{i + j\sqrt{D}}{k} \qquad \text{with } i, j, k, D \text{ integers and } D > 0.$$

Returning to our original number $A = [a_1, a_2, \ldots, a_\ell, \overline{b_1, b_2, \ldots, b_m}\,]$, we can write A as

$$A = a_1 + \cfrac{1}{a_2 + \cfrac{1}{\ddots + \cfrac{1}{a_\ell + \cfrac{1}{B}}}} = a_1 + \cfrac{1}{a_2 + \cfrac{1}{\ddots + \cfrac{1}{a_\ell + \cfrac{1}{\frac{i + j\sqrt{D}}{k}}}}}.$$

Again we repeatedly flip, combine, and simplify, which eventually yields an expression for A of the form

$$A = \frac{e + f\sqrt{D}}{g + h\sqrt{D}}, \qquad \text{where } e, f, g, h \text{ are integers.}$$

Finally, we multiply both numerator and denominator by $g - h\sqrt{D}$. This rationalizes the denominator and expresses A as a number of the form

$$A = \frac{r + s\sqrt{D}}{t}, \qquad \text{where } r, s, t \text{ are integers.}$$

We have completed the proof of part (a) of the Periodic Continued Fraction Theorem 40.2. The proof of part (b) is left to you as (challenging) Exercise 40.10. □

The Continued Fraction of $\sqrt{D}$ and Pell's Equation

The convergents to a continued fraction form a list of rational numbers that get closer and closer to the original number. For example, the number $\sqrt{71}$ has continued fraction

$$\sqrt{71} = [8, \overline{2, 2, 1, 7, 1, 2, 2, 16}\,]$$

and its first few convergents are

$$\frac{17}{2}, \ \frac{42}{5}, \ \frac{59}{7}, \ \frac{455}{54}, \ \frac{514}{61}, \ \frac{1483}{176}, \ \frac{3480}{413}, \ \frac{57163}{6784}, \ \frac{117806}{13981}, \ \frac{292775}{34746}.$$

If p/q is a convergent to $\sqrt{D}$, then

$$\frac{p}{q} \approx \sqrt{D}, \qquad \text{so} \qquad \frac{p^2}{q^2} \approx D.$$

p	q	$p^2 - 17q^2$
17	2	5
42	5	−11
59	7	2
455	54	−11
514	61	5
1483	176	−7
3480	413	1
57163	6784	−7
117806	13981	5
292775	34746	−11

Table 40.2: Convergents p/q to $\sqrt{17}$

Multiplying by q^2, this means that we would expect p^2 to be fairly close to Dq^2. Table 40.2 lists the value of the difference $p^2 - Dq^2$ for the first few convergents to $\sqrt{71}$.

Among the many striking features of the data in Table 40.2, we pick out the seemingly mundane appearance of the number 1 in the final column. This occurs on the seventh row and reflects that fact that

$$3480^2 - 71 \cdot 413^2 = 1.$$

Thus the convergent $3480/413$ to the number $\sqrt{71}$ provides a solution $(3480, 413)$ to the Pell equation

$$x^2 - 71y^2 = 1.$$

This suggests a connection between the convergents to $\sqrt{D}$ and Pell's equation $x^2 - Dy^2 = 1$.

In Chapters 30 and 32 we carefully and completely proved that Pell's equation

$$x^2 - Dy^2 = 1$$

always has a solution. But if you look back at Chapter 32, you will see that our proof does not provide an efficient way to actually find a solution. It would thus be very useful if the convergents to $\sqrt{D}$ could be used to efficiently compute a solution to Pell's equation.

The continued fraction of $\sqrt{71}$,

$$\sqrt{71} = [8, \overline{2, 2, 1, 7, 1, 2, 2, 16}],$$

has period 8, and the convergent that gives the solution to Pell's equation is

$$[8, 2, 2, 1, 7, 1, 2, 2] = \frac{3480}{413}.$$

A brief examination of Table 40.1 shows that the continued fractions of square roots $\sqrt{D}$ have many special features.[1] Here are some further examples with moderately large periods.

$$\sqrt{73} = [8, \overline{1, 1, 5, 5, 1, 1, 16}], \qquad \text{Period} = 7.$$
$$\sqrt{89} = [9, \overline{2, 3, 3, 2, 18}], \qquad \text{Period} = 5.$$
$$\sqrt{97} = [9, \overline{1, 5, 1, 1, 1, 1, 1, 1, 5, 1, 18}], \qquad \text{Period} = 11.$$

For $D = 71$, the convergent that solved Pell's equation was the one obtained by removing the overline and dropping the last entry. Let's try doing the same for $D = 73$, $D = 89$, and $D = 97$. The results are shown in Table 40.3.

$\sqrt{D}$	$[a, b_1, b_2, \dots, b_{m-1}] = \frac{p}{q}$	$p^2 - Dq^2$
$\sqrt{71}$	$[8, 2, 2, 1, 7, 1, 2, 2] = \frac{3480}{413}$	$3480^2 - 71 \cdot 413^2 = 1$
$\sqrt{73}$	$[8, 1, 1, 5, 5, 1, 1] = \frac{1068}{125}$	$1068^2 - 73 \cdot 125^2 = -1$
$\sqrt{79}$	$[8, 1, 7, 1] = \frac{80}{9}$	$80^2 - 79 \cdot 9^2 = 1$
$\sqrt{97}$	$[9, 1, 5, 1, 1, 1, 1, 1, 1, 5, 1] = \frac{5604}{569}$	$5604^2 - 97 \cdot 569^2 = -1$

Table 40.3: Convergents to $\sqrt{D}$ and Pell's equation

This looks very promising. We did not get solutions to Pell's equation in all cases, but we either found a solution to Pell's equation $p^2 - Dq^2 = 1$ or a solution to the similar equation $p^2 - Dq^2 = -1$. Furthermore, we obtain a plus sign when the period of $\sqrt{D}$ is even and a minus sign when the period of $\sqrt{D}$ is odd. We summarize our observations in the following wonderful theorem.

Theorem 40.3. *Let D be a positive integer that is not a perfect square. Write the continued fraction of $\sqrt{D}$ as*

$$\sqrt{D} = [a, \overline{b_1, b_2, \dots, b_{m-1}, b_m}] \qquad \textit{and let} \qquad \frac{p}{q} = [a, b_1, b_2, \dots, b_{m-1}].$$

[1]Exercise 40.9 describes various special properties of the continued fraction for $\sqrt{D}$, but before you look at that exercise, you should try to discover some for yourself.

Then (p, q) is the smallest solution in positive integers to the equation

$$p^2 - Dq^2 = (-1)^m.$$

We do not give the proof of Theorem 40.3, since it is time to wrap up our discussion of continued fractions and move on to other topics. If you are interested in reading the proof, you will find it in Chapter 4 of Davenport's *The Higher Arithmetic* and in many other number theory textbooks. Instead, we conclude with one final observation and one Brobdingnagian[2] example.

Our observation has to do with the problem of solving $x^2 - Dy^2 = 1$ when Theorem 40.3 happens to give a solution to $x^2 - Dy^2 = -1$. In other words, what can we do when $\sqrt{D} = [a, \overline{b_1, b_2, \ldots, b_{m-1}, b_m}\,]$ and m is odd? The answer is provided by our earlier work. Recall that Pell's Equation Theorem 30.1 says that if (x_1, y_1) is the smallest solution to $x^2 - Dy^2 = 1$ in positive integers, then every other solution (x_k, y_k) can be computed from the smallest solution via the formula

$$x_k + y_k\sqrt{D} = \left(x_1 + y_1\sqrt{D}\right)^k, \qquad k = 1, 2, 3, \ldots. \tag{$*$}$$

The reason that this formula works is because

$$\begin{aligned} x_k^2 - Dy_k^2 &= \left(x_k + y_k\sqrt{D}\right)\left(x_k - y_k\sqrt{D}\right) \\ &= \left(x_1 + y_1\sqrt{D}\right)^k \left(x_1 - y_1\sqrt{D}\right)^k \\ &= (x_1^2 - Dy_1^2)^k \\ &= 1 \qquad \text{since we have assumed that } x_1^2 - Dy_1^2 = 1. \end{aligned}$$

Suppose instead that (x_1, y_1) is a solution to $x^2 - Dy^2 = -1$ and that we compute (x_k, y_k) using formula $(*)$. Then we get

$$x_k^2 - Dy_k^2 = (x_1^2 - Dy_1^2)^k = (-1)^k.$$

So if k is even, then we get a solution to Pell's equation $x^2 - Dy^2 = 1$.

Do you see how this solves our problem? Suppose that m is odd in Theorem 40.3, so (p, q) satisfies $p^2 - Dq^2 = -1$. Then we simply compute the square

$$(p + q\sqrt{D})^2 = (p^2 + q^2D) + 2pq\sqrt{D}$$

to find the desired solution $(p^2 + q^2D, 2pq)$ to $x^2 - Dy^2 = 1$. This finally gives us an efficient way to solve Pell's equation in all cases.

[2] "The Learning of this People [the Brobdingnags] is very defective, consisting only in Morality, History, Poetry, and Mathematicks, wherein they must be allowed to excel." (*Gulliver's Travels*, Chapter II:7, Jonathan Swift)

Theorem 40.4 (Continued Fractions and Pell's Equation Theorem). *Write the continued fraction of $\sqrt{D}$ as*

$$\sqrt{D} = [a, \overline{b_1, b_2, \ldots, b_{m-1}, b_m}] \qquad \textit{and let} \qquad \frac{p}{q} = [a, b_1, b_2, \ldots, b_{m-1}].$$

Then the smallest solution in positive integers to Pell's equation

$$x^2 - Dy^2 = 1 \qquad \textit{is given by} \qquad (x_1, y_1) = \begin{cases} (p, q) & \textit{if } m \textit{ is even,} \\ (p^2 + q^2 D, 2pq) & \textit{if } m \textit{ is odd.} \end{cases}$$

All other solutions are given by the formula

$$x_k + y_k\sqrt{D} = \left(x_1 + y_1\sqrt{D}\right)^k, \qquad k = 1, 2, 3, \ldots.$$

We conclude our exploration of the world of continued fractions by solving the seemingly innocuous Pell equation

$$x^2 - 313y^2 = 1.$$

The continued fraction of $\sqrt{313}$ is

$$\sqrt{313} = [17, \overline{1, 2, 4, 11, 1, 1, 3, 2, 2, 3, 1, 1, 11, 4, 2, 1, 34}].$$

Following the procedure laid out by Theorem 40.4, we discard the last number in the periodic part, which in this case is the number 34, and compute the fraction

$$\frac{126862368}{7170685} = [17, 1, 2, 4, 11, 1, 1, 3, 2, 2, 3, 1, 1, 11, 4, 2, 1].$$

The period m is equal to 17, so the pair $(p, q) = (126862368, 7170685)$ gives a solution to

$$126862368^2 - 313 \cdot 7170685^2 = -1.$$

To find the smallest solution to Pell's equation, Theorem 40.4 tells us to compute

$$p^2 + q^2 D = 126862368^2 + 7170685^2 \cdot 313 = 32188120829134849$$
$$2pq = 2 \cdot 126862368 \cdot 7170685 = 1819380158564160$$

Thus the smallest solution[3] to

$$x^2 - 313y^2 = 1 \quad \text{is} \quad (x, y) = (32188120829134849, 1819380158564160).$$

[3]As noted in Chapter 30, this is the solution found by Brouncker in 1657. Now you know how someone could find such a large solution back in the days before computers!

And if we desire the next smallest solution, we simply square

$$32188120829134849 + 1819380158564160\sqrt{313}$$

and read off the answer

$$(x, y) = (2072150245021969438104715652505601,$$
$$1171248567559874056477817168236680).$$

Exercises

40.1. Find the value of each of the following periodic continued fractions. Express your answer in the form $\frac{r+s\sqrt{D}}{t}$, where r, s, t, D are integers, just as we did in the text when we computed the value of $[1, 2, 3, \overline{4, 5}\,]$ to be $\frac{80-\sqrt{30}}{52}$.

(a) $[\,\overline{1, 2, 3}\,] = [1, 2, 3, 1, 2, 3, 1, 2, 3, 1, 2, 3, \ldots]$
(b) $[1, 1, \overline{2, 3}\,] = [1, 1, 2, 3, 2, 3, 2, 3, 2, 3, 2, 3, 2, 3, 2, \ldots]$
(c) $[1, 1, 1, \overline{3, 2}\,] = [1, 1, 1, 3, 2, 3, 2, 3, 2, 3, 2, 3, 2, \ldots]$
(d) $[3, \overline{2, 1}\,] = [3, 2, 1, 2, 1, 2, 1, 2, 1, 2, 1, 2, 1, 2, \ldots]$
(e) $[\,\overline{1, 3, 5}\,] = [1, 3, 5, 1, 3, 5, 1, 3, 5, 1, 3, 5, 1, 3, 5, \ldots]$
(f) $[1, 2, \overline{1, 3, 4}\,] = [1, 2, 1, 3, 4, 1, 3, 4, 1, 3, 4, 1, 3, 4, \ldots]$

40.2. For each of the following numbers, find their (periodic) continued fraction. What is the period?

(a) $\dfrac{16 - \sqrt{3}}{11}$ (b) $\dfrac{1 + \sqrt{293}}{2}$ (c) $\dfrac{3 + \sqrt{5}}{7}$ (d) $\dfrac{1 + 2\sqrt{5}}{3}$

40.3. During the proof of the Periodic Continued Fraction Theorem 40.2, we simplified the continued fraction $[b_1, b_2, B]$ and found that it equals $\frac{(b_1 b_2+1)B+b_1}{b_2 B+1}$.

(a) Do a similar calculation for $[b_1, b_2, b_3, B]$ and write it as

$$[b_1, b_2, b_3, B] = \frac{uB + v}{wB + z},$$

where u, v, w, z are given by formulas that involve b_1, b_2, and b_3.

(b) Repeat (a) for $[b_1, b_2, b_3, b_4, B]$.

(c) Look at your answers in (a) and (b). Do the expressions for u, v, w, z look familiar? (*Hint*. Compare them to the fractions $[b_1, b_2]$, $[b_1, b_2, b_3]$, and $[b_1, b_2, b_3, b_4]$. These are convergents to $[b_1, b_2, b_3, \ldots]$. Also look at Table 39.2.)

(d) More generally, when the continued fraction $[b_1, b_2, \ldots, b_m, B]$ is simplified as

$$[b_1, b_2, b_3, \ldots, b_m, B] = \frac{u_m B + v_m}{w_m B + z_m},$$

explain how the numbers u_m, v_m, w_m, z_m can be described in terms of the convergents $[b_1, b_2, b_3, \ldots, b_{m-1}]$ and $[b_1, b_2, b_3, \ldots, b_m]$. Prove that your description is correct.

40.4. Proposition 40.1 describes the number with continued fraction expansion $[a, \overline{b}]$.

(a) Do a similar computation to find the number whose continued fraction expansion is $[a, \overline{b, c}]$.

(b) If you let $b = c$ in your formula, do you get the same result as described in Proposition 40.1? (If your answer is "No," then you made a mistake in (a)!)

(c) For which values of a, b, c does the number in (a) have the form $\frac{s\sqrt{D}}{t}$ for integers s, t, D?

(d) For which values of a, b, c is the number in (a) equal to the square root $\sqrt{D}$ of some integer D?

40.5. Theorem 40.3 tells us that if the continued fraction of $\sqrt{D}$ has odd period we can find a solution to $x^2 - Dy^2 = -1$.

(a) Among the numbers $2 \le D \le 20$ with D not a perfect square, which $\sqrt{D}$ have odd period and which have even period. Do you see a pattern?

(b) Same question for $\sqrt{p}$ for primes $2 \le p \le 40$. (See Table 40.1.)

(c) Write down infinitely many positive integers D so that $\sqrt{D}$ has odd period. For each of your D values, give a solution to the equation $x^2 - Dy^2 = -1$. (*Hint.* Look at Proposition 40.1.)

(d) Write down infinitely many positive integers D so that $\sqrt{D}$ has even period. (*Hint.* Use your solution to Exercise 40.4(d).)

40.6. (a) Write a program that takes as input a positive integer D and returns as output a list of numbers $[a, b_1, \ldots, b_m]$ so that the continued fraction expansion of $\sqrt{D}$ is $[a, \overline{b_1, \ldots, b_m}]$. Use your program to print a table of continued fractions of $\sqrt{D}$ for all nonsquare D between 2 and 50.

(b) Generalize (a) by writing a program that takes as input integers r, s, t, D with $t > 0$ and $D > 0$ and returns as output a list of numbers

$$[a_1, \ldots, a_\ell, b_1, \ldots, b_m] \qquad \text{so that} \qquad \frac{r + s\sqrt{D}}{t} = [a_1, \ldots, a_\ell, \overline{b_1, \ldots, b_m}].$$

Use your program to print a table of continued fractions of $(3 + 2\sqrt{D})/5$ for all nonsquare D between 2 and 50.

40.7. (a) Write a program that takes as input a list $[b_1, \ldots, b_m]$ and returns the value of the purely periodic continued fraction $[\overline{b_1, b_2, \ldots, b_m}]$. The output should be in the form (r, s, t, D), where the value of the continued fraction is $(r + s\sqrt{D})/t$.

(b) Use your program from (a) to compute the values of each of the following continued fractions:

$$[\overline{1}], \quad [\overline{1, 2}], \quad [\overline{1, 2, 3}], \quad [\overline{1, 2, 3, 4}], \quad [\overline{1, 2, 3, 4, 5}], \quad [\overline{1, 2, 3, 4, 5, 6}].$$

(c) Extend your program in (a) to handle periodic continued fractions that are not purely periodic. In other words, take as input two lists $[a_1, \ldots, a_\ell]$ and $[b_1, \ldots, b_m]$ and return the value of $[a_1, \ldots, a_\ell, \overline{b_1, b_2, \ldots, b_m}]$.

(d) Use your program from (c) to compute the values of each of the following continued fractions:

$$[6,5,4,3,2,\overline{1}],\ [6,5,4,3,\overline{1,2}],\ [6,5,4,\overline{1,2,3}],\ [6,5,\overline{1,2,3,4}],\ [6,\overline{1,2,3,4,5}].$$

40.8. Write a program to solve Pell's equation $x^2 - Dy^2 = 1$ using the method of continued fractions. If it turns out that there is a solution to $x^2 - Dy^2 = -1$, list a solution to this equation also.

(a) Use your program to solve Pell's equation for all nonsquare values of D between 2 and 20. Check your answers against Table 30.1 (page 220).

(b) Use your program to extend the table by solving Pell's equation for all nonsquare values of D between 76 and 99.

40.9. (hard problem) Let D be a positive integer that is not a perfect square.

(a) Prove that the continued fraction of $\sqrt{D}$ is periodic.

(b) More precisely, prove that the continued fraction of $\sqrt{D}$ looks like

$$\sqrt{D} = [a, \overline{b_1, b_2, \ldots, b_m}].$$

(c) Prove that $b_m = 2a$.

(d) Prove that the list of numbers $b_1, b_2, \ldots, b_{m-1}$ is symmetric; that is, it's the same left to right as it is right to left.

40.10. (hard problem) Let r, s, t, D be integers with $D > 0$ and $t \neq 0$ and let

$$A = \frac{r + s\sqrt{D}}{t}.$$

Prove that the continued fraction of A is periodic. (This is part (b) of the Periodic Continued Fraction Theorem 40.2.)

Chapter 41

Generating Functions

Aptitude tests, intelligence tests, and those ubiquitous grade school math worksheets teem with questions such as:[1]

What is the next number in the sequence
23, 27, 28, 32, 36, 37, 38, 39, 41, 43, 47, 49, 50, 51, 52, 53, 56, 58, 61, 62, 77, 78, ___?

Number theory abounds with interesting sequences. We've seen lots of them in our excursions, including, for example,[2]

Natural Numbers	$0, 1, 2, 3, 4, 5, 6, \ldots$
Square Numbers	$0, 1, 4, 9, 16, 25, 36, \ldots$
Fibonacci Numbers	$0, 1, 1, 2, 3, 5, 8, 13, 21, \ldots$

We have also seen that sequences can be described in various ways, for example by a formula such as

$$s_n = n^2$$

or by a recursion such as

$$F_n = F_{n-1} + F_{n-2}.$$

Both of these methods of describing a sequence are useful, but since a sequence consists of an infinitely long list of numbers, it would be nice to have a way of bundling the entire sequence into a single package. We will build these containers out of power series.

[1]This problem is for baseball fans only. Answer given at the end of the chapter.

[2]For this chapter, it is convenient to start these interesting sequences with 0, rather than with 1.

For example, we can package the sequence 0, 1, 2, 3, ... of natural numbers into the power series

$$0 + 1 \cdot x + 2x^2 + 3x^3 + 4x^4 + 5x^5 + \cdots,$$

and we can package the sequence 0, 1, 1, 2, 3, 5, 8, ... of Fibonacci numbers into the power series

$$0 + 1 \cdot x + 1 \cdot x^2 + 2x^3 + 3x^4 + 5x^5 + 8x^6 + \cdots.$$

In general, any sequence

$$a_0, a_1, a_2, a_3, a_4, a_5, \ldots$$

can be packaged into a power series

$$A(x) = a_0 + a_1 x + a_2 x^2 + a_3 x^3 + a_4 x^4 + a_5 x^5 + \cdots$$

that is called the *generating function* for the sequence $a_0, a_1, a_2, a_3, \ldots$.

What good are generating functions, other than providing a moderately inconvenient way to list the terms of a sequence? The answer lies in that powerful word *function.* A generating function $A(x)$ is a function of the variable x; that is, we can substitute in a value for x and (if we're lucky) get back a value for $A(x)$. We say "if we're lucky" because, as you know if you have studied calculus, a power series need not converge for every value of x.

To illustrate these ideas, we start with the seemingly uninteresting sequence

$$1, 1, 1, 1, 1, 1, 1, 1, 1, \ldots$$

consisting of all ones.[3] Its generating function, which we call $G(x)$, is

$$G(x) = 1 + x + x^2 + x^3 + x^4 + x^5 + \cdots.$$

The ratio test[4] from calculus says that this series converges provided that

$$1 > \rho = \lim_{n\to\infty} \left| \frac{x^{n+1}}{x^n} \right| = |x|.$$

---

[3]If intelligence tests asked for the next term in sequences like this one, we could all have an IQ of 200!

[4]Recall that the ratio test says that a series $b_0 + b_1 + b_2 + \cdots$ converges if the limiting ratio $\rho = \lim_{n\to\infty} |b_{n+1}/b_n|$ satisfies $\rho < 1$.

You've undoubtedly already recognized that $G(x)$ is the *geometric series* and you probably also remember its value, but in case you've forgotten, here is the elegant method used to evaluate the geometric series.

$$\begin{aligned} xG(x) &= x(1 + x + x^2 + x^3 + x^4 + x^5 + \cdots) \\ &= x + x^2 + x^3 + x^4 + x^5 + x^6 + \cdots \\ &= (1 + x + x^2 + x^3 + x^4 + x^5 + x^6 + \cdots) - 1 \\ &= G(x) - 1. \end{aligned}$$

Thus $xG(x) = G(x) - 1$, and we can solve this equation for $G(x)$ to obtain the formula

$$G(x) = \frac{1}{1-x}.$$

This proves the following formula.

Geometric Series Formula

$$1 + x + x^2 + x^3 + x^4 + x^5 + \cdots = \frac{1}{1-x} \quad \text{valid for } |x| < 1$$

The sequence 1, 1, 1, 1,... is rather dull, so let's move on to the sequence of natural numbers 0, 1, 2, 3,... whose generating function is

$$N(x) = x + 2x^2 + 3x^3 + 4x^4 + 5x^5 + 6x^6 + \cdots.$$

The ratio test tells us that $N(x)$ converges provided that

$$1 > \rho = \lim_{n\to\infty} \left| \frac{(n+1)x^{n+1}}{nx^n} \right| = |x|.$$

We would like to find a simple formula for $N(x)$, similar to the formula we found for $G(x)$. The way we do this is to start with the Geometric Series Formula

$$1 + x + x^2 + x^3 + x^4 + x^5 + \cdots = \frac{1}{1-x}$$

and use a little bit of calculus. If we differentiate both sides of this formula, we get

$$\underbrace{0 + 1 + 2x + 3x^2 + 4x^3 + 5x^4 + \cdots}_{\text{When multiplied by } x \text{, this becomes } N(x).} = \frac{d}{dx}\left(\frac{1}{1-x}\right) = \frac{1}{(1-x)^2}.$$

Multiplying both sides of this equation by x gives us the formula

$$N(x) = x + 2x^2 + 3x^3 + 4x^4 + 5x^5 + 6x^6 + \cdots = \frac{x}{(1-x)^2}.$$

If we differentiate again and multiply both sides by x, we get a formula for the generating function

$$S(x) = x + 4x^2 + 9x^3 + 16x^4 + 25x^5 + 36x^6 + \cdots$$

for the sequence of squares 0, 1, 4, 9, 16, 25, ... Thus

$$\begin{aligned} x\frac{dN(x)}{dx} &= x\frac{d}{dx}(x + 2x^2 + 3x^3 + 4x^4 + \cdots) &&= x\frac{d}{dx}\left(\frac{x}{(1-x)^2}\right) \\ S(x) &= x + 4x^2 + 9x^3 + 16x^4 + 25x^5 + \cdots &&= \frac{x+x^2}{(1-x)^3}. \end{aligned}$$

We now turn to the Fibonacci sequence 0, 1, 1, 2, 3, 5, 8, ... and its generating function

$$\begin{aligned} F(x) &= F_1x + F_2x^2 + F_3x^3 + F_4x^4 + F_5x^5 + F_6x^6 + \cdots \\ &= x + x^2 + 2x^3 + 3x^4 + 5x^5 + 8x^6 + 13x^7 + \cdots. \end{aligned}$$

How can we find a simple expression for $F(x)$? The differentiation trick we used earlier doesn't seem to help, so instead we make use of the recursive formula

$$F_n = F_{n-1} + F_{n-2}.$$

Thus we can replace F_3 with $F_2 + F_1$, and we can replace F_4 with $F_3 + F_2$, and so on, which means we can write $F(x)$ as

$$\begin{aligned} F(x) &= F_1x + F_2x^2 + F_3x^3 + F_4x^4 + F_5x^5 + \cdots \\ &= F_1x + F_2x^2 + (F_2 + F_1)x^3 + (F_3 + F_2)x^4 + (F_4 + F_3)x^5 + \cdots. \end{aligned}$$

Ignoring the first two terms for the moment, we regroup the other terms in the following manner:

Group these terms together

$$(F_2 + F_1)x^3 + (F_3 + F_2)x^4 + (F_4 + F_3)x^5 + (F_5 + F_4)x^6 + \cdots.$$

Group these terms together

This gives the formula

$$\begin{aligned} F(x) = F_1x + F_2x^2 &+ \{F_1x^3 + F_2x^4 + F_3x^5 + F_4x^6 + \cdots\} \\ &+ \{F_2x^3 + F_3x^4 + F_4x^5 + F_5x^6 + \cdots\}. \end{aligned}$$

Now observe that the series between the first set of braces is almost equal to the generating function $F(x)$ with which we started; more precisely, it is equal to $x^2F(x)$. Similarly, the series between the second set of braces is equal to $xF(x)$ except that it is missing the initial F_1x^2 term. In other words,

$$\begin{aligned} F(x) &= F_1x + F_2x^2 + \{F_1x^3 + F_2x^4 + \cdots\} + \{F_2x^3 + F_3x^4 + \cdots\} \\ &= F_1x + F_2x^2 + x^2\{\underbrace{F_1x + F_2x^2 + \cdots}_{\text{equals } F(x)}\} + x\{\underbrace{F_2x^2 + F_3x^3 + \cdots}_{\text{equals } F(x) - F_1x}\}. \end{aligned}$$

If we use the values $F_1 = 1$ and $F_2 = 1$, this gives us the formula

$$\begin{aligned} F(x) &= x + x^2 + x^2F(x) + x\big(F(x) - x\big) \\ &= x + x^2F(x) + xF(x). \end{aligned}$$

This gives us an equation that we can solve for $F(x)$ to obtain the following beautiful formula.

Fibonacci Generating Function Formula

$$x + x^2 + 2x^3 + 3x^4 + 5x^5 + 8x^6 + \cdots = \frac{x}{1 - x - x^2}$$

We can use the formula for the Fibonacci generating function together with the method of partial fractions that you learned in calculus to rederive Binet's formula for the nth Fibonacci number (Theorem 37.1). The first step is to use the quadratic formula to find the roots of the polynomial $1 - x - x^2$. The roots are $\frac{-1\pm\sqrt{5}}{2}$, which are the reciprocals of the two numbers[5]

$$\alpha = \frac{1+\sqrt{5}}{2} \qquad \text{and} \qquad \beta = \frac{1-\sqrt{5}}{2}.$$

This lets us factor the polynomial as

$$1 - x - x^2 = (1 - \alpha x)(1 - \beta x).$$

The idea of partial fractions is to take the function $\dfrac{x}{1 - x - x^2}$ and split it up into the sum of two pieces

$$\frac{x}{1 - x - x^2} = \frac{A}{1 - \alpha x} + \frac{B}{1 - \beta x},$$

[5]Notice how the Golden Ratio (page 303) has suddenly appeared! This should not be surprising, since we saw in Chapter 37 that the Fibonacci sequence and the Golden Ratio are intimately related to one another.

where we need to find the correct values for A and B. To do this, we clear denominators by multiplying both sides by $1 - x - x^2$ to get

$$x = A(1 - \beta x) + B(1 - \alpha x).$$

[Remember that $1 - x - x^2 = (1 - \alpha x)(1 - \beta x)$.] Rearranging this relation yields

$$x = (A + B) - (A\beta + B\alpha)x.$$

We're looking for values of A and B that make the polynomial x on the left equal to the polynomial on the right, so we must choose A and B to satisfy

$$\begin{aligned} 0 &= A + B \\ 1 &= -A\beta - B\alpha. \end{aligned}$$

It is easy to solve these two equations for the unknown quantities A and B (remember that α and β are particular numbers). We find that

$$A = \frac{1}{\alpha - \beta} \qquad \text{and} \qquad B = \frac{1}{\beta - \alpha},$$

and using the values $\alpha = (1 + \sqrt{5})/2$ and $\beta = (1 - \sqrt{5})/2$ gives

$$A = \frac{1}{\sqrt{5}} \qquad \text{and} \qquad B = -\frac{1}{\sqrt{5}}.$$

To recapitulate, we have found the partial fraction decomposition

$$\frac{x}{1 - x - x^2} = \frac{1}{\sqrt{5}}\left(\frac{1}{1 - \alpha x}\right) - \frac{1}{\sqrt{5}}\left(\frac{1}{1 - \beta x}\right).$$

This may not seem like progress, but it is, because we have replaced the complicated function $x/(1 - x - x^2)$ with the sum of two simpler expressions. If this were a calculus textbook, I would now ask you to compute the indefinite integral $\int \frac{x}{1-x-x^2}\,dx$ and you would use the partial fraction formula to compute

$$\begin{aligned} \int \frac{x}{1 - x - x^2}\,dx &= \frac{1}{\sqrt{5}}\int \frac{dx}{1 - \alpha x} - \frac{1}{\sqrt{5}}\int \frac{dx}{1 - \beta x} \\ &= \frac{-1}{\sqrt{5}\,\alpha}\log|1 - \alpha x| + \frac{1}{\sqrt{5}\,\beta}\log|1 - \beta x| + C. \end{aligned}$$

However, our subject is not calculus, it is number theory, so we instead observe that the two pieces of the partial fraction decomposition can be expanded using the geometric series as

$$\frac{1}{1-\alpha x} = 1 + \alpha x + (\alpha x)^2 + (\alpha x)^3 + (\alpha x)^4 + \cdots,$$
$$\frac{1}{1-\beta x} = 1 + \beta x + (\beta x)^2 + (\beta x)^3 + (\beta x)^4 + \cdots.$$

This lets us write the function $x/(1-x-x^2)$ as a power series

$$\begin{aligned}\frac{x}{1-x-x^2} &= \frac{1}{\sqrt{5}}\left(\frac{1}{1-\alpha x}\right) - \frac{1}{\sqrt{5}}\left(\frac{1}{1-\beta x}\right) \\ &= \frac{\alpha-\beta}{\sqrt{5}}x + \frac{\alpha^2-\beta^2}{\sqrt{5}}x^2 + \frac{\alpha^3-\beta^3}{\sqrt{5}}x^3 + \cdots.\end{aligned}$$

But we know that $x/(1-x-x^2)$ is the generating function for the Fibonacci sequence,

$$\frac{x}{1-x-x^2} = F_1x + F_2x^2 + F_3x^3 + F_4x^4 + F_5x^5 + F_6x^6 + \cdots,$$

so matching the two series for $x/(1-x-x^2)$, we find that

$$F_1 = \frac{\alpha-\beta}{\sqrt{5}}, \quad F_2 = \frac{\alpha^2-\beta^2}{\sqrt{5}}, \quad F_3 = \frac{\alpha^3-\beta^3}{\sqrt{5}}, \ldots.$$

Substituting the values of α and β, we again obtain Binet's formula (Theorem 37.1) for the nth Fibonacci number.

Binet's Formula

$$F_n = \frac{1}{\sqrt{5}}\left\{\left(\frac{1+\sqrt{5}}{2}\right)^n - \left(\frac{1-\sqrt{5}}{2}\right)^n\right\}$$

The two numbers appearing in the Binet's Formula are approximately equal to

$$\alpha = \frac{1+\sqrt{5}}{2} \approx 1.618034 \quad \text{and} \quad \beta = \frac{1-\sqrt{5}}{2} \approx -0.618034.$$

Notice that $|\beta| < 1$, so if we raise β to a large power, it becomes very small. In particular,

$$\begin{aligned}F_n &= \text{Closest integer to } \frac{1}{\sqrt{5}}\left(\frac{1+\sqrt{5}}{2}\right)^n \\ &\approx (0.447213\ldots) \times (1.61803\ldots)^n.\end{aligned}$$

For example,

$$F_{10} \approx 55.003636\ldots \quad \text{and} \quad F_{25} \approx 75024.999997334\ldots,$$

which are indeed extremely close to the correct values $F_{10} = 55$ and $F_{25} = 75025$.

Exercises

41.1. (a) Find a simple formula for the generating function $E(x)$ for the sequence of even numbers 0, 2, 4, 6, 8, ...

(b) Find a simple formula for the generating function $O(x)$ for the sequence of odd numbers 1, 3, 5, 7, 9, ...

(c) What does $E(x^2) + xO(x^2)$ equal? Why?

41.2. Find a simple formula for the generating function of the sequence of numbers

$$a, \quad a+m, \quad a+2m, \quad a+3m, \quad a+4m, \ldots.$$

(If $0 \le a < m$, then this is the sequence of nonnegative numbers that are congruent to a modulo m.)

41.3. (a) Find a simple formula for the generating function of the sequence whose n^{th} term is n^3, that is, the sequence 0, 1, 8, 27, 64, ...

(b) Repeat (a) for the generating function of the sequence 0, 1, 16, 81, 256, ... (This is the sequence whose n^{th} term is n^4.)

(c) If you have access to a computer that does symbolic differentiation or if you enjoy length calculations with paper and pencil, find the generating function for the sequence whose n^{th} term is n^5.

(d) Repeat (c) for the sequence whose n^{th} term is n^6.

41.4. Let $G(x) = 1 + x + x^2 + x^3 + \cdots$ be the generating function of the sequence $1, 1, 1, \ldots$.

(a) Compute the first five coefficients of the power series $G(x)^2$.

(b) Prove that the power series $G(x)^2 - G(x)$ is equal to some other power series that we studied in this chapter.

41.5. Let $T(x) = x + 3x^2 + 6x^3 + 10x^4 + \cdots$ be the generating function for the sequence $0, 1, 3, 6, 10, \ldots$ of triangular numbers. Find a simple expression for $T(x)$.

41.6. This question investigates the generating functions of certain sequences whose terms are binomial coefficients (see Chapter 36).

(a) Find a simple expression for the generating function of the sequence whose n^{th} term is $\binom{n}{1}$.

(b) Same question for the sequence whose n^{th} term is $\binom{n}{2}$.

(c) Same question for the sequence whose n^{th} term is $\binom{n}{3}$.

(d) For a fixed number k, make a conjecture giving a simple expression for the generating function of the sequence whose n^{th} term is $\binom{n}{k}$.

(e) Prove that your conjecture in (d) is correct.

41.7. Let $k \ge 0$ be an integer and let $D_k(x)$ be the generating function of the sequence $0^k, 1^k, 2^k, 3^k, 4^k, \ldots$. In this chapter we computed

$$D_0(x) = \frac{1}{1-x}, \qquad D_1(x) = \frac{x}{(1-x)^2}, \qquad D_2(x) = \frac{x+x^2}{(1-x)^3},$$

and in Exercise 41.3 you computed further examples. These computations suggest that $D_k(x)$ looks like

$$D_k(x) = \frac{P_k(x)}{(1-x)^{k+1}}$$

for some polynomial $P_k(x)$.

(a) Prove that there is a polynomial $P_k(x)$ so that $D_k(x)$ has the form $P_k(x)/(1-x)^{k+1}$. (*Hint*. Use induction on k.)

(b) Make a list of values of $P_k(0)$ for $k = 0, 1, 2, \ldots$ and make a conjecture. Prove that your conjecture is correct.

(c) Same as (b) for the values of $P_k(1)$.

(d) Repeat (b) and (c) for the values of the derivative $P_k'(0)$ and $P_k'(1)$.

(e) What other patterns can you find in the $P_k(x)$ polynomials?

41.8. Let ϕ be Euler's phi function (see Chapter 11), and let p be a prime number. Find a simple formula for the generating function of the sequence $\phi(1), \phi(p), \phi(p^2), \phi(p^3), \ldots$.

41.9. The *Lucas sequence* is the sequence of numbers L_n given by the rules $L_1 = 1$, $L_2 = 3$, and $L_n = L_{n-1} + L_{n-2}$.

(a) Write down the first 10 terms of the Lucas sequence.

(b) Find a simple formula for the generating function of the Lucas sequence.

(c) Use the partial fraction method to find a simple formula for L_n, similar to Binet's Formula for the Fibonacci number F_n.

41.10. Write down the first few terms in each of the following recursively defined sequences, and then find a simple formula for the generating function.

(a) $a_1 = 1$, $a_2 = 2$, and $a_n = 5a_{n-1} - 6a_{n-2}$ for $n = 3, 4, 5, \ldots$

(b) $b_1 = 1$, $b_2 = 3$, and $b_n = 2b_{n-1} - 2b_{n-2}$ for $n = 3, 4, 5, \ldots$

(c) $c_1 = 1$, $c_2 = 1$, $c_3 = 1$, and $c_n = 4c_{n-1} + 11c_{n-2} - 30c_{n-3}$ for $n = 4, 5, 6, \ldots$

41.11. Use generating functions and the partial fraction method to find a simple formula for the nth term of each of the following sequences similar to the formula we found in the text for the nth term of the Fibonacci sequence. (Note that these are the same sequences as in the previous exercise.) Be sure to check your answer for the first few values of n.

(a) $a_1 = 1$, $a_2 = 2$, and $a_n = 5a_{n-1} - 6a_{n-2}$ for $n = 3, 4, 5, \ldots$

(b) $b_1 = 1$, $b_2 = 3$, and $b_n = 2b_{n-1} - 2b_{n-2}$ for $n = 3, 4, 5, \ldots$ (*Hint*. You may need to use complex numbers!)

(c) $c_1 = 1$, $c_2 = 1$, $c_3 = 1$, and $c_n = 4c_{n-1} + 11c_{n-2} - 30c_{n-3}$ for $n = 4, 5, 6, \ldots$

41.12. (a) Fix an integer $k \geq 0$, and let $H(x)$ be the generating function of the sequence whose n^{th} term is $h_n = n^k$. Use the ratio test to find the interval of convergence of the generating function $H(x)$.

(b) Use the ratio test to find the interval of convergence of the generating function $F(x)$ of the Fibonacci sequence $0, 1, 1, 2, 3, 5, \ldots$.

41.13. Sequences $a_0, a_1, a_2, a_3, \ldots$ are also sometimes packaged in an *exponential generating function*

$$a_0 + a_1\frac{x}{1!} + a_2\frac{x^2}{2!} + a_3\frac{x^3}{3!} + a_4\frac{x^4}{4!} + a_5\frac{x^5}{5!} + \cdots.$$

(a) What is the exponential generating function for the sequence 1, 1, 1, 1, ...? (*Hint.* Your answer explains why the word *exponential* is used in the name of this type of generating function.)

(b) What is the exponential generating function for the sequence 0, 1, 2, 3, ... of natural numbers?

41.14. Let $f(x)$ be the exponential generating function of the Fibonacci sequence

$$f(x) = F_0 + F_1\frac{x}{1!} + F_2\frac{x^2}{2!} + F_3\frac{x^3}{3!} + F_4\frac{x^4}{4!} + F_5\frac{x^5}{5!} + \cdots.$$

(a) Find a simple relation satisfied by $f(x)$ and its derivatives $f'(x)$ and $f''(x)$.

(b) Find a simple formula for $f(x)$.

41.15. Fix an integer N and create a sequence of numbers $a_0, a_1, a_2, \ldots$ in the following way:

$$\begin{aligned}
a_0 &= 1^0 + 2^0 + 3^0 + \cdot + N^0 \\
a_1 &= 1^1 + 2^1 + 3^1 + \cdot + N^1 \\
a_2 &= 1^2 + 2^2 + 3^2 + \cdot + N^2 \\
a_3 &= 1^3 + 2^3 + 3^3 + \cdot + N^3 \\
&\vdots \qquad\qquad \vdots
\end{aligned}$$

Compute the exponential generating function of this sequence. (We will study these power sums further in Chapter 42.)

Solution to Sequence on page 355. The next four terms in the sequence

$$23, 27, 28, 32, 36, 37, 38, 39, 41, 43, 47, 49, 50, 51, 52, 53, 56, 58, 61, 62, 77, 78$$

given at the beginning of this chapter are 96, 98, 99, and 00, as is obvious to those who know that the New York Yankees won the World Series in the years 1923, 1927, 1928, ..., 1977, 1978, 1996, 1998, 1999, and 2000. Those who are not Yankee fans might prefer to complete the shorter sequence 03, 12, 15, 16, 18, ___. (*Hint.* There is a gap of 86 years before the final entry.)

Chapter 42

Sums of Powers

The nth triangular number

$$T_n = 1 + 2 + 3 + 4 + \cdots + n$$

is the sum of the first n integers. In Chapter 1 we used geometry to find a simple formula for T_n,

$$T_n = \frac{n(n+1)}{2}.$$

This formula was extremely useful in Chapter 29, where we described all numbers that are simultaneously triangular and square.

The reason that the formula for T_n is so helpful is because it expresses a sum of n numbers as a simple polynomial in the variable n. To say this another way, let $F(X)$ be the polynomial

$$F(X) = \frac{1}{2}X^2 + \frac{1}{2}X.$$

Then the sum

$$1 + 2 + 3 + 4 + \cdots + n,$$

which at first glance requires us to add n numbers, can be computed very simply as the value $F(n)$.

Now suppose that instead of adding the first n integers, we instead add the first n squares,

$$R_n = 1 + 4 + 9 + 16 + \cdots + n^2.$$

We make a short table of the first few values and look for patterns.

n	1	2	3	4	5	6	7	8	9	10
R_n	1	5	14	30	55	91	140	204	285	385

The numbers $R_1, R_2, R_3, \ldots$ are increasing fairly rapidly, but they don't seem to obey any simple pattern. It isn't easy to see how we might get our hands on these numbers. We use a tool called the method of *telescoping sums*. To illustrate this technique, we first look at the following easier problem. Suppose we want to compute the value of the sum

$$S_n = \frac{1}{1 \cdot 2} + \frac{1}{2 \cdot 3} + \frac{1}{3 \cdot 4} + \frac{1}{4 \cdot 5} + \cdots + \frac{1}{(n-1) \cdot n}.$$

For this sum, if we compute the first few values, it's easy to see the pattern:

n	2	3	4	5	6	7	8	9	10
S_n	$\frac{1}{2}$	$\frac{2}{3}$	$\frac{3}{4}$	$\frac{4}{5}$	$\frac{5}{6}$	$\frac{6}{7}$	$\frac{7}{8}$	$\frac{8}{9}$	$\frac{9}{10}$

So we guess that S_n is probably equal to $\frac{n-1}{n}$, but how can we prove that this is true? The key is to observe that the first few terms of the sum can be written as

$$\frac{1}{1 \cdot 2} = 1 - \frac{1}{2} \qquad \text{and} \qquad \frac{1}{2 \cdot 3} = \frac{1}{2} - \frac{1}{3} \qquad \text{and} \qquad \frac{1}{3 \cdot 4} = \frac{1}{3} - \frac{1}{4}$$

and so on. More generally, the ith term of the sum is equal to

$$\frac{1}{i \cdot (i+1)} = \frac{1}{i} - \frac{1}{i+1}.$$

Hence the sum S_n is equal to

$$\begin{aligned} S_n &= \frac{1}{1 \cdot 2} + \frac{1}{2 \cdot 3} + \frac{1}{3 \cdot 4} + \cdots + \frac{1}{(n-1) \cdot n} \\ &= \left(1 - \frac{1}{2}\right) + \left(\frac{1}{2} - \frac{1}{3}\right) + \left(\frac{1}{3} - \frac{1}{4}\right) + \cdots + \left(\frac{1}{n-1} - \frac{1}{n}\right). \end{aligned}$$

Now look what happens when we add the terms on this last line. We start with 1. Next we get $+\frac{1}{2}$ followed by $-\frac{1}{2}$, so these two terms cancel. Then we get $+\frac{1}{3}$, which is followed by $-\frac{1}{3}$, so these two terms also cancel. Notice how the sum "telescopes" (imagine how the tubes of a telescope fold into one another), with only the first term and the last term remaining at the end. This proves the formula

$$S_n = 1 - \frac{1}{n} = \frac{n-1}{n}.$$

Now we return to the problem of computing the sum of squares

$$R_n = 1^2 + 2^2 + 3^2 + \cdots + n^2.$$

For reasons that will become apparent in a moment, we look at the following telescoping sum involving cubes:

$$(n+1)^3 = 1^3 + (2^3 - 1^3) + (3^3 - 2^3) + \cdots + ((n+1)^3 - n^3).$$

Using summation notation, we can write this as

$$(n+1)^3 = 1 + \sum_{i=1}^{n} ((i+1)^3 - i^3).$$

Next we expand the expression $(i+1)^3$ using the Binomial Formula (see Chapter 36)

$$(i+1)^3 = i^3 + 3i^2 + 3i + 1.$$

Substituting this into the telescoping sum gives (notice that the i^3 terms cancel)

$$(n+1)^3 = 1 + \sum_{i=1}^{n} (3i^2 + 3i + 1).$$

Now we split the sum into three pieces and add each piece individually,

$$\begin{aligned}(n+1)^3 &= 1 + 3\sum_{i=1}^{n} i^2 + 3\sum_{i=1}^{n} i + \sum_{i=1}^{n} 1 \\ &= 1 + 3R_n + 3T_n + n.\end{aligned}$$

But we already know that $T_n = \sum_{i=1}^{n} i$ is equal to $(n^2+n)/2$, so we can solve for R_n,

$$\begin{aligned}R_n &= \frac{(n+1)^3 - n - 1}{3} - T_n \\ &= \frac{n^3 + 3n^2 + 2n}{3} - \frac{n^2+n}{2} \\ &= \frac{2n^3 + 3n^2 + n}{6}.\end{aligned}$$

That was a lot of algebra, but we are amply rewarded for our efforts by the beautiful formula

$$1^2 + 2^2 + 3^2 + \cdots + n^2 = \frac{2n^3 + 3n^2 + n}{6}.$$

Notice how nifty this formula is. If we want to compute the value of

$$1^2 + 2^2 + 3^2 + 4^2 + \cdots + 9999^2 + 10000^2,$$

we could add up 10000 terms, but using the formula for R_n, we only need to compute

$$R_{10000} = \frac{2 \cdot 10000^3 + 3 \cdot 10000^2 + 10000}{6} = 333{,}383{,}335{,}000.$$

Now take a deep breath, because we next tackle the problem of adding sums of kth powers for higher values of k. We write

$$F_k(n) = 1^k + 2^k + 3^k + \cdots + n^k$$

for the sum of the first n numbers, each raised to the kth power.

The telescoping sum method that worked so well computing sums of squares works just as well for higher powers. We begin with the telescoping sum

$$(n+1)^k = 1^k + (2^k - 1^k) + (3^k - 2^k) + \cdots + \left((n+1)^k - n^k\right),$$

which, using summation notation, becomes

$$(n+1)^k = 1 + \sum_{i=1}^{n} \left((i+1)^k - i^k\right).$$

Just as before, we expand $(i+1)^k$ using the Binomial Formula (Chapter 36),

$$(i+1)^k = \sum_{j=0}^{k} \binom{k}{j} i^j.$$

The last term (i.e., the $j = k$ term) is i^k, so it cancels the i^k in the telescoping sum, leaving

$$(n+1)^k = 1 + \sum_{i=1}^{n} \sum_{j=0}^{k-1} \binom{k}{j} i^j.$$

Now switch the order of the two sums, and lo and behold, we find the power sums $F_0(n), F_1(n), \ldots, F_{k-1}(n)$ appearing,

$$(n+1)^k = 1 + \sum_{j=0}^{k-1} \binom{k}{j} \sum_{i=1}^{n} i^j = 1 + \sum_{j=0}^{k-1} \binom{k}{j} F_j(n).$$

What good is a formula like this, which seems to involve all sorts of quantities that we don't know? The answer is that it relates each of $F_0(n), F_1(n), F_2(n), \ldots$

with the previous ones. To make this clearer, we pull off the last term in the sum, that is, the term with $j = k - 1$, and we move all the other terms to the other side,

$$\binom{k}{k-1} F_{k-1}(n) = (n+1)^k - 1 - \sum_{j=0}^{k-2} \binom{k}{j} F_j(n).$$

Now $\binom{k}{k-1} = k$, so dividing by k gives the recursive formula

$$F_{k-1}(n) = \frac{(n+1)^k - 1}{k} - \frac{1}{k} \sum_{j=0}^{k-2} \binom{k}{j} F_j(n).$$

We call this a recursive formula for the F_k's, because it expresses each F_k in terms of the previous ones. It is thus similar in some ways to the recursive formula used to describe the Fibonacci sequence (Chapter 37), although this formula is obviously much more complicated than the Fibonacci formula.

Let's use the recursive formula to find a new power-sum formula. Taking $k = 4$ in the recursive formula gives

$$F_3(n) = \frac{(n+1)^4 - 1}{4} - \frac{1}{4} \left\{ \binom{4}{0} F_0(n) + \binom{4}{1} F_1(n) + \binom{4}{2} F_2(n) \right\}.$$

We already know from our earlier work that

$$F_1(n) = T_n = \frac{n^2 + n}{2} \qquad \text{and} \qquad F_2(n) = R_n = \frac{2n^3 + 3n^2 + n}{6},$$

while the value of $F_0(n)$ is clearly equal to

$$F_0(n) = 1^0 + 2^0 + \cdots + n^0 = \underbrace{1 + 1 + \cdots + 1}_{n \text{ terms}} = n.$$

Substituting in these values for $F_0(n)$, $F_1(n)$, and $F_2(n)$ yields

$$\begin{aligned} F_3(n) &= \frac{(n+1)^4 - 1}{4} - \frac{1}{4} \left\{ \binom{4}{0} F_0(n) + \binom{4}{1} F_1(n) + \binom{4}{2} F_2(n) \right\} \\ &= \frac{n^4 + 4n^3 + 6n^2 + 4n + 1 - 1}{4} \\ &\qquad - \frac{1}{4} \left\{ n + 4\frac{n^2 + n}{2} + 6\frac{2n^3 + 3n^2 + n}{6} \right\} \\ &= \frac{n^4 + 2n^3 + n^2}{4}. \end{aligned}$$

Thus

$$F_3(n) = 1^3 + 2^3 + 3^3 + \cdots + n^3 = \frac{n^4 + 2n^3 + n^2}{4}.$$

For example,

$$\begin{aligned} 1^3 + 2^3 + 3^3 + 4^3 + \cdots + 10000^3 &= \frac{10000^4 + 2 \cdot 10000^3 + 10000^2}{4} \\ &= 2{,}500{,}500{,}025{,}000{,}000. \end{aligned}$$

The recursive formula for power sums is very beautiful, so we record our discovery in the form of a theorem.

Theorem 42.1 (Sum of Powers Theorem). *Let $k \geq 0$ be an integer. There is a polynomial $F_k(X)$ of degree $k+1$ so that*

$$F_k(n) = 1^k + 2^k + 3^k + \cdots + n^k \quad \textit{for every value of } n = 1, 2, 3, \ldots.$$

These polynomials can be computed using the recurrence formula

$$F_{k-1}(X) = \frac{(X+1)^k - 1}{k} - \frac{1}{k} \sum_{i=0}^{k-2} \binom{k}{i} F_i(X).$$

Proof. We proved above that the power sums can be computed by the recurrence formula. It is also clear from the recurrence formula that the power sums are polynomials, since each successive power sum is simply the polynomial $\frac{(X+1)^k - 1}{k}$ adding to some multiples of the previous power sums.

All that remains is to prove that $F_k(X)$ has degree $k+1$. We use induction. To start the induction, we observe that $F_0(X) = X$ has the correct degree. Now suppose that we know that $F_k(X)$ has degree $k+1$ for $k = 0, 1, 2, \ldots, m-1$. In other words, suppose we've finished the proof for all values of k less than m. We use the recurrence formula with $k = m+1$ to compute

$$F_m(X) = \frac{(X+1)^{m+1} - 1}{m+1} - \frac{1}{m+1} \sum_{i=0}^{m-1} \binom{m+1}{i} F_i(X).$$

The first part looks like

$$\frac{(X+1)^{m+1} - 1}{m+1} = \frac{1}{m+1} X^{m+1} + \cdots.$$

On the other hand, by the induction hypothesis we know that $F_i(X)$ has degree $i+1$ for each $i = 0, 1, \ldots, m-1$, so the polynomials in the sum have degree at most m. This proves that the X^{m+1} coming from the first part isn't canceled by any of the other terms, so $F_m(X)$ has degree $m+1$. This completes our induction proof. □

We've computed the power-sum polynomials

$$F_1(X) = \frac{1}{2}(X^2 + X),$$
$$F_2(X) = \frac{1}{6}(2X^3 + 3X^2 + X),$$
$$F_3(X) = \frac{1}{4}(X^4 + 2X^3 + X^2).$$

Now it's your turn to compute the next few power-sum polynomials, so turn to Exercise 42.2 and use the recursive formula to compute $F_4(X)$ and $F_5(X)$. Be sure to check your answers.

Lest you feel that I've done the easy computations and you've been stuck with the hard ones, here are the next few power-sum polynomials.

$$F_6(X) = \frac{1}{42}(6X^7 + 21X^6 + 21X^5 - 7X^3 + X)$$
$$F_7(X) = \frac{1}{24}(3X^8 + 12X^7 + 14X^6 - 7X^4 + 2X^2)$$
$$F_8(X) = \frac{1}{90}(10X^9 + 45X^8 + 60X^7 - 42X^5 + 20X^3 - 3X)$$
$$F_9(X) = \frac{1}{20}(2X^{10} + 10X^9 + 15X^8 - 14X^6 + 10X^4 - 3X^2)$$
$$F_{10}(X) = \frac{1}{66}(6X^{11} + 33X^{10} + 55X^9 - 66X^7 + 66X^5 - 33X^3 + 5X)$$

You can use this list to look for patterns and to test conjectures.

Three-Dimensional Number Shapes

In our work on number theory and geometry, we have studied various sorts of number shapes, such as triangular numbers and square numbers (Chapters 1 and 29) and even pentagonal numbers (Exercise 29.4). Triangles, squares, and pentagons are plane figures; that is, they lie on a flat surface. We, on the other hand, live in three-dimensional space, so it's about time we looked at three-dimensional number shapes. We'll build pyramids with triangular bases, as illustrated in Figure 42.1.

The fancy mathematical term for this sort of solid shape is a *tetrahedron*. We define the nth *Tetrahedral Number* to be the number of dots in a tetrahedron with n layers, and we let

$$\mathbb{T}_n = \text{ the } n^{\text{th}} \text{ Tetrahedral Number.}$$

Looking at Figure 42.1, we see that

$$\mathbb{T}_1 = 1, \qquad \mathbb{T}_2 = 4, \qquad \mathbb{T}_3 = 10, \qquad \text{and} \qquad \mathbb{T}_4 = 20.$$

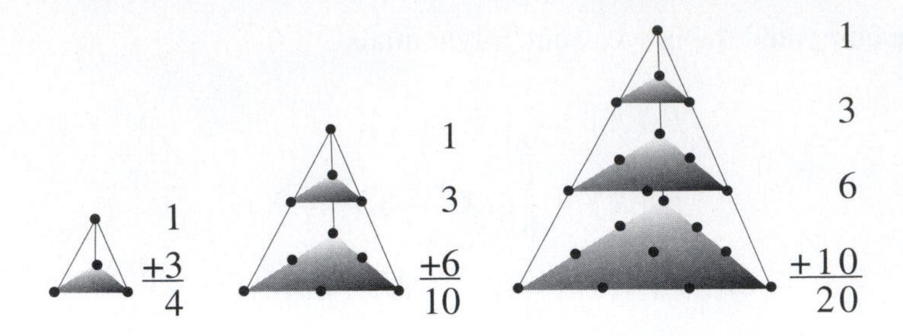

Figure 42.1: The Tetrahedral Numbers $\mathbb{T}_2 = 4$, $\mathbb{T}_3 = 10$, and $\mathbb{T}_4 = 20$

The pictures illustrate how tetrahedral numbers are formed,

$$\begin{aligned} \mathbb{T}_1 &= 1, \\ \mathbb{T}_2 &= 4 \ \ = 1 + 3, \\ \mathbb{T}_3 &= 10 = 1 + 3 + 6, \\ \mathbb{T}_4 &= 20 = 1 + 3 + 6 + 10. \end{aligned}$$

To form the fifth tetrahedral number, we need to add another triangle onto the bottom of the previous tetrahedron. In other words, we need to add the next triangular number to the previous tetrahedral number. If this isn't clear, notice how $\mathbb{T}_4$ is formed by adding the first four triangular numbers: 1, 3, 6, and 10. So to get $\mathbb{T}_5$, we take $\mathbb{T}_4$ and add on the fifth triangular number $T_5 = 15$ to get

$$\mathbb{T}_5 = \mathbb{T}_4 + T_5 = (1 + 3 + 6 + 10) + 15 = 35.$$

In general, the nth tetrahedral number is equal to the sum of the first n triangular numbers,

$$\mathbb{T}_n = T_1 + T_2 + T_3 + \cdots + T_n.$$

We know that the nth triangular number T_n is given by

$$T_n = \frac{n^2 + n}{2},$$

so we can find a formula for $\mathbb{T}_n$ by adding

$$\mathbb{T}_n = \sum_{j=1}^{n} T_j = \sum_{j=1}^{n} \frac{j^2 + j}{2} = \frac{1}{2}\sum_{j=1}^{n} j^2 + \frac{1}{2}\sum_{j=1}^{n} j.$$

To finish the calculation, we use our power-sum formulas, in particular our formula for the sum of the first n squares, to compute

$$\mathbb{T}_n = \frac{1}{2}\left(\frac{2n^3+3n^2+n}{6}\right) + \frac{1}{2}\left(\frac{n^2+n}{2}\right) = \frac{n^3+3n^2+2n}{6}.$$

It is interesting to observe that the tetrahedral polynomial factors as

$$\mathbb{T}_n = \frac{n(n+1)(n+2)}{6}.$$

Thus the nth triangular number and the nth tetrahedral number can be expressed using binomial coefficients

$$T_n = \binom{n+1}{2} \quad \text{and} \quad \mathbb{T}_n = \binom{n+2}{3}.$$

In other words, the nth two-dimensional pyramid (triangle) has $\binom{n+1}{2}$ dots, and the nth three-dimensional pyramid (tetrahedron) has $\binom{n+2}{3}$ dots. How many dots do you think it takes to fill up a four-dimensional pyramid?

Exercises

42.1. In the text we used a telescoping sum to prove that the quantity $S_n = \frac{1}{1\cdot 2} + \frac{1}{2\cdot 3} + \frac{1}{3\cdot 4} + \frac{1}{4\cdot 5} + \cdots + \frac{1}{(n-1)\cdot n}$ is equal to $\frac{n-1}{n}$. Use induction to give a different proof of this formula.

42.2. (a) Use the recursive formula to compute the polynomial $F_4(X)$. Be sure to check your answer by computing $F_4(1)$, $F_4(2)$, and $F_4(3)$ and verifying that they equal 1, $1+2^4=17$, and $1+2^4+3^4=98$, respectively.

(b) Find the polynomial $F_5(X)$ and check your answer as in (a).

42.3. (a) Prove that the leading coefficient of $F_k(X)$ is $\frac{1}{k+1}$. In other words, prove that $F_k(X)$ looks like

$$F_k(X) = \frac{1}{k+1}X^{k+1} + aX^k + bX^{k-1} + \cdots.$$

(b) Try to find a similar formula for the next coefficient (that is, the coefficient of X^k) in the polynomial $F_k(X)$.

(c) Find a formula for the coefficient of X^{k-1} in the polynomial $F_k(X)$.

42.4. (a) What is the value of $F_k(0)$?

(b) What is the value of $F_k(-1)$?

(c) Use (b) to prove that if p is a prime number then

$$1^k + 2^k + \cdots + (p-1)^k \equiv 0 \pmod{p}.$$

(d) What is the value of $F_k(-1/2)$? More precisely, try to find a large collection of k's for which you can guess (and prove correct) the value of $F_k(-1/2)$.

42.5. Prove the remarkable fact that

$$(1 + 2 + 3 + \cdots + n)^2 = 1^3 + 2^3 + 3^3 + \cdots + n^3.$$

42.6. The coefficients of the polynomial $F_k(N)$ are rational numbers. We would like to multiply by some integer to clear all the denominators. For example,

$$F_1(X) = \tfrac{1}{2}X^2 + \tfrac{1}{2}X \quad \text{and} \quad F_2(X) = \tfrac{1}{3}X^3 + \tfrac{1}{2}X^2 + \tfrac{1}{6}X,$$

so $2 \cdot F_1(X)$ and $6 \cdot F_2(X)$ have coefficients that are integers.

(a) Prove that

$$(k+1)! \cdot F_k(X)$$

has integer coefficients.

(b) It is clear from the examples in this chapter that $(k+1)!$ is usually much larger than necessary for clearing the denominators of the coefficients of $F_k(X)$. Can you find any sort of patterns in the actual denominator?

42.7. A pyramid with a square base of side n requires $F_2(n)$ dots, so $F_2(n)$ is the nth *Square Pyramid Number*. In Chapter 29 we found infinitely many numbers that are both triangular and square. Search for numbers that are both tetrahedral and square pyramid numbers. Do you think there are finitely many, or infinitely many, such numbers?

42.8. (a) Find a simple expression for the sum

$$\mathbb{T}_1 + \mathbb{T}_2 + \mathbb{T}_3 + \cdots + \mathbb{T}_n$$

of the first n tetrahedral numbers.

(b) Express your answer in (a) as a single binomial coefficient.

(c) Try to understand and explain the following statement: "The number $\mathbb{T}_1 + \mathbb{T}_2 + \cdots + \mathbb{T}_n$ is the number of dots needed to form a pyramid shape in four-dimensional space."

42.9. The nth triangular number T_n equals the binomial coefficient $\binom{n+1}{2}$, and the nth tetrahedral number $\mathbb{T}_n$ equals the binomial coefficient $\binom{n+2}{3}$. This means that the formula $\mathbb{T}_n = T_1 + T_2 + \cdots + T_n$ can be written using binomial coefficients as

$$\binom{2}{2} + \binom{3}{2} + \binom{4}{2} + \cdots + \binom{n+1}{2} = \binom{n+2}{3}.$$

(a) Illustrate this formula for $n = 5$ by taking Pascal's Triangle (see Chapter 36), circling the numbers $\binom{2}{2}, \binom{3}{2}, \ldots, \binom{6}{2}$, and putting a box around their sum $\binom{7}{3}$.

(b) Write the formula $1+2+3+\cdots+n = T_n$ using binomial coefficients and illustrate your formula for $n = 5$ using Pascal's Triangle as in (a). [*Hint.* $\binom{n}{1} = n$.]

(c) Generalize these formulas to write a sum of binomial coefficients $\binom{r}{r}, \binom{r+1}{r}, \ldots$ in terms of a binomial coefficient.

(d) Prove that your formula in (c) is correct.

42.10 (For students who know calculus). Let $P_0(x)$ be the polynomial

$$P_0(x) = 1 + x + x^2 + x^3 + \cdots + x^{n-1}.$$

Next let

$$P_1(x) = \frac{d}{dx}(xP_0(x)), \quad \text{and} \quad P_2(x) = \frac{d}{dx}(xP_1(x)), \quad \text{and so on.}$$

(a) What does $P_k(x)$ look like? What is the value of $P_k(1)$? (*Hint.* The answer has something to do with the material in this chapter.)

(b) The polynomial $P_0(x)$ is the geometric sum that we used in Chapter 14. Recall that the formula for the geometric sum is $P_0(x) = (x^n - 1)/(x - 1)$, at least provided that $x \neq 1$. Compute the limit

$$\lim_{x \to 1} \frac{x^n - 1}{x - 1}$$

and check that it gives the same value as $P_0(1)$. (*Hint.* Use L'Hôpital's rule.)

(c) Find a formula for $P_1(x)$ by differentiating,

$$P_1(x) = \frac{d}{dx}\left(x\frac{x^n - 1}{x - 1}\right).$$

(d) Compute the limit of your formula in (c) as $x \to 1$. Explain why this gives a new proof for the value of $1 + 2 + \cdots + n$.

(e) Starting with your formula in (c), repeat (c) and (d) to find a formula for $P_2(x)$ and for the limit of $P_2(x)$ as $x \to 1$.

(f) Starting with your formula in (e), repeat (c) and (d) to find a formula for $P_3(x)$ and for the limit of $P_3(x)$ as $x \to 1$.

42.11. Fix an integer $k \geq 0$ and let $F_k(n) = 1^k + 2^k + \cdots + n^k$ be the sum of powers studied in this chapter. Let

$$\begin{aligned} A(x) &= \text{generating function of the sequence } F_k(0), F_k(1), F_k(2), F_k(3), \ldots \\ &= F_k(1)x + F_k(2)x^2 + F_k(3)x^3 + \cdots, \\ B(x) &= \text{generating function of the sequence } 0^k, 1^k, 2^k, 3^k, \ldots \\ &= x + 2^k x^2 + 3^k x^3 + 4^k x^4 + \cdots. \end{aligned}$$

Find a simple formula relating $A(x)$ and $B(x)$.

Chapter 43

Cubic Curves and Elliptic Curves

We have now studied solutions to several different sorts of polynomial equations, including

$X^2 + Y^2 = Z^2$	Pythagorean Triples Equation (Chapters 2 & 3)
$x^4 + y^4 = z^4$	Fermat's Equation of Degree 4 (Chapter 28)
$x^2 - Dy^2 = 1$	Pell's Equation (Chapters 30, 32 & 40)

These are all examples of what are known as *Diophantine Equations*. A Diophantine equation is a polynomial equation in one or more variables for which we are to find solutions in either integers or in rational numbers. For example, in Chapter 2 we showed that every solution in (relatively prime) integers to the Pythagorean triples equation is given by the formulas

$$X = st, \qquad Y = \frac{s^2 - t^2}{2}, \qquad Z = \frac{s^2 + t^2}{2}.$$

We reached a very different conclusion in Chapter 28 concerning Fermat's equation of degree 4, where we showed that there were no solutions in integers with $xyz \neq 0$. Pell's equation, on the other hand, has infinitely many solutions in integers, and we showed in Chapter 32 that every solution can be obtained by taking a single basic solution and raising it to powers.

In the next few chapters we discuss a new kind of Diophantine equation, one given by a polynomial of degree 3. We are especially interested in the rational number solutions, but we also discuss solutions in integers and solutions "modulo p." Diophantine equations of degree 2 are fairly well understood by mathematicians today, but equations of degree 3 already pose enough difficulties to be topics of current research. Also, surprisingly, it is by using equations of degree 3 that Andrew Wiles proved that Fermat's equation $x^n + y^n = z^n$ has no solutions in integers with $xyz \neq 0$ for all degrees $n \geq 3$.

The degree 3 equations that we weill study are called *elliptic curves*.[1] Elliptic curves are given by equations of the form

$$y^2 = x^3 + ax^2 + bx + c.$$

The numbers a, b, and c are fixed, and we are looking for pairs of numbers (x, y) that solve the equation. Here are three sample elliptic curves:

$$E_1 : y^2 = x^3 + 17,$$
$$E_2 : y^2 = x^3 + x,$$
$$E_3 : y^2 = x^3 - 4x^2 + 16.$$

The graphs of E_1, E_2, and E_3 are shown in Figure 43.1. We will return to these three examples many times in the ensuing chapters to illustrate the general theory.

As already mentioned, we will be studying solutions in rational numbers, in integers, and modulo p. Each of our three examples has solutions in integers, for example

E_1 has the solutions $(-2, 3)$, $(-1, 4)$, and $(2, 5)$,
E_2 has the solution $(0, 0)$,
E_3 has the solutions $(0, 4)$ and $(4, 4)$.

We found these solutions by trial and error. In other words, we plugged in small values for x and checked to see if $x^3 + ax^2 + bx + c$ turned out to be a perfect square. Similarly, checking a few small rational values for x, we discover the rational solution $(1/4, 33/8)$ to E_1. How might we go about creating more solutions?

A principal theme of this chapter is the interplay between geometry and number theory. We've already seen this idea at work in Chapter 3, where we used the geometry of lines and circles to find Pythagorean triples. Briefly, in Chapter 3 we took a line through the point $(-1, 0)$ on the unit circle and looked at the other point where the line intersected the circle. By taking lines whose slope was a rational number, we found that the second intersection point had x, y-coordinates that were rational numbers. In this way we used lines through the one point $(-1, 0)$ to create lots of new points with rational coordinates. We want to use the same sort of method to find lots of points with rational coordinates on elliptic curves.

[1]Contrary to popular opinion, an elliptic curve is not an ellipse. You may recall that an ellipse looks like a squashed circle. This is not at all the shape of the elliptic curves illustrated in Figure 43.1. Elliptic curves first arose when mathematicians tried to compute the circumference of an ellipse, whence their somewhat unfortunate moniker. A more accurate, but less euphonious, name for elliptic curves is *abelian varieties of dimension one*.

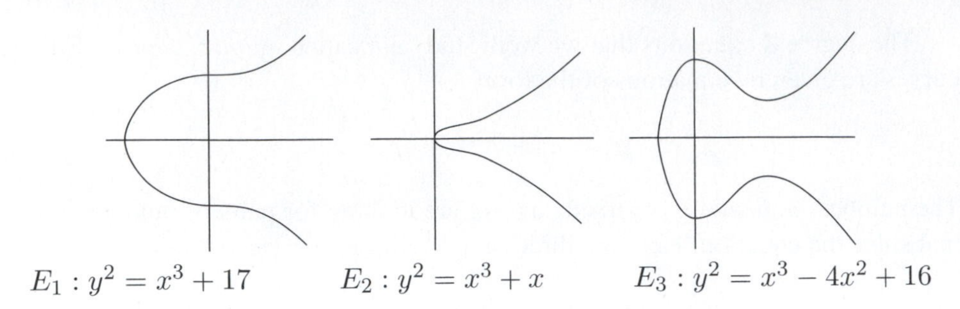

Figure 43.1: Graphs of Three Representative Elliptic Curves

Let's try the exact same idea using the elliptic curve

$$E_1 : y^2 = x^3 + 17.$$

We draw lines through the point $P = (-2, 3)$ and see what other points we find. For example, suppose we try the line with slope 1,

$$y - 3 = x + 2.$$

To find the intersection of this line with E_1, we substitute $y = x + 5$ into the equation for E_1 and solve for x. Thus,

$$\begin{aligned} y^2 &= x^3 + 17 && \text{The equation for } E_1. \\ (x+5)^2 &= x^3 + 17 && \text{Substitute in the equation of the line.} \\ 0 &= x^3 - x^2 - 10x - 8 && \text{Multiply out and combine terms.} \end{aligned}$$

You probably don't know how to find the roots of cubic polynomials,[2] but in this instance we already know one of the solutions. The elliptic curve E_1 and the line both go through the point $P = (-2, 3)$, so $x = -2$ must be a root. This allows us to factor the cubic polynomial as

$$x^3 - x^2 - 10x - 8 = (x + 2)(x^2 - 3x - 4).$$

[2]There actually is a cubic formula, although it is considerably more complicated than its cousin, the quadratic formula. The first step in finding the roots of $x^3 + Ax^2 + Bx + C = 0$ is to make the substitution $x = t - A/3$. After some work, the equation for t looks like $t^3 + pt + q = 0$. A root of this equation is then given by Cardano's formula

$$t = \sqrt[3]{-q/2 + \sqrt{q^2/4 + p^3/27}} + \sqrt[3]{-q/2 - \sqrt{q^2/4 + p^3/27}}.$$

There is a yet more complicated quartic formula for the roots of fourth-degree polynomials, but that is where the story ends. In the early 1800s, Niels Abel and Evariste Galois showed that there are no similar formulas giving the roots of polynomials of degree 5 or greater. This result is one of the great triumphs of modern mathematics, and the tools that were developed to prove it are still of fundamental importance in algebra and number theory.

Now we can use the quadratic formula to find the roots $x = -1$ and $x = 4$ of $x^2 - 3x - 4$. Substituting these values into the equation of the line $y = x + 5$ then gives the y-coordinates of our new points $(-1, 4)$ and $(4, 9)$. You should check that these points do indeed satisfy the equation $y^2 = x^3 + 17$.

FOXTROT ©Bill Amend. Reprinted with permission of UNIVERSAL SYNDICATE. All rights reserved

This looks good, but before we become overconfident, we should try (at least) one more example. Suppose we take the line through $P = (-2, 3)$ having slope 3. This line has equation

$$y - 3 = 3(x + 2),$$

which, after rearranging, becomes

$$y = 3x + 9.$$

We substitute $y = 3x + 9$ into the equation for E_1 and compute.

$y^2 = x^3 + 17$	The equation for E_1.
$(3x + 9)^2 = x^3 + 17$	Substitute $y = 3x + 9$.
$0 = x^3 - 9x^2 - 54x - 64$	Expand and combine terms.
$0 = (x + 2)(x^2 - 11x - 32)$	Factor out the known root.

Just as before, we can use the quadratic formula to find the roots of $x^2 - 11x - 32$, but unfortunately what we find are the two values

$$x = \frac{11 \pm \sqrt{249}}{2}.$$

This is obviously not the sort of answer we were hoping for, since we are looking for points on E_1 having rational coordinates.

What causes the problem? Suppose that we draw the line L of slope m through the point $P = (-2, 3)$ and find its intersection with E_1. The line L is given by the equation

$$L : y - 3 = m(x + 2).$$

To find the intersection of L and E_1, we substitute $y = m(x + 2) + 3$ into the equation for E_1 and solve for x. When we do this, we get the following formidable-looking cubic equation to solve:

$$\begin{aligned} y^2 &= x^3 + 17 \\ \bigl(m(x+2)+3\bigr)^2 &= x^3 + 17 \\ 0 &= x^3 - m^2x^2 - (4m^2 + 6m)x - (4m^2 + 12m - 8). \end{aligned}$$

Of course, we do know one root is $x = -2$, so the equation factors as

$$0 = (x + 2)(x^2 - (m^2 + 2)x - (2m^2 + 6m - 4)).$$

Unfortunately for our plans, the other two roots are unlikely to be rational numbers.

It looks like the idea of using lines through known points to produce new points has hit a brick wall. As is so often the case in mathematics (and in life?), stepping back and taking a slightly wider view reveals a way to clamber over, squeeze under, or just plain walk around the wall. In this case, our problem is that we have a cubic polynomial, and we know that one of the roots is a rational number; but this leaves the other two roots as solutions of a quadratic polynomial whose roots may not be rational. How can we compel that quadratic polynomial to have rational roots? Harking back to our work in Chapter 3, we see that if a quadratic polynomial has one rational root then the other root will also be rational. In other words, we really want to force the original cubic polynomial to have two rational roots, and then the third one will be rational, too.

This brings us to the crux of the problem. The original cubic polynomial had one rational root because we chose a line going through the point $P = (-2, 3)$, thereby ensuring that $x = -2$ is a root. To force the cubic polynomial to have two rational roots, we should choose a line that already goes through two rational points on the elliptic curve E_1.

An example illustrates this idea. We start with the two points $P = (-2, 3)$ and $Q = (2, 5)$ on the elliptic curve

$$E_1 : y^2 = x^3 + 17.$$

The line connecting P and Q has slope $(5 - 3)/(2 - (-2)) = 1/2$, so its equation is

$$y = \frac{1}{2}x + 4.$$

Substituting this into the equation for E_1 gives

$$\begin{aligned} y^2 &= x^3 + 17 \\ \left(\frac{1}{2}x + 4\right)^2 &= x^3 + 17 \\ 0 &= x^3 - \frac{1}{4}x^2 - 4x + 1. \end{aligned}$$

This must have $x = -2$ and $x = 2$ as roots, so it factors as

$$0 = (x - 2)(x + 2)\left(x - \frac{1}{4}\right).$$

Notice that the third root is indeed a rational number, $x = 1/4$, and substituting this value into the equation of the line gives the corresponding y-coordinate, $y = 33/8$. In summary, by taking the line through the two known solutions $(-2, 3)$ and $(2, 5)$, we have found the rational solution $(1/4, 33/8)$ for our elliptic curve. This procedure is illustrated in Figure 43.2.

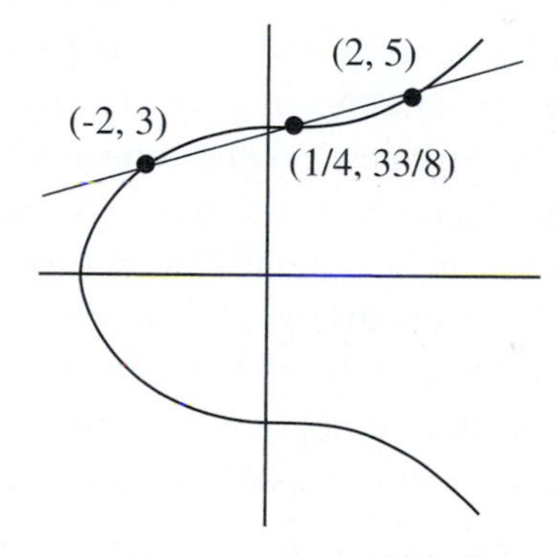

Figure 43.2: Using Two Known Points to Find a New Point

Suppose that we try to repeat this procedure with the new solution $(1/4, 33/8)$. If we draw the line through $(-2, 3)$ and $(1/4, 33/8)$, say, we know that the third intersection point with E_1 is the point $(2, 5)$. So we end up back where we started. Again it seems we are stuck, but again a simple observation sets us moving again. This simple observation is that if (x, y) is a point on the elliptic curve E_1 then the point $(x, -y)$ is also a point on E_1. This is clear from the symmetry of E_1 about the x-axis (see Figure 40.2). So what we do is take the new point $(1/4, 33/8)$, replace it with $(1/4, -33/8)$, and then repeat the above procedure using the line through $(1/4, -33/8)$ and $(-2, 3)$. This line has slope $-19/6$ and is given by the equation $y = -19x/6 - 10/3$. Substituting into the equation for E_1, we end up having to find the roots of

$$x^3 - \frac{361}{36}x^2 - \frac{190}{9}x + \frac{53}{9}.$$

Two of the roots are $1/4$ and -2, so we can divide this cubic polynomial by $(x - 1/4)(x + 2)$ to find the other root,

$$x^3 - \frac{361}{36}x^2 - \frac{190}{9}x + \frac{53}{9} = \left(x - \frac{1}{4}\right)(x + 2)\left(x - \frac{106}{9}\right).$$

This gives $x = 106/9$, and substituting this value of x into the equation of the line gives $y = -1097/27$. So we have found a new point $(106/9, -1097/27)$ satisfying the equation

$$E_1 : y^2 = x^3 + 17.$$

Continuing in this fashion, we find lots and lots of points. In fact, just as with Pell's equation, we get infinitely many points with rational coordinates. For Pell's equation, we showed that all solutions can be obtained by taking powers of a single smallest solution. It turns out that every point on E_1 with rational coordinates can be found by starting with the two points P and Q, connecting them by a line to find a new point, reflecting about the x-axis, drawing more lines through the known points to find new points, reflecting again, and repeating the process over and over. The important point to observe here is that every point on E_1 with rational coordinates can be obtained by starting with just two points and repeatedly applying a simple geometric procedure, just as every solution to Pell's equation was obtained by starting with one basic solution and repeatedly applying a simple rule. The fact that the infinitely many rational solutions to E_1 can be created from a finite generating set is a special case of a famous theorem.

Theorem 43.1 (Mordell's Theorem). (L.J. Mordell, 1922) *Let E be an elliptic curve given by the equation*

$$E : y^2 = x^3 + ax^2 + bx + c,$$

where a, b, c are integers such that the discriminant

$$\Delta(E) = -4a^3c + a^2b^2 - 4b^3 - 27c^2 + 18abc$$

is not zero.[3] *Then there is a finite list of solutions*

$$P_1 = (x_1, y_1),\ P_2 = (x_2, y_2),\ \ldots,\ P_r = (x_r, y_r),$$

[3]If $\Delta(E) = 0$, then the cubic polynomial $x^3 + ax^2 + bx + c$ has a double or triple root, and the curve E either crosses itself or has a sharp point. (See Exercise 43.7.) The discriminant $\Delta(E)$ will appear in several guises as we continue our study of elliptic curves.

with rational coordinates so that every rational solution to E can be obtained starting from these r points and repeatedly taking lines through pairs of points, intersecting with E, and reflecting to create new points.

Mordell proved his theorem in 1922. Unfortunately, the proof is too complicated for us to give in detail, but the following outline of Mordell's proof shows that it is nothing more than a fancy version of Fermat's descent method:

(1) The first step is to make a list $P_1, P_2, \ldots, P_r$ of "small" points on E having rational coordinates.

(2) The next step is to show that, if Q is any point with rational coordinates that is not in the list, then it is possible to choose one of the P_i's so that the line through P_i and Q intersects E in a third point Q' that is "smaller" than Q.

(3) Repeating this process, we get a list of points $Q, Q', Q'', Q''', \ldots$ of decreasing size, and we show that eventually the size gets so small that we end up with one of the P_i's in our original list.

Notice the similarity to our work on Pell's equation, where we showed that any large solution is always the product of a smaller solution and the smallest solution. Of course, it's not even clear what "larger" and "smaller" mean for points with rational coordinates on an elliptic curve E. This is one of the many ideas that Mordell had to work out before his proof was complete.

Let's take a look at some of the rational solutions to E_1. We start with $P_1 = (-2, 3)$ and $P_2 = (-1, 4)$. The line through P_1 and P_2 intersects E_1 in a third point, which we reflect about the x-axis and call P_3. Next we take the line through the points P_1 and P_3, intersect it with E_1 and reflect across the x-axis to get P_4. Using the line through P_1 and P_4, we similarly get P_5, and so on. The following are the first few P_n's. As you can see, the numbers get complicated with frightening rapidity.

$$P_1 = (-2, 3), \quad P_2 = (-1, 4), \quad P_3 = (4, -9), \quad P_4 = (2, 5),$$

$$P_5 = \left(\frac{1}{4}, \frac{-33}{8}\right), \quad P_6 = \left(\frac{106}{9}, \frac{1097}{27}\right), \quad P_7 = \left(\frac{-2228}{961}, \frac{-63465}{29791}\right),$$

$$P_8 = \left(\frac{76271}{289}, \frac{-21063928}{4913}\right), \quad P_9 = \left(\frac{-9776276}{6145441}, \frac{54874234809}{15234548239}\right),$$

$$P_{10} = \left(\frac{3497742218}{607770409}, \frac{-215890250625095}{14983363893077}\right).$$

We would like a quantitative way to measure the "size" of these points. One way to do this is to look at the numerator and denominator of the x-coordinates. In other words, if we write the coordinates of P_n as

$$P_n = \left(\frac{A_n}{B_n}, \frac{C_n}{D_n}\right)$$

in lowest terms, we might define the *size of* P_n to be[4]

$$\text{size}(P_n) = \text{maximum of } |A_n| \text{ and } |B_n|.$$

For example,

$$\text{size}(P_1) = \max\{|-2|, |1|\} = 2$$

and

$$\text{size}(P_7) = \max\{|-2228|, |961|\} = 2228.$$

The first 20 P_n's together with their sizes are listed in Table 43.1.

Do you see any sort of pattern in Table 43.1? The actual numbers don't seem to follow a pattern, but try moving back a little bit and squinting while staring at the table. Imagine that the numbers are solid black boxes and look at the curve that separates the black area from the white area. Does it look familiar? If not, look at Table 43.2, which extends Table 43.1 up to $n \le 50$ with the digits replaced by black boxes.

The curve separating the black region from the white region looks very much like a parabola (lying on its side). What this means is that the number of digits in $\text{size}(P_n)$ looks like cn^2 for some constant c. Using more advanced methods, it is possible to show that c is approximately 0.1974.[5] In other words, for large values of n, the size of P_n looks like

$$\# \text{ of digits in size}(P_n) \approx 0.1974n^2,$$
$$\text{size}(P_n) \approx 10^{0.1974n^2} \approx (1.574)^{n^2}.$$

It is instructive to compare this with the solutions to Pell's equation that we found in Chapter 30. We showed there that the size of the nth solution (x_n, y_n) to Pell's equation $x^2 - 2y^2 = 1$ is approximately

$$x_n \approx \frac{1}{2}(5.82843)^n.$$

The exponential growth rate for Pell's equation is quite rapid, but it pales in comparison to the speed with which the points on an elliptic curve grow.

---

[4]The mathematical term for what we are calling the size is the *height*.

[5]The value of c is computed with the theory of canonical heights developed by André Néron and John Tate in the 1960s. Using this theory, we can show that the ratio $\ln(\text{size}(P_n))/n^2$ gets closer and closer to $0.4546168651\ldots$ as n gets larger and larger.

n	size(P_n)
1	2
2	1
3	4
4	2
5	4
6	106
7	2228
8	76271
9	9776276
10	3497742218
11	1160536538401
12	1610419388060961
13	43923749623043363812
14	102656671584861356692801
15	18531846858359807878734515284
16	370183335711420357564604634095918
17	125067940343620957546805016634617881761
18	14803896396546295880463242120819717253248409
19	41495337621274074603425488675302807756680196997372
20	83094719816361303226380666143399722139698613105279866991

Table 43.1: The Size of Points P_n on E_1

Table 43.2: The Size of Points P_n on E_1

Exercises

43.1. For each of the following pairs of points on the elliptic curve $E_1 : y^2 = x^3 + 17$, use the line connecting the points to find a new point with rational coordinates on E_1.

(a) The points $(-1, 4)$ and $(2, 5)$

(b) The points $(43, 282)$ and $(52, -375)$

(c) The points $(-2, 3)$ and $(19/25, 522/125)$

43.2. The elliptic curve

$$E : y^2 = x^3 + x - 1$$

has the points $P = (1, 1)$ and $Q = (2, -3)$ with rational coordinates.

(a) Use the line connecting P and Q to find a new point R on E having rational coordinates.

(b) Let R' be the point obtained by reflecting R through the x-axis. [That is, if $R = (x, y)$, then $R' = (x, -y)$.] Use the line through P and R' to find a new point S with rational coordinates on E.

(c) Same as (b), but use the line through Q and R' to find a new point T.

(d) Let S be the point you found in (b), and let S' be the point obtained by reflecting S through the x-axis. What point do you get if you use the line through P and S' to find a new point on E?

43.3. Suppose that $Q_1, Q_2, Q_3, \ldots$ is a list of points with rational coordinates on an elliptic curve E, and suppose that their sizes are strictly decreasing,

$$\text{size}(Q_1) > \text{size}(Q_2) > \text{size}(Q_3) > \text{size}(Q_4) > \cdots.$$

Explain why the list must stop after a finite number of points. In other words, explain why a list of points with strictly decreasing sizes must be a finite list. Do you see why this makes the size a good tool for proofs by descent?

43.4. Write a short biography of Girolamo Cardano, including especially a description of his publication of the solution to the cubic equation and the ensuing controversy.

43.5 (This exercise is for people who have taken some calculus.). There is another way to find points with rational coordinates on elliptic curves that involves using tangent lines. This exercise explains the method for the curve

$$E : y^2 = x^3 - 3x + 7.$$

(a) The point $P = (2, 3)$ is a point on E. Find an equation for the tangent line L to the elliptic curve E at the point P. (*Hint.* Use implicit differentiation to find the slope dy/dx at P.)

(b) Find where the tangent line L intersects the elliptic curve E by substituting the equation for L into E and solving. You should discover a new point Q with rational coordinates on E. (Notice that $x = 2$ is a double root of the cubic equation you need to solve. This reflects the fact that L is tangent to E at the point where $x = 2$.)

(c) Let R be the point you get by reflecting Q across the x-axis. [In other words, if $Q = (x_1, y_1)$, let $R = (x_1, -y_1)$.] Take the line through P and R and intersect it with E to find a third point with rational coordinates on E.

43.6. Let L be the line $y = m(x+2)+3$ of slope m going through the point $(-2, 3)$. This line intersects the elliptic curve $E_1 : y^2 = x^3 + 17$ in the point $(-2, 3)$ and in two other points. If all three of these points have rational coordinates, show that the quantity

$$m^4 + 12m^2 + 24m - 12$$

must be the square of a rational number. Substitute in values of m between -10 and 10 to find which ones make this quantity a square, and use the values you find to obtain rational solutions to $y^2 = x^3 + 17$.

43.7. The discriminant of each of the curves

$$C_1 : y^2 = x^3 \qquad \text{and} \qquad C_2 : y^2 = x^3 + x^2$$

is zero. Graph these two curves and explain in what way your graphs are different from each other and different from the graphs of the elliptic curves illustrated in Figure 43.1

43.8. Let E be the elliptic curve

$$E : y^2 = x^3 + ax^2 + bx + c,$$

and let $P_1 = (x_1, y_1)$ and $P_2 = (x_2, y_2)$ be two points on E.

(a) Let L be the line connecting P_1 and P_2. Write a program to compute the third point $P_3 = (x_3, y_3)$ where the line L intersects E. (If L is a vertical line, then there won't be a real third intersection point, so your program should return a warning message.) You should keep track of the coordinates as rational numbers; if your computer language won't let you work with rational numbers directly, you'll have to store a rational number A/B as a pair (A, B), in which case you should always cancel $\gcd(A, B)$.

(b) Modify your program so that the output is the reflected point $(x_3, -y_3)$. We denote this point with the suggestive notation $P_1 \oplus P_2$, since it is a sort of "addition" rule for the points of E.

(c) Let E be the elliptic curve

$$E : y^2 = x^3 + 3x^2 - 7x + 3,$$

and consider the points $P = (2, -3)$, $Q = (37/36, 53/216)$, and $R = (3, 6)$. Use your program to compute

$$P \oplus Q, \qquad Q \oplus R, \qquad \text{and} \qquad P \oplus R.$$

Next compute

$$(P \oplus Q) \oplus R \qquad \text{and} \qquad P \oplus (Q \oplus R).$$

Are the answers the same regardless of the order in which you "add" the points? Do you find this surprising? (If not, try proving that the corresponding fact is true for every elliptic curve.)

Chapter 44

Elliptic Curves with Few Rational Points

The elliptic curve E_1 with equation $y^2 = x^3 + 17$ has lots of points with rational coordinates. On the other hand, the elliptic curve E_2 with equation $y^2 = x^3 + x$ appears to have very few such points. In fact, the only point that is immediately visible is the point $(0, 0)$. We show that this is indeed the only rational point on E_2.

Theorem 44.1. *The only point with rational coordinates on the elliptic curve*

$$E_2 : y^2 = x^3 + x$$

is the point $(x, y) = (0, 0)$.

Verification. Suppose that $(A/B, C/D)$ is a point on E_2 with rational coordinates, where we write the fractions A/B and C/D in lowest terms. In particular, we take the denominators B and D to be positive. Our task is to show that $A = 0$ and $C = 0$. Substituting $x = A/B$ and $y = C/D$ into the equation for E_2 and clearing denominators, we get the equation

$$C^2B^3 = A^3D^2 + AB^2D^2. \qquad (*)$$

Any solution in integers to this equation (with B and D not zero) gives a rational point on E_2.

The equation $(*)$ contains a lot of divisibility information from which we draw numerous conclusions. For example, factoring the right-hand side of $(*)$ gives

$$C^2B^3 = D^2A(A^2 + B^2),$$

so D^2A divides C^2B^3. However, we know that $\gcd(C, D) = 1$, so D^2 must divide B^3. Similarly, rearranging $(*)$ and factoring gives

$$A^3D^2 = C^2B^3 - AB^2D^2 = B^2(C^2B - AD^2),$$

so B^2 divides A^3D^2. Since $\gcd(A, B) = 1$, it follows that B^2 must divide D^2, which of course means that B divides D. We have verified that

$$D^2|B^3 \qquad \text{and} \qquad B|D.$$

Let $v = D/B$, so we know that v is an integer. Substituting $D = Bv$ into the relation $D^2|B^3$ tells us that $B^2v^2|B^3$, so $v^2|B$. In other words, we can write B as $B = v^2z$ for some integer z. Notice $D = Bv = v^3z$. Substituting $B = v^2z$ and $D = v^3z$ into the equation $(*)$ yields

$$\begin{aligned} C^2B^3 &= A^3D^2 + AB^2D^2 \\ C^2(v^2z)^3 &= A^3(v^3z)^2 + A(v^2z)^2(v^3z)^2 \\ C^2z &= A^3 + Av^4z^2, \end{aligned}$$

and rearranging gives

$$A^3 = C^2z - Av^4z^2 = z(C^2 - Av^4z).$$

Thus, z divides A^3. However, z also divides B and $\gcd(A, B) = 1$, so we must have $z = \pm 1$. On the other hand, $B = v^2z$ and we know that B is positive, so in fact $z = 1$. We now know that

$$B = v^2 \qquad \text{and} \qquad D = v^3,$$

so our original point $(A/B, C/D)$ on E_2 looks like $(A/v^2, C/v^3)$, and the equation $(*)$ becomes

$$C^2 = A^3 + Av^4.$$

Factoring the right-hand side, we see that

$$C^2 = A(A^2 + v^4).$$

This is a very interesting equation, because it expresses the perfect square C^2 as the product of the two numbers A and $A^2 + v^4$. I claim that these two numbers have no common factors. Do you see why? Well, if A and $A^2 + v^4$ were to have a common factor, say they were both divisible by some prime p, then v would also have to be divisible by p. However, A and v can't both be divisible by p, since the fraction $A/B = A/v^2$ is written in lowest terms.

So we now know that A and $A^2 + v^4$ have no common factors and that their product is a perfect square. The only way that this can happen is if each of them individually is a square. (Does this reasoning look familiar? We used it long ago in Chapter 2 to we derive a formula for Pythagorean triples.) In other words, we can find integers u and w so that

$$A = u^2 \qquad \text{and} \qquad A^2 + v^4 = w^2.$$

Substituting the value $A = u^2$ into the second equation gives

$$u^4 + v^4 = w^2.$$

Let's recapitulate our progress. We began with some solution to the elliptic curve E_2, which we wrote as $(A/B, C/D)$ in lowest terms. Starting from this solution, we showed that there must be integers u, v, and w satisfying the equation

$$u^4 + v^4 = w^2.$$

Furthermore, given such integers u, v, w, we can recover the solution to E_2 from the formulas $A/B = u^2/v^2$ and $C/D = uw/v^3$. Do you recognize this u, v, w equation? It should look familiar, since it is exactly the equation that we studied in Chapter 28, where we showed that the only solutions are those with either $u = 0$ or $v = 0$. Since $u = 0$ leads to $(A/B, C/D) = (0, 0)$ and $v = 0$ leads to zeros in the denominator, it follows that the only point with rational coordinates on E_2 is the point $(0, 0)$. This completes our verification. □

We now turn to our third representative elliptic curve

$$E_3 : y^2 = x^3 - 4x^2 + 16.$$

A brief search reveals four points on E_3,

$$P_1 = (0, 4), \quad P_2 = (4, 4), \quad P_3 = (0, -4), \quad P_4 = (4, -4).$$

What happens if we use these four points and play the same game that we played on E_1? The line connecting P_1 and P_2 has equation $y = 4$. To find where this line intersects E_3, we substitute $y = 4$ into the equation of E_3 and solve for x:

$$\begin{aligned} 4^2 &= x^3 - 4x^2 + 16, \\ 0 &= x^3 - 4x^2 = x^2(x - 4). \end{aligned}$$

Notice that $x = 0$ is a double root, so the line connecting P_1 and P_2 intersects E_3 only at P_1 and P_2. The game is lost; we've failed to find any new points. A similar thing happens if we choose any two of P_1, P_2, P_3, P_4, connect them with a line, and compute the intersection of the line with E_3. We never find any new points. In fact, it turns out that the only points on E_3 with rational coordinates are the four points P_1, P_2, P_3, P_4. (Unfortunately, the proof is too long for us to give.)

More generally, a finite collection of points

$$P_1, P_2, \ldots, P_t$$

(with $t \geq 3$) on an elliptic curve

$$E : y^2 = x^3 + ax^2 + bx + c$$

is called a *torsion collection* if, whenever you draw a line L through two of the P_i's, all intersection points of L and E are already in the collection. Another way to say this is that a torsion collection cannot be enlarged using the geometric method of taking lines and intersections. For example, E_3 has the torsion collection consisting of the four points $(0, \pm 4), (4, \pm 4)$. The following important theorem describes torsion collections.

Theorem 44.2 (Torsion Theorem). *Let $E : y^2 = x^3 + ax^2 + bx + c$ be an elliptic curve with integer coefficients a, b, c, and let $P_1, P_2, \ldots, P_t$ be a torsion collection on E consisting of points whose coordinates are rational numbers. Also let*

$$\Delta(E) = -4a^3c + a^2b^2 - 4b^3 - 27c^2 + 18abc$$

be the discriminant of E, and suppose that $\Delta(E) \neq 0$.

(a) (Nagell–Lutz Theorem, 1935/37) *Write the coordinates of each P_i as $P_i = (x_i, y_i)$. Then all the x_i's and y_i's are integers. Furthermore, if $y_i \neq 0$, then $y_i^2 | 16\Delta(E)$.*

(b) (Mazur's Theorem, 1977) *A torsion collection can contain at most 15 points.*

The Nagell–Lutz portion of the Torsion Theorem says that points in a torsion collection have integer coordinates. We've also seen examples of points with integer coordinates that do not lie in a torsion collection, such as the point $(-2, 3)$ on the curve $E_1 : y^2 = x^3 + 17$. Our investigations have unearthed quite a few points with integer coordinates on E_1, including

$$(-2, \pm 3), \quad (-1, \pm 4), \quad (2, \pm 5), \quad (4, \pm 9),$$
$$(8, \pm 23), \quad (43, \pm 282), \quad (52, \pm 375).$$

We know that the curve E_1 has infinitely many points with rational coordinates, so there's no reason why it shouldn't be possible to extend this list indefinitely. Continuing our search, we soon find another integer point on E_1,

$$(5234, \pm 378661),$$

but after that we find no others, even if we search up to $x < 10^{100}$. Eventually, we begin to suspect that there are no other integer points on E_1. This turns out to be true; it is a special case of the following fundamental result.

Theorem 44.3 (Siegel's Theorem). (C.L. Siegel, 1926) *Let E be an elliptic curve*

$$E : y^2 = x^3 + ax^2 + bx + c$$

given by an equation whose coefficients a, b, and c are integers and with discriminant $\Delta(E) \neq 0$. Then there are only finitely many solutions in integers x and y.

Siegel actually gave two very different proofs of his theorem. The first, published in 1926 in the *Journal of the London Mathematical Society*,[1] works directly with the equation for E and uses factorization methods. The second proof, published in 1929, begins with Mordell's theorem and uses the geometric method for generating new points from old points. Ultimately, however, both proofs rely on the theory of Diophantine approximation (Chapter 31), specifically on advanced results which say that certain numbers cannot be closely approximated by rational numbers.

Exercises

44.1. A Pythagorean triple (a, b, c) describes a right triangle whose sides have lengths that are integers. We will call such a triangle a Pythagorean triangle. Find all Pythagorean triangles whose area is twice a perfect square.

44.2. (a) Let E be the elliptic curve $E : y^2 = x^3 + 1$. Show that the points $(-1, 0)$, $(0, 1)$, $(0, -1)$, $(2, 3)$, $(2, -3)$ form a torsion collection on E.

(b) Let E be the elliptic curve $E : y^2 = x^3 - 43x + 166$. The four points $(3, 8)$, $(3, -8)$, $(-5, 16)$, and $(-5, -16)$ form part of a torsion collection on E. Draw lines through pairs of these points and intersect the lines with E to construct the full torsion collection.

(c) Let E be an elliptic curve given by an equation

$$y^2 = (x - \alpha)(x - \beta)(x - \gamma).$$

Verify that the set of points $(\alpha, 0)$, $(\beta, 0)$, $(\gamma, 0)$ is a torsion collection.

[1] In both England and Germany in the 1920s, there was still a great deal of lingering bitterness from World War I, so Siegel published his article using the pseudonym "X".

44.3. How many integer solutions can you find on the elliptic curve

$$y^2 = x^3 - 16x + 16?$$

44.4. This exercise guides you in proving that the elliptic curve

$$E : y^2 = x^3 + 7$$

has no solutions in integers x and y. (This special case of Siegel's Theorem was originally proven by V.A. Lebesgue in 1869.)

(a) Suppose that (x, y) is a solution in integers. Show that x must be odd.

(b) Show that $y^2 + 1 = (x + 2)(x^2 - 2x + 4)$.

(c) Show that $x^2 - 2x + 4$ must be congruent to 3 modulo 4. Explain why $x^2 - 2x + 4$ must be divisible by some prime q satisfying $q \equiv 3 \pmod 4$.

(d) Reduce the original equation $y^2 = x^3 + 7$ modulo q, and use the resulting congruence to show that -1 is a quadratic residue modulo q. Explain why this is impossible, thereby proving that $y^2 = x^3 + 7$ has no solutions in integers.

44.5. The elliptic curve $E : y^2 = x^3 - 2x + 5$ has the four integer points $P = (-2, \pm 1)$ and $Q = (1, \pm 2)$.

(a) Find four more integer points by plugging in $x = 2, 3, 4, \ldots$ and seeing if $x^3 - 2x + 5$ is a square.

(b) Use the line through P and Q to find a new point R having rational coordinates. Reflect R across the x-axis to get a point R'. Now take the line through Q and R' and intersect it with E to find a point with rather large integer coordinates.

44.6. (a) Show that the equation $y^2 = x^3 + x^2$ has infinitely many solutions in integers x, y. (*Hint.* Try substituting $y = tx$.)

(b) Does your answer in (a) mean that Siegel's Theorem is incorrect? Explain.

(c) Show that the equation $y^2 = x^3 - x^2 - x + 1$ has infinitely many solutions in integers x, y.

44.7. Let $E : y^2 = x^3 + ax^2 + bx + c$ be an elliptic curve with a, b, and c integers. Suppose that $P = \left(\frac{A}{B}, \frac{C}{D}\right)$ is a point on E whose coordinates are rational numbers, written in lowest terms with B and D positive. Prove that there is an integer v so that $B = v^2$ and $D = v^3$.

44.8. Write a program to search for all points on the elliptic curve

$$E : y^2 = x^3 + ax^2 + bx + c$$

such that x is an integer and $|x| < H$. Do this by trying all possible x values and checking if $x^3 + ax^2 + bx + c$ is a perfect square.

44.9. (a) Write a program to search for points on the elliptic curve

$$E : y^2 = x^3 + ax^2 + bx + c$$

such that x and y are rational numbers. Exercise 44.7 says that any such point must look like $(x, y) = (A/D^2, B/D^3)$, so the user should input an upper bound H and your program should loop through all integers $|A| \leq H$ and $1 \leq D \leq \sqrt{H}$ and check if

$$A^3 + aA^2D^2 + bAD^4 + cD^6$$

is a perfect square. If it equals B^2, then you've found the point $(A/D^2, B/D^3)$.

(b) Use your program to find all points on the elliptic curve

$$y^2 = x^3 - 2x^2 + 3x - 2$$

whose x-coordinate has the form $x = A/D^2$ with $|A| \leq 1500$ and $1 \leq D \leq 38$.

Chapter 45

Points on Elliptic Curves Modulo p

It can be quite difficult to solve a Diophantine equation. So rather than trying to solve using integers or rational numbers, we treat the Diophantine equation as a congruence and try to find solutions modulo p. This is a far easier task. To see why, consider the following example.

How might we find all solutions "modulo 7" to the equation

$$x^2 + y^2 = 1?$$

In other words, what are the solutions to the congruence

$$x^2 + y^2 \equiv 1 \pmod 7?$$

This question is easy; we can just try each pair (x, y) with $0 \le x, y \le 6$ and see which ones make the congruence true. Thus, $(1, 0)$ and $(2, 2)$ are solutions, while $(1, 2)$ and $(3, 2)$ are not solutions. The full set of solutions is

$$(0,1), (0,6), (1,0), (2,2), (2,5), (5,2), (5,5), (6,0).$$

We conclude that the equation $x^2 + y^2 = 1$ has 8 solutions modulo 7. Similarly, there are 12 solutions modulo 11:

$$(0,1), (0,10), (1,0), (3,5), (3,6), (5,3), (5,8),$$
$$(6,3), (6,8), (8,5), (8,6), (10,0).$$

Now we look at some elliptic curves and count how many points they have modulo p for various primes p. We begin with the curve

$$E_2 : y^2 = x^3 + x$$

p	Points modulo p on $E_2 : y^2 = x^3 + x$	N_p
2	$(0,0), (1,0)$	2
3	$(0,0), (2,1), (2,2)$	3
5	$(0,0), (2,0), (3,0)$	3
7	$(0,0), (1,3), (1,4), (3,3), (3,4), (5,2), (5,5)$	7
11	$(0,0), (5,3), (5,8), (7,3), (7,8), (8,5), (8,6),$ $(9,1), (9,10), (10,3), (10,8)$	11
13	$(0,0), (2,6), (2,7), (3,2), (3,11), (4,4), (4,9),$ $(5,0), (6,1), (6,12), (7,5), (7,8), (8,0), (9,6),$ $(9,7), (10,3), (10,10), (11,4), (11,9)$	19
17	$(0,0), (1,6), (1,11), (3,8), (3,9),$ $(4,0), (6,1), (6,16), (11,4), (11,13),$ $(13,0), (14,2), (14,15), (16,7), (16,10)$	15
19	$(0,0), (3,7), (3,12), (4,7), (4,12), (5,4), (5,15),$ $(8,8), (8,11), (9,4), (9,15), (12,7), (12,12),$ $(13,5), (13,14), (17,3), (17,16), (18,6), (18,13)$	19

Table 45.1: Points Modulo p on E_2

whose lone rational point is $(0,0)$. However, as Table 45.1 indicates, E_2 tends to have lots of points modulo p. In the last column of Table 45.1 we have listed N_p, the number of points modulo p.

The number of points modulo p on an elliptic curve exhibits many wonderful and subtle patterns. Look closely at Table 45.1. Do you see any patterns? If not, maybe some more data would help. Table 45.2 gives the number of solutions modulo p without bothering to list the actual solutions.

One partial pattern that immediately strikes the eye is that there are many primes for which N_p is equal to p. This occurs for the primes

$$p = 2, 3, 7, 11, 19, 23, 31, 43, 47, 59, 67, \text{ and } 71,$$

which is surely too often to be entirely random. Indeed, aside from the initial entry 2, this list is precisely the set of primes (less than 71) that are congruent to 3 modulo 4. So we are led to make the following guess:

> *Guess.* If $p \equiv 3 \pmod 4$, then the elliptic curve $E_2 : y^2 = x^3 + x$ has exactly $N_p = p$ points modulo p.

p	2	3	5	7	11	13	17	19	23	29
N_p	2	3	3	7	11	19	15	19	23	19
p	31	37	41	43	47	53	59	61	67	71
N_p	31	35	31	43	47	67	59	51	67	71

Table 45.2: The Number N_p of Points Modulo p on E_2

What about the other primes, those congruent to 1 modulo 4? The N_p's in this case look fairly random. Sometimes N_p is less than p, such as for $p = 5$ and 17, while sometimes N_p is greater than p, such as for $p = 13$ and 53. However, it also seems to be true that as p gets larger N_p also becomes larger. In fact, N_p is usually found hovering in the general neighborhood of p. A little thought suggests why this is very reasonable.

In general, if we are trying to find the solutions modulo p to an elliptic curve

$$y^2 = x^3 + ax^2 + bx + c,$$

we substitute $x = 0, 1, 2, \ldots, p-1$ and check, for each x, whether

$$x^3 + ax^2 + bx + c$$

turns out to be a square. It is reasonable to suppose that the values we get for $x^3 + ax^2 + bx + c$ are essentially randomly distributed, so we would expect the values to be squares about half the time and to be nonsquares about half the time. This follows from the fact, proved in Chapter 23, that half the numbers from 1 to $p-1$ are quadratic residues and the other half are nonresidues. We also observe that if $x^3 + ax^2 + bx + c$ happens to be a square, say it is congruent to $t^2 \pmod{p}$, then there are two possible values for y: $y = t$ and $y = -t$. In summary, approximately half of the values of x give two solutions modulo p, and about half give no solutions modulo p, so we would expect to find approximately $2 \times \frac{1}{2}p = p$ solutions. Of course, this argument doesn't prove that there are always exactly p solutions; it merely gives a hint why the number of solutions should be more or less in the neighborhood of p.

All this suggests that it might be interesting to investigate the difference between p and N_p. We write this difference as

$$a_p = p - N_p$$

p	5	13	17	29	37	41	53	61	73	89
N_p	3	19	15	19	35	31	67	51	79	79
a_p	2	-6	2	10	2	10	-14	10	-6	10

p	97	101	109	113	137	149	157	173	181	193
N_p	79	99	115	127	159	163	179	147	163	207
a_p	18	2	-6	-14	-22	-14	-22	26	18	-14

Table 45.3: The p-Defect $a_p = p - N_p$ for E_2

and call it the *p-defect of E_2*.[1] Table 45.3 lists the p-defects for the elliptic curve E_2. This table exhibits a subtle pattern that is closely related to a topic that we studied earlier. Take a few minutes to see if you can discover the pattern yourself before reading on.

During our number theoretic investigations, we found that the primes that are congruent to 1 modulo 4 display many interesting properties. One of the most striking was our discovery in Chapter 26 that these are the primes that can be written as a sum of two squares $p = A^2 + B^2$. For example,

$$5 = 1^2 + 2^2, \quad 13 = 3^2 + 2^2, \quad 17 = 1^2 + 4^2, \quad \text{and} \quad 29 = 5^2 + 2^2.$$

Furthermore, Legendre's theorem in Chapter 34 tells us that if we require A to be odd and A and B both positive, then there is only one choice for A and B. [In the notation of Theorem 34.5, $R(p) = 8(D_1 - D_3) = 8$, where the 8 is accounted for by switching A and B and/or changing their signs.] Compare these formulas with the values $a_5 = 2$, $a_{13} = -6$, $a_{17} = 2$, and $a_{29} = 10$. Now do you see a pattern? It looks as if, when we write $p = A^2 + B^2$ with A positive and odd, a_p is either $2A$ or $-2A$. Another way to say this is that it appears that the quantity $p - (a_p/2)^2$ is always a perfect square. We check this for a few more values of p:

$$53 - (a_{53}/2)^2 = 2^2, \quad 73 - (a_{73}/2)^2 = 8^2, \quad 193 - (a_{193}/2)^2 = 12^2.$$

Amazingly, the pattern continues to hold.

[1] The actual mathematical name for the quantity a_p is the *trace of Frobenius*, a terminology whose full explication is unfortunately beyond the scope of this book. However, if you wish to impress your mathematical friends or kill a conversation at a cocktail party, try casually venturing a remark concerning "the trace of Frobenius acting on the ℓ-adic cohomology of an elliptic curve."

p	5	13	17	29	37	41	53	61	73	89
$a_p/2$	1	-3	1	5	1	5	-7	5	-3	5

p	97	101	109	113	137	149	157	173	181	193
$a_p/2$	9	1	-3	-7	-11	-7	-11	13	9	-7

Table 45.4: The Value of $a_p/2$ for E_2

One question remains. When does $a_p = 2A$ and when does $a_p = -2A$? Looking at Table 45.3, we see that

$$a_p = 2A \quad \text{for} \quad p = 5, 17, 29, 37, 41, 61, 89, 97, 101, 173, 181,$$
$$a_p = -2A \quad \text{for} \quad p = 13, 53, 73, 109, 113, 137, 149, 157, 193.$$

These two lists don't seem to follow any orderly pattern. However, if we look at the values of $a_p/2$ listed in Table 45.4, a pattern emerges.

Every $a_p/2$ value is congruent to 1 modulo 4. So if we write $p = A^2 + B^2$ with A positive and odd, then $a_p = 2A$ if $A \equiv 1 \pmod 4$ and $a_p = -2A$ if $A \equiv 3 \pmod 4$. The following statement summarizes all our conclusions.

Theorem 45.1 (The Number of Points Modulo p on $E_2 : y^2 = x^3 + x$). *Let p be an odd prime, and let N_p denote the number of points modulo p on the elliptic curve $E_2 : y^2 = x^3 + x$.*

(a) *If $p \equiv 3 \pmod 4$, then $N_p = p$.*

(b) *If $p \equiv 1 \pmod 4$, write $p = A^2 + B^2$ with A positive and odd. (We know from Chapter 26 that this is always possible.)* Then $N_p = p \pm 2A$, where the sign is chosen negative if $A \equiv 1 \pmod 4$ and positive if $A \equiv 3 \pmod 4$.

The first part is comparatively easy to verify, but we omit the proof because we will be proving a similar result later. The second part is considerably more difficult, so we are content to illustrate it with one more example. The prime $p = 130657$ is congruent to 1 modulo 4. Using trial and error, a computer, or the method described in Chapter 26, we write $130657 = 111^2 + 344^2$ as a sum of two squares. Now $111 \equiv 3 \pmod 4$, so we conclude that E_2 has $130657 + 2 \cdot 111 = 130879$ points modulo 130657.

Next we look at our old friend, the elliptic curve

$$E_1 : y^2 = x^3 + 17.$$

p	2	3	5	7	11	13	17	19	23	29
N_p	2	3	5	12	11	20	17	26	23	29
a_p	0	0	0	-5	0	-7	0	-7	0	0

p	31	37	41	43	47	53	59	61	67	71
N_p	42	48	41	56	47	53	59	48	62	71
a_p	-11	-11	0	-13	0	0	0	13	5	0

p	73	79	83	89	97	101	103	107	109	113
N_p	63	75	83	89	102	101	110	107	111	113
a_p	10	4	0	0	-5	0	-7	0	-2	0

Table 45.5: The Number of Points Modulo p and Defect a_p for E_1

Just as we did for E_2, we make a table giving the number N_p of points on E_2 modulo p and the defect $a_p = p - N_p$. The values are listed in Table 45.5.

Again there are many primes for which the defect a_p is zero:

$$p = 2, 3, 5, 11, 17, 23, 29, 41, 47, 53, 59, 71, 83, 89, 101, 107, 113.$$

These primes don't follow any pattern modulo 4, but they do follow a pattern modulo 3. Aside from 3 itself, they are all congruent to 2 modulo 3. So we might guess that if $p \equiv 2 \pmod 3$, then $N_p = p$. We can use primitive roots to verify that this guess is correct.

Theorem 45.2. *If $p \equiv 2 \pmod 3$, then the number of points N_p on the elliptic curve*

$$E_1 : y^2 = x^3 + 17 \qquad \textit{modulo } p$$

satisfies $N_p = p$.

Verification. Before trying to give a proof, let's look at an example. We take the prime $p = 11$. To find the points modulo 11 on E_1, we substitute $x = 0, 1, \ldots, 10$ into $x^3 + 17$ and check if the value is a square modulo 11. Here's what happens when we substitute:

$x \pmod{11}$	0	1	2	3	4	5	6	7	8	9	10
$x^3 \pmod{11}$	0	1	8	5	9	4	7	2	6	3	10
$x^3 + 17 \pmod{11}$	6	7	3	0	4	10	2	8	1	9	5

Notice that the numbers $x^3 \pmod{11}$ are just the numbers $0, 1, \ldots, 10$ rearranged, and the same for the numbers $x^3 + 17 \pmod{11}$. So when we look for solutions to

$$y^2 \equiv 0^3 + 17 \pmod{11}, \quad y^2 \equiv 1^3 + 17 \pmod{11}, \quad y^2 \equiv 2^3 + 17 \pmod{11},$$
$$y^2 \equiv 3^3 + 17 \pmod{11}, \qquad \ldots \qquad y^2 \equiv 10^3 + 17 \pmod{11},$$

we're really just looking for solutions to

$$y^2 \equiv 0 \pmod{11}, \qquad y^2 \equiv 1 \pmod{11}, \qquad y^2 \equiv 2 \pmod{11},$$
$$y^2 \equiv 3 \pmod{11}, \qquad \ldots \qquad y^2 \equiv 10 \pmod{11},$$

The first congruence, $y^2 \equiv 0 \pmod{11}$ has one solution, $y \equiv 0 \pmod{11}$. As for the other 10 congruences, we know from Chapter 23 that half of the numbers from 1 to 10 are quadratic residues modulo 11, and the other half are nonresidues. So half of the congruences $y^2 \equiv a \pmod{11}$ have two solutions (remember that if b is a solution then so is $p - b$), and half of them have no solutions. So overall there are $1 + 2 \cdot 5 = 11$ solutions.

If you try a few more examples, you'll find that the same phenomenon occurs. Of course, you must stick with primes $p \equiv 2 \pmod{3}$; the situation is entirely different for primes $p \equiv 1 \pmod{3}$, as you can check for yourself by computing $x^3 + 17 \pmod{7}$ for $x = 0, 1, 2, \ldots, 6$.

So we try to show that if $p \equiv 2 \pmod{3}$ then the numbers

$$0^3 + 17, \quad 1^3 + 17, \quad 2^3 + 17, \ldots, (p-1)^3 + 17 \pmod{p}$$

are the same as the numbers

$$0, 1, 2, \ldots, p-1 \pmod{p}$$

in some order. Notice that each list contains exactly p numbers. So all that we need to do is show that the numbers in the first list are distinct, since that will imply that they hit all the numbers in the second list.

Suppose we take two numbers from the first list, say $b_1^3 + 17$ and $b_2^3 + 17$, and suppose that they are equal modulo p. In other words,

$$b_1^3 + 17 \equiv b_2^3 + 17 \pmod{p}, \qquad \text{so} \qquad b_1^3 \equiv b_2^3 \pmod{p}.$$

We want to prove that $b_1 = b_2$. If $b_1 \equiv 0 \pmod{p}$, then $b_2 \equiv 0 \pmod{p}$, and vice-versa, so we may as well assume that $b_1 \not\equiv 0 \pmod{p}$ and $b_2 \not\equiv 0 \pmod{p}$.

We would like to take the cube root of both sides of the congruence

$$b_1^3 \equiv b_2^3 \pmod{p},$$

but how? The answer is to apply Fermat's Little Theorem $b^{p-1} \equiv 1 \pmod{p}$. We also make use of the assumption that $p \equiv 2 \pmod{3}$, which tells us that 3 does not divide $p-1$. Thus 3 and $p-1$ are relatively prime, so the Linear Equation Theorem (Chapter 6) says that we can find a solution to the equation

$$3u - (p-1)v = 1.$$

In fact, it's easy to write down a solution, $u = (2p-1)/3$ and $v = 2$. Of course, $(2p-1)/3$ is an integer because $p \equiv 2 \pmod{3}$.

Notice that $3u \equiv 1 \pmod{p-1}$, so in some sense raising to the uth power is the same as raising to the $1/3$ power (i.e., taking a cube root; you may recognize that we developed this idea in a more general setting in Chapter 17). So we raise both sides of the congruence $b_1^3 \equiv b_2^3 \pmod{p}$ to the uth power and use Fermat's Little Theorem to compute

$$\begin{aligned}
\left(b_1^3\right)^u &\equiv \left(b_2^3\right)^u \pmod{p} \\
b_1^{3u} &\equiv b_2^{3u} \pmod{p} \\
b_1^{1+(p-1)v} &\equiv b_2^{1+(p-1)v} \pmod{p} \\
b_1 \cdot \left(b_1^{p-1}\right)^v &\equiv b_2 \cdot \left(b_2^{p-1}\right)^v \pmod{p} \\
b_1 &\equiv b_2 \pmod{p}.
\end{aligned}$$

This proves that the numbers $0^3 + 17$, $1^3 + 17, \ldots, (p-1)^3 + 17$ are all different modulo p, so they must equal $0, 1, \ldots, p-1$ in some order.

To recapitulate, we have shown that if we substitute

$$x = 0, 1, 2, \ldots, p-1$$

into $x^3 + 17 \pmod{p}$, we get back precisely the numbers

$$0, 1, 2, \ldots p-1 \pmod{p}.$$

The congruence $y^2 \equiv 0 \pmod{p}$ has one solution: $y \equiv 0 \pmod{p}$. On the other hand, half of the congruences

$$y^2 \equiv 1 \pmod{p}, \quad y^2 \equiv 2 \pmod{p}, \quad y^2 \equiv 3 \pmod{p}, \ldots,$$
$$y^2 \equiv p-2 \pmod{p}, \quad y^2 \equiv p-1 \pmod{p}$$

have two solutions, and the other half have no solutions, since half of the numbers are quadratic residues and the other half are nonresidues (see Chapter 23). Hence, the Diophantine equation $y^2 = x^3 + 17$ has exactly

$$N_p = 1 + 2 \cdot \left(\frac{p-1}{2}\right) = p$$

solutions modulo p. □

p	2	3	5	7	11	13	17	19	23	29
N_p	2	4	4	9	10	9	19	19	24	29
a_p	0	-1	1	-2	1	4	-2	0	-1	0

p	31	37	41	43	47	53	59	61	67	71
N_p	24	34	49	49	39	59	54	49	74	74
a_p	7	3	-8	-6	8	-6	5	12	-7	-3

p	73	79	83	89	97	101	103	107	109	113
N_p	69	89	89	74	104	99	119	89	99	104
a_p	4	-10	-6	15	-7	2	-16	18	10	9

Table 45.6: The Number of Points Modulo p and Defect a_p for E_3

We now understand what happens on E_1 modulo p for primes $p \equiv 2 \pmod 3$. Exercise 45.3 asks you to discover a far more subtle pattern lurking in the a_p's when $p \equiv 1 \pmod 3$.

Let's pause to review the patterns we've discovered. For the elliptic curves E_1 and E_2, we've found that the p-defect a_p equals 0 for about half of all primes, and we've been able to describe quite precisely these 0-defect primes. For the other primes we've seen that the a_p's satisfy a more subtle pattern involving squares; that is, $p - (a_p/2)^2$ is a perfect square for E_2, and something similar for E_1 (see Exercise 45.3). Of course, E_1 and E_2 are only two elliptic curves among the myriad, so having discovered common patterns for E_1 and E_2, we should certainly investigate at least one or two more examples. Table 45.6 gives the number of points modulo p and the p-defects for the elliptic curve

$$E_3 : y^2 = x^3 - 4x^2 + 16.$$

Alas and alack, it seems that there are very few primes for which the p-defect a_p is 0. Even if we extend our Table 45.6, we find that the only primes $p < 5000$ with $a_p = 0$ are

$$p = 2, 19, 29, 199, 569, 809, 1289, 1439, 2539, 3319, 3559, 3919.$$

These primes do all happen to be congruent to 9 modulo 10, but unfortunately there are lots of 9 mod 10 primes, such as 59, 79, 89, and 109, that are not in the list.

There doesn't appear to be a simple pattern governing which primes are in the list, and indeed no one has been able to find a pattern. It wasn't until 1987 that Noam Elkies was able to show that there are always infinitely many primes p for which $a_p = 0$.

Lacking primes with $a_p = 0$, we might try looking for patterns involving squares; but again we search in vain, and no pattern emerges. In fact, what we find if we look at other elliptic curves is that most of them are like E_3, with very few a_p's being 0 and no patterns involving squares. The elliptic curves E_1 and E_2 are of a very special type; they are elliptic curves with *complex multiplication*.[2] We do not give the precise definition, but only say that elliptic curves with complex multiplication have half of their a_p's equal to 0, while elliptic curves without complex multiplication have very few of their a_p's equal to 0.

Exercises

45.1. (a) For each prime number p, let M_p be the number of solutions modulo p to the equation $x^2 + y^2 = 1$. Figure out the values of M_3, M_5, M_{13}, and M_{17}. (*Hint.* Here's an efficient way to do this computation. First, make a list of all of the squares modulo p. Second, substitute in each $0 \le y < p$ and check if $1 - y^2$ is a square modulo p.)

(b) Use your data from (a) and the values $M_7 = 8$ and $M_{11} = 12$ that we computed earlier to make a conjecture about the value of M_p. Test your conjecture by computing M_{19}. According to your conjecture, what is the value of M_{1373}? of M_{1987}?

(c) Prove that your conjecture in (b) is correct. (*Hint.* Formulas in Chapter 3 might be helpful.)

45.2. (a) Find all solutions to the Diophantine equation $y^2 = x^5 + 1$ modulo 7. How many solutions are there?

(b) Find all solutions to the Diophantine equation $y^2 = x^5 + 1$ modulo 11. How many solutions are there?

(c) Let p be a prime with the property that $p \not\equiv 1 \pmod 5$. Prove that the Diophantine equation $y^2 = x^5 + 1$ has exactly p solutions modulo p.

45.3. For each prime $p \equiv 1 \pmod 3$ in the table for E_1, compute the quantity $4p - a_p^2$. Do the numbers you compute have some sort of special form?

45.4. Write a program to count the number of solutions of the congruence

$$E : y^2 \equiv x^3 + ax^2 + bx + c \pmod p$$

using one of the following methods:

[2] An elliptic curve has complex multiplication if its equation satisfies a certain special sort of transformation property. For example, if (x, y) is a solution to the equation $E_2 : y^2 = x^3 + x$, then the pair $(-x, iy)$ will also be a solution. The presence of numbers such as $i = \sqrt{-1}$ in these formulas led to the name "complex multiplication."

(i) First make a list of the squares modulo p, then substitute $x = 0, 1, \ldots, p-1$ into $x^3 + ax^2 + bx + c$ and look at the remainder modulo p. If it is a nonzero square, add 2 to your list, if it is zero, add 1 to your list, and if is is not a square, ignore it.

(ii) For each $x = 0, 1, \ldots, p-1$, compute the Legendre symbol $\left(\frac{x^3+ax^2+bx+c}{p}\right)$. If it is $+1$, add 2 to your list; if it is -1, ignore it. [And if $x^3+ax^2+bx+c \equiv 0 \pmod{p}$, then just add 1 to your list.]

Use your program to compute the number of points N_p and the p-defect $a_p = p - N_p$ for each of the following curves and for all primes $2 \le p \le 100$. Which curve(s) do you think have complex multiplication?

(a) $y^2 = x^3 + x^2 - 3x + 11$ (c) $y^2 = x^3 + 4x^2 + 2x$
(b) $y^2 = x^3 - 595x + 5586$ (d) $y^2 = x^3 + 2x - 7$

45.5. In this exercise you will discover the pattern of the p-defects for the elliptic curve $E : y^2 = x^3 + 1$. To assist you, I offer the following list.

p	2	3	5	7	11	13	17	19	23	29
a_p	0	0	0	−4	0	2	0	8	0	0

p	31	37	41	43	47	53	59	61	67	71
a_p	−4	−10	0	8	0	0	0	14	−16	0

p	73	79	83	89	97	101	103	107	109	113
a_p	−10	−4	0	0	14	0	20	0	2	0

The Defect a_p for the Elliptic Curve $E : y^2 = x^3 + 1$

(a) Make a conjecture as to which primes have defect $a_p = 0$, and prove that your conjecture is correct.

(b) For those primes with $a_p \neq 0$, compute the value of $4p - a_p^2$ and discover what is special about these numbers.

(c) For every prime $p < 113$ with $p \equiv 1 \pmod{3}$, find all pairs of integers (A, B) that satisfy $4p = A^2 + 3B^2$. (Note that there may be several solutions. For example, $4 \cdot 7 = 28$ equals $5^2 + 3 \cdot 1^2$ and $4^2 + 3 \cdot 2^2$. An efficient way to find the solutions is to compute $4p - 3B^2$ for all $B < \sqrt{4p/3}$ and pick out those values for which $4p - 3B^2$ is a perfect square.)

(d) Compare the values of A and B with the values of a_p given in the table. Make as precise a conjecture as you can as to how they are related.

(e) For each of the following primes p, I have given you the pairs (A, B) satisfying $4p = A^2 + 3B^2$. Use your conjecture in (d) to guess the value of a_p.

(i) $p = 541$ $(A, B) = (46, 4), (29, 21), (17, 25)$
(ii) $p = 2029$ $(A, B) = (79, 25), (77, 27), (2, 52)$
(iii) $p = 8623$ $(A, B) = (173, 39), (145, 67), (28, 106)$

Chapter 46

Torsion Collections Modulo p and Bad Primes

In the last chapter we found simple patterns for the p-defects of E_1 and E_2, but there did not seem to be any similar pattern for E_3. However, the N_p's for E_3 do exhibit a pattern that you may have already noticed. If not, take a moment now to look back at Table 45.6 and try to discover the pattern for yourself before reading on.

It appears that the N_p's for E_3 have the property that

$$N_p \equiv 4 \pmod{5} \quad \text{for all primes except } p = 2 \text{ and } p = 11.$$

Although we won't give a full verification of this property, we can at least give some idea why it is true. Recall from Chapter 44 that E_3 has a torsion collection consisting of the four points

$$P_1 = (0,4), \quad P_2 = (0,-4), \quad P_3 = (4,4), \quad P_4 = (4,-4).$$

This means that the lines connecting any two of these points do not intersect E_3 in any additional points. The method of taking pairs of points on an elliptic curve, connecting them with a line, and intersecting with the curve can all be done using equations without any reference to geometry. This means that we can use the same method to find points modulo p!

Let's look at an example. The point $Q = (1,8)$ is a solution to

$$y^2 \equiv x^3 - 4x^2 + 16 \pmod{17}.$$

The line through Q and $P_1 = (0,4)$ is $y = 4x + 4$. Substituting the equation of the line into the equation of the elliptic curve gives

$$\begin{aligned}(4x+4)^2 &\equiv x^3 - 4x^2 + 16 \pmod{17}\\ x^3 - 3x^2 + 2x &\equiv 0 \pmod{17}\\ x(x-1)(x-2) &\equiv 0 \pmod{17}.\end{aligned}$$

So we get the two known points Q and P_1 with x-coordinates $x = 0$ and $x = 1$, and we also get a new point with $x = 2$. Subsituting $x = 2$ into the equation of the line gives $y = 12$, so we have found a new solution $(2, 12)$ for E_3 modulo 17.

If we try the same idea with the points $Q = (1, 8)$ and $P_3 = (4, 4)$, we get the line $y = -(4/3)x + 28/3$. How can we make sense of this line modulo 17? Well, the fraction $-4/3$ is just the solution to the equation $3u = -4$. So the number "$-4/3$ modulo 17" is the solution to the congruence $-3u \equiv 4 \pmod{17}$. We know how to solve such congruences; in this case, the answer is $u = 10$. Similarly, "$28/3$ modulo 17" is 15, so the line through $Q = (1, 8)$ and $P_3 = (4, 4)$ modulo 17 is $y = 10x + 15$. Now we substitute into the equation of E_3 and solve as before to find a new solution $(14, 2)$ on E_3 modulo 17.

We can also do the same thing with Q and P_2, giving the solution $(11, 9)$, and with Q and P_4, giving the solution $(15, 3)$. Thus, starting with the single solution Q, we used the four points in the torsion packet to find four more solutions.

Now consider the curve E_3 modulo p for any prime p. It already has the four points P_1, P_2, P_3, P_4. Each time we find another point Q on E_3 modulo p, we can take the line L_i connecting Q to each of the P_i's. Each line L_i intersects E_3 in a new point Q_i. In this way we get four additional points Q_1, Q_2, Q_3, Q_4 to go with the original point Q. Thus, points on E_3 modulo p come in bundles of five, except that there are only four P_i's. Hence

$$\left\{\begin{matrix}\text{Solutions to}\\ E_3 \text{ modulo } p\end{matrix}\right\} = \left\{\begin{matrix}\text{The 4 solutions}\\ P_1, P_2, P_3, P_4\end{matrix}\right\} + \left\{\begin{matrix}\text{Bundles containing}\\ \text{5 solutions each}\end{matrix}\right\}.$$

Therefore, the total number of solutions to E_3 modulo p is equal to 4 plus a multiple of 5; that is, $N_p \equiv 4 \pmod{5}$. This is true for all primes except $p = 2$ and $p = 11$. (For $p = 2$ and $p = 11$, some of the bundles of 5 points contain repetitions.)

The congruence $N_p \equiv 4 \pmod{5}$ also explains our earlier observation about the primes with $a_p = 0$. To see why, suppose that $a_p = 0$. Then

$$p = N_p \equiv 4 \pmod{5}.$$

Furthermore, p is odd, so $p \equiv 9 \pmod{10}$. This shows that if $a_p = 0$ then p is 9 modulo 10; but it does not say that every 9 modulo 10 prime has $a_p = 0$. This is an important distinction that stands in sharp contrast to our results for E_1 and E_2.

The preceding argument is fine, but what about the primes $p = 2$ and $p = 11$ that do not follow the pattern? It turns out that 2 and 11 are somewhat special for the elliptic curve $E_3 : y^2 = x^3 - 4x^2 + 16$. The reason they are special is because they are the only primes for which the polynomial $x^3 - 4x^2 + 16$ has a double or triple root modulo p. Thus,

$$x^3 - 4x^2 + 16 \equiv x^3 \pmod{2} \quad \text{has a triple root } x = 0\text{, and}$$
$$x^3 - 4x^2 + 16 \equiv (x+1)^2(x+5) \pmod{11} \quad \text{has a double root } x = -1.$$

In general, we say that p is a *bad prime* for an elliptic curve

$$E : y^2 = x^3 + ax^2 + bx + c$$

if the polynomial $x^3 + ax^2 + bx + c$ has a double or triple root modulo p. It is not hard to find the bad primes for E, since one can show that they are exactly the primes that divide the *discriminant of* E,[1]

$$\Delta(E) = -4a^3c + a^2b^2 - 4b^3 - 27c^2 + 18abc.$$

For example,

$$\Delta(E_1) = -7803 = -3^3 \cdot 17^2,$$
$$\Delta(E_2) = -4 = -2^2,$$
$$\Delta(E_3) = -2816 = -2^8 \cdot 11.$$

Exercises

46.1. Suppose that the elliptic curve E has a torsion collection consisting of the t points $P_1, P_2, \ldots, P_t$. Explain why the number of solutions to E modulo p should satisfy

$$N_p \equiv t \pmod{t+1}.$$

46.2. Exercise 44.2(c) says that the elliptic curve $E : y^2 = x^3 - x$ has a torsion collection $\{(0,0), (1,0), (-1,0)\}$ containing three points.

(a) Find the number of points on E modulo p for $p = 2, 3, 5, 7, 11$. Which ones satisfy $N_p \equiv 3 \pmod{4}$?

(b) Find the solutions to E modulo 11, other than the solutions in the torsion collection, and group them into bundles of four solutions each by drawing lines through the points in the torsion collection.

---

[1] We've cheated a little bit in our description of the bad primes, since for various technical reasons the prime 2 is always bad for our elliptic curves. However, it is sometimes possible to turn a bad prime into a good prime by using an equation for E that includes an xy term or a y term.

46.3. This exercise investigates the values of a_p for the bad primes.

(a) Find the bad primes for each of the following elliptic curves.

(i) $E : y^2 = x^3 + x^2 - x + 2$
(ii) $E : y^2 = x^3 + 3x + 4$
(iii) $E : y^2 = x^3 + 2x^2 + x + 3$

(b) For each curve in (a), compute the p-defects a_p for its bad primes.

(c) Here are a few more sample elliptic curves, together with a list of the p-defects for their bad primes.

E	$\Delta(E)$	a_p for bad primes
$y^2 = x^3 + 2x + 3$	$-5^2 \cdot 11$	$a_5 = -1, \quad a_{11} = -1$
$y^2 = x^3 + x^2 + 2x + 3$	$-5^2 \cdot 7$	$a_5 = 0, \quad a_7 = 1$
$y^2 = x^3 + 5$	$-3^3 \cdot 5^2$	$a_3 = 0, \quad a_5 = 0$
$y^2 = x^3 + 2x^2 - 7x + 3$	$11 \cdot 43$	$a_{11} = -1, \quad a_{43} = 1$
$y^2 = x^3 + 21x^2 + 37x + 42$	$-31 \cdot 83 \cdot 239$	$a_{31} = -1, \quad a_{83} = 1,$ $a_{239} = -1$

Several patterns, of varying degrees of subtlety, are exhibited by the p-defects of bad primes. Describe as many as you can.

46.4. For this exercise, p is a prime greater than 3.

(a) Check that the elliptic curve $y^2 = x^3 + p$ has p as a bad prime. Figure out the value of a_p. Prove that your guess is correct.

(b) Check that the elliptic curve $y^2 = x^3 + x^2 + p$ has p as a bad prime. Figure out the value of a_p. Prove that your guess is correct.

(c) Check that the elliptic curve $y^2 = x^3 - x^2 + p$ has p as a bad prime. Figure out the value of a_p. Prove that your guess is correct. [*Hint*. For (c), the value of a_p will depend on p.]

Chapter 47

Defect Bounds and Modularity Patterns

Chapters 45 and 46, despite their length, have barely begun to scratch the surface of the wonderful patterns lurking in elliptic curves modulo p. In this chapter we continue the investigation.

We have already indicated why the number of points N_p on an elliptic curve modulo p should be approximately equal to p, and we have found many patterns for the p-defect $a_p = p - N_p$. How might we quantify the statement that "N_p is approximately p"? We could say that "a_p tends to be small," but this just raises the question of how small. Looking at the tables for E_1, E_2, and E_3 in Chapter 45, it seems that a_p can get fairly large when p is large. One thing we might do is study the relative size of p and a_p. Table 47.1 lists those primes p for which the p-defect on E_3 seems to be particularly large, either positively or negatively. For comparison purposes, we have also listed the values of $\sqrt{p}$, $\sqrt[3]{p}$, and $\log(p)$.

It is clear from Table 47.1 that although the a_p's are indeed much smaller than p, they can grow to be much larger than $\sqrt[3]{p}$ and $\log(p)$. The a_p's are also larger than $\sqrt{p}$, but as you will observe, they are never twice as large. In other words, it appears that $|a_p|$ is never more than $2\sqrt{p}$.

Theorem 47.1 (Hasse's Theorem). (H. Hasse, 1933) *Let N_p be the number of points modulo p on an elliptic curve, and let $a_p = p - N_p$ be the p-defect. Then*

$$|a_p| < 2\sqrt{p}.$$

In other words, the number of points N_p on an elliptic curve modulo p is approximately equal to p, with an error of no more than $2\sqrt{p}$. This beautiful result was conjectured by Emil Artin in the 1920s and proved by Helmut Hasse during

a_p	p	$\sqrt{p}$	$\sqrt[3]{p}$	$\log(p)$
−30	239	15.45962	6.20582	2.37840
40	439	20.95233	7.60014	2.64246
44	593	24.35159	8.40140	2.77305
50	739	27.18455	9.04097	2.86864
53	797	28.23119	9.27156	2.90146
−52	827	28.75761	9.38646	2.91751
68	1327	36.42801	10.98897	3.12287
−72	1367	36.97296	11.09829	3.13577
−68	1381	37.16181	11.13605	3.14019
−70	1429	37.80212	11.26360	3.15503
−71	1453	38.11824	11.32631	3.16227
78	1627	40.33609	11.76149	3.21139
84	2053	45.31004	12.70953	3.31239
89	2083	45.63989	12.77114	3.31869
−86	2113	45.96738	12.83216	3.32490
−91	2143	46.29255	12.89261	3.33102
93	2267	47.61302	13.13663	3.35545
−98	2551	50.50743	13.66376	3.40671
−103	3221	56.75385	14.76829	3.50799
114	3733	61.09828	15.51265	3.57206
−123	4051	63.64747	15.94119	3.60756
129	4733	68.79680	16.78980	3.67514
−132	4817	69.40461	16.88854	3.68278
132	5081	71.28113	17.19160	3.70595
138	5407	73.53231	17.55168	3.73296
−146	5693	75.45197	17.85584	3.75534
−138	5711	75.57116	17.87464	3.75671
−147	6317	79.47956	18.48575	3.80051
−146	6373	79.83107	18.54021	3.80434
164	7043	83.92258	19.16840	3.84776
153	7187	84.77618	19.29816	3.85655
162	7211	84.91761	19.31962	3.85800

Table 47.1: Large Values of a_p for the curve $E_3 : y^2 = x^3 - 4x^2 + 16$

the 1930s. A generalized version was proved by André Weil in the 1940s, and this was again vastly generalized by Pierre Deligne in the 1970s.

The proof of Hasse's Theorem for general elliptic curves is beyond our present means, but we can at least indicate why it is true for the elliptic curve E_1 with equation $y^2 = x^3 + x$. Recall from Chapter 45 that the defect for this curve is given by the rules

$$\begin{aligned} a_p &= 0 && \text{if } p \equiv 3 \pmod 4, \text{ and} \\ a_p &= \pm 2A && \text{if } p \equiv 1 \pmod 4, \text{ where we write } p = A^2 + B^2. \end{aligned}$$

If $a_p = 0$, there is little more to be said. On the other hand, if $p \equiv 1 \pmod 4$, then we can estimate

$$|a_p| = 2A = 2\sqrt{p - B^2} < 2\sqrt{p},$$

which is exactly Hasse's Theorem.

The final a_p pattern that we discuss is so unexpected and unusual that you may be amazed that anyone noticed it at all. Indeed, it took many years and indications from many sources before mathematicians finally began to realize that this remarkable *Modularity Pattern* might be universally true. Although we are not able to give a full explanation of exactly what a Modularity Pattern is, we can convey the flavor by examining our representative elliptic curve

$$E_3 : y^2 = x^3 - 4x^2 + 16.$$

The other quantity that we look at is the following product:

$$\begin{aligned} \Theta = T\{(1-T)(1-T^{11})\}^2\{(1-T^2)(1-T^{22})\}^2 \\ \times \{(1-T^3)(1-T^{33})\}^2\{(1-T^4)(1-T^{44})\}^2 \cdots . \end{aligned}$$

This product is meant to continue indefinitely, but if we multiply out the first factors, we find that the beginning terms stabilize and don't change when we multiply by additional factors. For example, if we multiply out all the factors up to $\{(1-T^{23})(1-T^{253})\}^2$, then we get

$$\begin{aligned} \Theta = T - 2T^2 - T^3 + 2T^4 + T^5 + 2T^6 - 2T^7 - 2T^9 - 2T^{10} + T^{11} \\ - 2T^{12} + 4T^{13} + 4T^{14} - T^{15} - 4T^{16} - 2T^{17} + 4T^{18} \\ + 2T^{20} + 2T^{21} - 2T^{22} - T^{23} + \cdots , \end{aligned}$$

and these first 23 terms won't change if we multiply by more factors.

At this point you are probably wondering what this product Θ could possibly have to do with the elliptic curve E_3. To answer your question, here again is a list of the p-defects for E_3 for all primes up to 23:

$$a_2 = 0, \quad a_3 = -1, \quad a_5 = 1, \quad a_7 = -2, \quad a_{11} = 1,$$
$$a_{13} = 4, \quad a_{17} = -2, \quad a_{19} = 0, \quad a_{23} = -1.$$

Ignoring a_2, can you see a relation between these a_p's and the product Θ? When we write the product Θ as a sum, it appears that the coefficient of T^p is equal to a_p. Amazingly, this pattern continues for all primes.

Theorem 47.2 (Modularity Theorem for E_3). *Let E_3 be the elliptic curve*

$$E_3 : y^2 = x^3 - 4x^2 + 16,$$

and let Θ be the product

$$\Theta = T\{(1-T)(1-T^{11})\}^2\{(1-T^2)(1-T^{22})\}^2$$
$$\times \{(1-T^3)(1-T^{33})\}^2\{(1-T^4)(1-T^{44})\}^2 \cdots .$$

Multiply out Θ and write it as a sum

$$\Theta = c_1T + c_2T^2 + c_3T^3 + c_4T^4 + c_5T^5 + \cdots .$$

Then for every prime $p \geq 3$, the p-defect of E_3 satisfies $a_p = c_p$.

In the 1950s, Yutaka Taniyama made a sweeping conjecture concerning modularity patterns, and during the 1960s Goro Shimura refined Taniyama's conjecture to the assertion that every elliptic curve should exhibit a modularity pattern. André Weil then proved a Converse Theorem that helped the conjecture of Shimura and Taniyama to gain widespread acceptance.

Conjecture 47.3 (Modularity Conjecture). (Shimura, Taniyama) *Every elliptic curve E is modular. That is, the p-defects of E exhibit a modularity pattern.*

What does it mean to say that the p-defects of an elliptic curve E "exhibit a Modularity Pattern?" It means that there is a series

$$\Theta = c_1T + c_2T^2 + c_3T^3 + c_4T^4 + c_5T^5 + \cdots$$

so that for (most) primes p, the coefficient c_p equals the p-defect a_p of E, and so that Θ has certain wonderful transformation properties that are unfortunately too complicated for us to describe precisely.[1] Despite this lack of precision, I hope that the Θ for E_3 helps to convey the flavor of what modularity means.

Exercises

47.1. In this exercise you will look for further patterns in the coefficients of the product Θ described in the Modularity Theorem for E_3. If we write Θ as a sum,

$$\Theta = c_1 T + c_2 T^2 + c_3 T^3 + c_4 T^4 + c_5 T^5 + \cdots,$$

the Modularity Theorem says that for primes $p \geq 3$ the pth coefficient c_p is equal to the p-defect a_p of E_3. Use the following table, which lists the c_n coefficients of Θ for all $n \leq 100$, to formulate conjectures.

n	1	2	3	4	5	6	7	8	9	10	11	12	13	14	15	16	17
c_n	1	−2	−1	2	1	2	−2	0	−2	−2	1	−2	4	4	−1	−4	−2

n	18	19	20	21	22	23	24	25	26	27	28	29	30	31	32	33	34
c_n	4	0	2	2	−2	−1	0	−4	−8	5	−4	0	2	7	8	−1	4

n	35	36	37	38	39	40	41	42	43	44	45	46	47	48	49	50	51
c_n	−2	−4	3	0	−4	0	−8	−4	−6	2	−2	2	8	4	−3	8	2

n	52	53	54	55	56	57	58	59	60	61	62	63	64	65	66	67	68
c_n	8	−6	−10	1	0	0	0	5	−2	12	−14	4	−8	4	2	−7	−4

n	69	70	71	72	73	74	75	76	77	78	79	80	81	82	83	84	85
c_n	1	4	−3	0	4	−6	4	0	−2	8	−10	−4	1	16	−6	4	−2

n	86	87	88	89	90	91	92	93	94	95	96	97	98	99	100
c_n	12	0	0	15	4	−8	−2	−7	−16	0	−8	−7	6	−2	−8

(a) Find a relationship between c_m, c_n, and c_{mn} when $\gcd(m, n) = 1$.

(b) Find a relationship between c_p and c_{p^2} for primes p. To assist you, here are the values of c_{p^2} for $p \leq 37$.

$$c_{2^2} = 2, \quad c_{3^2} = -2, \quad c_{5^2} = -4, \quad c_{7^2} = -3,$$
$$c_{11^2} = 1, \quad c_{13^2} = 3, \quad c_{17^2} = -13, \quad c_{19^2} = -19,$$
$$c_{23^2} = -22, \quad c_{29^2} = -29, \quad c_{31^2} = 18, \quad c_{37^2} = -28$$

(*Hint*. The prime $p = 11$ is a bad prime for E_3, so you may want to treat c_{11^2} as experimental error and ignore it!)

[1] For those who have had some complex analysis, here is the main part of the modularity condition. We think of Θ as being a function of T, and we set $f(z) = \Theta(e^{2\pi i z})$. Then there is an integer $N \geq 1$ so that, if A, B, C, D are any integers satisfying $AD - BCN = 1$, then the function $f(z)$ satisfies

$$f\left(\frac{Az + B}{CNz + D}\right) = \frac{1}{(Cz + D)^2} f(z)$$

for all complex numbers $z = x + iy$ with $y > 0$.

(c) Generalize (b) by finding a relationship between various c_{p^k}'s for primes p. To assist you, here are the values of c_{p^k} for $p = 3$ and 5 and $1 \le k \le 8$.

$$\begin{array}{llll} c_{3^1} = -1, & c_{3^2} = -2, & c_{3^3} = 5, & c_{3^4} = 1, \\ c_{3^5} = -16, & c_{3^6} = 13, & c_{3^7} = 35, & c_{3^8} = -74, \\ c_{5^1} = 1, & c_{5^2} = -4, & c_{5^3} = -9, & c_{5^4} = 11, \\ c_{5^5} = 56, & c_{5^6} = 1, & c_{5^7} = -279, & c_{5^8} = -284. \end{array}$$

(d) Use the relationships you have discovered to compute the following c_m values: (i) c_{400} (ii) c_{289} (iii) c_{1521} (iv) c_{16807}.

47.2. In this exercise we look at the modularity pattern for the elliptic curve

$$E : y^2 = x^3 + 1.$$

The p-defects for E are listed in Exercise 45.5. Consider the product

$$\Theta = T(1 - T^k)^4(1 - T^{2k})^4(1 - T^{3k})^4(1 - T^{4k})^4 \cdots .$$

(a) Multiply out the first few factors of Θ,

$$\Theta = c_1T + c_2T^2 + c_3T^3 + c_4T^4 + c_5T^5 + c_6T^6 + \cdots .$$

Try to guess what value of k makes the c_p's equal to the a_p's of E.

(b) Using your chosen value of k from (a), find the values of $c_1, c_2, \ldots, c_{18}$.

(c) If you're using a computer, find the values of $c_1, c_2, \ldots, c_{100}$. How is the value of c_{91} related to the values of c_7 and c_{13}? How is the value of c_{49} related to the value of c_7? Make a conjecture.

47.3. The product

$$f(X) = (1 - X)(1 - X^2)(1 - X^3)(1 - X^4)(1 - X^5) \cdots$$

is useful for describing modularity patterns. For example, the modularity pattern for the elliptic curve E_3 is given by $\Theta = T \cdot f(T)^2 \cdot f(T^{11})^2$. Now consider the elliptic curve

$$y^2 = x^3 - x^2 - 4x + 4.$$

It turns out that the modularity pattern for this curve looks like

$$\Theta = T \cdot f(T^j) \cdot f(T^k) \cdot f(T^m) \cdot f(T^n)$$

for certain positive integers j, k, m, n. Accumulate some data and try to figure out the correct values for j, k, m, n. (You'll probably need a computer to do this problem.)

Chapter 48

Elliptic Curves and Fermat's Last Theorem

Fermat's Last Theorem says that if $n \geq 3$ then the Diophantine equation

$$A^n + B^n = C^n$$

has no solutions in nonzero integers. We proved in Chapter 28 that there are no solutions when $n = 4$. We also observe that if $p|n$, say $n = pm$, and if $A^n + B^n = C^n$ then

$$(A^m)^p + (B^m)^p = (C^m)^p.$$

So if Fermat's equation has no solutions for prime exponents, then it won't have solutions for nonprime exponents either.

The history of Fermat's Last Theorem was briefly discussed in Chapter 4. It is probably fair to say that most of the deep work on Fermat's equation done prior to the 1980s was based on factorization techniques of one type or another. In 1986 Gerhard Frey suggested a connection between Fermat's Last Theorem and elliptic curves that he thought might give a new line of attack.

Frey's idea was to take a supposed solution (A, B, C) to Fermat's equation and look at the elliptic curve

$$E_{A,B} : y^2 = x(x + A^p)(x - B^p).$$

This elliptic curve is now called the *Frey curve* in his honor. The discriminant of the Frey curve turns out to be

$$\Delta(E_{A,B}) = A^{2p}B^{2p}(A^p + B^p)^2 = (ABC)^{2p},$$

a perfect $2p^{\text{th}}$-power. This would be, to say the least, a trifle unusual. In fact, it would be so unusual that Frey suggested such a curve could not exist at all. More precisely, he conjectured that $E_{A,B}$ would be so strange that its p-defects could not exhibit a Modularity Pattern. Frey's conjecture was put into a more refined form by Jean-Pierre Serre, and in 1986 Ken Ribet proved that a Frey curve coming from a solution to Fermat's equation would indeed violate the Modularity Conjecture. In other words, Ribet proved that if $A^p + B^p = C^p$ with $ABC \neq 0$ then the Frey curve $E_{A,B}$ is not modular.

Inspired by Ribet's work, Andrew Wiles devoted the next six years to proving that every (or at least most) elliptic curves exhibit a Modularity Pattern. Ultimately, he was able to prove that every semistable elliptic curve exhibits a Modularity Pattern, and this is enough because the Frey curves turn out to be semistable.[1] We can now proceed as follows:

Proof (Sketch) of Fermat's Last Theorem

(1) Let $p \geq 3$ be a prime, and suppose that there is a solution (A, B, C) to $A^p + B^p = C^p$ with A, B, C nonzero integers and $\gcd(A, B, C) = 1$.

(2) Let $E_{A,B}$ be the Frey curve $y^2 = x(x + A^p)(x - B^p)$.

(3) Wiles's Theorem tells us that $E_{A,B}$ is modular; that is, its p-defects a_p follow a Modularity Pattern.

(4) Ribet's Theorem tells us that $E_{A,B}$ is so strange that it cannot possibly be modular.

(5) The only way out of this seeming contradiction is the conclusion that the equation $A^p + B^p = C^p$ has no solutions in nonzero integers. □

It is here, at the successful resolution of this most famous problem in mathematics, that we end our voyage through the Seven Seas of Number Theory. I hope you have enjoyed the tour as much as I have enjoyed being your guide and that you have found much to admire and much to ponder in this most beautiful of subjects. Above all, I hope that you have gained a sense of mathematics as a living, growing enterprise, with many wonderful treasures already discovered, but with many others, even more wonderful, waiting just over the horizon for the person having the insight, the daring, and the perseverance to sail into the unknown.

[1]An elliptic curve is *semistable* if, for every bad prime $p \geq 3$, the p-defect a_p is equal to ± 1. There is also a more complicated condition if the prime $p = 2$ is bad, but luckily it turns out that the Frey curves can be transformed so that 2 becomes a good prime.

Further Reading

Here are some books to assist you in your continuing study of Number Theory.

The Higher Arithmetic, H. Davenport, Cambridge University Press, 1952 (7th edition, 1999).

A beautiful introduction to number theory, covering many of the same topics as this book, but written in a more rigorous style. Highly recommended.

The following four books are standard introductions to number theory. They each include more material than we have been able to cover. The book of Ireland and Rosen uses advanced methods from abstract algebra.

An Introduction to the Theory of Numbers, G.H. Hardy and E.M. Wright, Oxford University Press, 1938 (4th edition, 1960).

A Classical Introduction to Modern Number Theory, K. Ireland and M. Rosen, Springer-Verlag, 1982 (2nd edition, 1990).

An Introduction to the Theory of Numbers, I. Niven, H. Zuckerman, and H. Montgomery, John Wiley & Sons, 1960 (5th edition, 1991).

A Course in Number Theory, H.E. Rose, Clarendon Press, Oxford, 1988 (2nd edition, 1994).

The remaining volumes in our list cover specific topics in more depth.

The Little Book of Primes, P. Ribenboim, Springer-Verlag, NY, 1991.

A delightful compendium of primes of all sizes and shapes.

13 Lectures on Fermat's Last Theorem, P. Ribenboim, Springer-Verlag, NY, 1979.

Fermat's Last Theorem through centuries, up to, but not including, the breakthrough proof of Wiles.

Introduction to Analytic Number Theory, T. Apostol, Springer-Verlag, NY, 1976.

Studying number theory via analytic (i.e., calculus) methods.

Rational Points on Elliptic Curves, J.H. Silverman and J. Tate, Springer-Verlag, NY, 1992.

Number theory and elliptic curves, including proofs of special cases of the theorems of Mordell, Hasse, and Siegel.

Appendix A

Factorization of Small Composite Integers

The following table gives the factorization of small composite integers that are not divisible by 2, 3, or 5. To use this table, first divide your number by powers of 2, 3, and 5 until no such factors remain; then look it up in the table.

$49 = 7^2$	$77 = 7 \cdot 11$	$91 = 7 \cdot 13$	$119 = 7 \cdot 17$
$121 = 11^2$	$133 = 7 \cdot 19$	$143 = 11 \cdot 13$	$161 = 7 \cdot 23$
$169 = 13^2$	$187 = 11 \cdot 17$	$203 = 7 \cdot 29$	$209 = 11 \cdot 19$
$217 = 7 \cdot 31$	$221 = 13 \cdot 17$	$247 = 13 \cdot 19$	$253 = 11 \cdot 23$
$259 = 7 \cdot 37$	$287 = 7 \cdot 41$	$289 = 17^2$	$299 = 13 \cdot 23$
$301 = 7 \cdot 43$	$319 = 11 \cdot 29$	$323 = 17 \cdot 19$	$329 = 7 \cdot 47$
$341 = 11 \cdot 31$	$343 = 7^3$	$361 = 19^2$	$371 = 7 \cdot 53$
$377 = 13 \cdot 29$	$391 = 17 \cdot 23$	$403 = 13 \cdot 31$	$407 = 11 \cdot 37$
$413 = 7 \cdot 59$	$427 = 7 \cdot 61$	$437 = 19 \cdot 23$	$451 = 11 \cdot 41$
$469 = 7 \cdot 67$	$473 = 11 \cdot 43$	$481 = 13 \cdot 37$	$493 = 17 \cdot 29$
$497 = 7 \cdot 71$	$511 = 7 \cdot 73$	$517 = 11 \cdot 47$	$527 = 17 \cdot 31$
$529 = 23^2$	$533 = 13 \cdot 41$	$539 = 7^2 \cdot 11$	$551 = 19 \cdot 29$
$553 = 7 \cdot 79$	$559 = 13 \cdot 43$	$581 = 7 \cdot 83$	$583 = 11 \cdot 53$
$589 = 19 \cdot 31$	$611 = 13 \cdot 47$	$623 = 7 \cdot 89$	$629 = 17 \cdot 37$
$637 = 7^2 \cdot 13$	$649 = 11 \cdot 59$	$667 = 23 \cdot 29$	$671 = 11 \cdot 61$
$679 = 7 \cdot 97$	$689 = 13 \cdot 53$	$697 = 17 \cdot 41$	$703 = 19 \cdot 37$
$707 = 7 \cdot 101$	$713 = 23 \cdot 31$	$721 = 7 \cdot 103$	$731 = 17 \cdot 43$
$737 = 11 \cdot 67$	$749 = 7 \cdot 107$	$763 = 7 \cdot 109$	$767 = 13 \cdot 59$
$779 = 19 \cdot 41$	$781 = 11 \cdot 71$	$791 = 7 \cdot 113$	$793 = 13 \cdot 61$
$799 = 17 \cdot 47$	$803 = 11 \cdot 73$	$817 = 19 \cdot 43$	$833 = 7^2 \cdot 17$
$841 = 29^2$	$847 = 7 \cdot 11^2$	$851 = 23 \cdot 37$	$869 = 11 \cdot 79$
$871 = 13 \cdot 67$	$889 = 7 \cdot 127$	$893 = 19 \cdot 47$	$899 = 29 \cdot 31$
$901 = 17 \cdot 53$	$913 = 11 \cdot 83$	$917 = 7 \cdot 131$	$923 = 13 \cdot 71$
$931 = 7^2 \cdot 19$	$943 = 23 \cdot 41$	$949 = 13 \cdot 73$	$959 = 7 \cdot 137$

$961 = 31^2$	$973 = 7 \cdot 139$	$979 = 11 \cdot 89$	$989 = 23 \cdot 43$
$1001 = 7 \cdot 11 \cdot 13$	$1003 = 17 \cdot 59$	$1007 = 19 \cdot 53$	$1027 = 13 \cdot 79$
$1037 = 17 \cdot 61$	$1043 = 7 \cdot 149$	$1057 = 7 \cdot 151$	$1067 = 11 \cdot 97$
$1073 = 29 \cdot 37$	$1079 = 13 \cdot 83$	$1081 = 23 \cdot 47$	$1099 = 7 \cdot 157$
$1111 = 11 \cdot 101$	$1121 = 19 \cdot 59$	$1127 = 7^2 \cdot 23$	$1133 = 11 \cdot 103$
$1139 = 17 \cdot 67$	$1141 = 7 \cdot 163$	$1147 = 31 \cdot 37$	$1157 = 13 \cdot 89$
$1159 = 19 \cdot 61$	$1169 = 7 \cdot 167$	$1177 = 11 \cdot 107$	$1183 = 7 \cdot 13^2$
$1189 = 29 \cdot 41$	$1199 = 11 \cdot 109$	$1207 = 17 \cdot 71$	$1211 = 7 \cdot 173$
$1219 = 23 \cdot 53$	$1241 = 17 \cdot 73$	$1243 = 11 \cdot 113$	$1247 = 29 \cdot 43$
$1253 = 7 \cdot 179$	$1261 = 13 \cdot 97$	$1267 = 7 \cdot 181$	$1271 = 31 \cdot 41$
$1273 = 19 \cdot 67$	$1309 = 7 \cdot 11 \cdot 17$	$1313 = 13 \cdot 101$	$1331 = 11^3$
$1333 = 31 \cdot 43$	$1337 = 7 \cdot 191$	$1339 = 13 \cdot 103$	$1343 = 17 \cdot 79$
$1349 = 19 \cdot 71$	$1351 = 7 \cdot 193$	$1357 = 23 \cdot 59$	$1363 = 29 \cdot 47$
$1369 = 37^2$	$1379 = 7 \cdot 197$	$1387 = 19 \cdot 73$	$1391 = 13 \cdot 107$
$1393 = 7 \cdot 199$	$1397 = 11 \cdot 127$	$1403 = 23 \cdot 61$	$1411 = 17 \cdot 83$
$1417 = 13 \cdot 109$	$1421 = 7^2 \cdot 29$	$1441 = 11 \cdot 131$	$1457 = 31 \cdot 47$
$1463 = 7 \cdot 11 \cdot 19$	$1469 = 13 \cdot 113$	$1477 = 7 \cdot 211$	$1501 = 19 \cdot 79$
$1507 = 11 \cdot 137$	$1513 = 17 \cdot 89$	$1517 = 37 \cdot 41$	$1519 = 7^2 \cdot 31$
$1529 = 11 \cdot 139$	$1537 = 29 \cdot 53$	$1541 = 23 \cdot 67$	$1547 = 7 \cdot 13 \cdot 17$
$1561 = 7 \cdot 223$	$1573 = 11^2 \cdot 13$	$1577 = 19 \cdot 83$	$1589 = 7 \cdot 227$
$1591 = 37 \cdot 43$	$1603 = 7 \cdot 229$	$1631 = 7 \cdot 233$	$1633 = 23 \cdot 71$
$1639 = 11 \cdot 149$	$1643 = 31 \cdot 53$	$1649 = 17 \cdot 97$	$1651 = 13 \cdot 127$
$1661 = 11 \cdot 151$	$1673 = 7 \cdot 239$	$1679 = 23 \cdot 73$	$1681 = 41^2$
$1687 = 7 \cdot 241$	$1691 = 19 \cdot 89$	$1703 = 13 \cdot 131$	$1711 = 29 \cdot 59$
$1717 = 17 \cdot 101$	$1727 = 11 \cdot 157$	$1729 = 7 \cdot 13 \cdot 19$	$1739 = 37 \cdot 47$
$1751 = 17 \cdot 103$	$1757 = 7 \cdot 251$	$1763 = 41 \cdot 43$	$1769 = 29 \cdot 61$
$1771 = 7 \cdot 11 \cdot 23$	$1781 = 13 \cdot 137$	$1793 = 11 \cdot 163$	$1799 = 7 \cdot 257$
$1807 = 13 \cdot 139$	$1813 = 7^2 \cdot 37$	$1817 = 23 \cdot 79$	$1819 = 17 \cdot 107$
$1829 = 31 \cdot 59$	$1837 = 11 \cdot 167$	$1841 = 7 \cdot 263$	$1843 = 19 \cdot 97$
$1849 = 43^2$	$1853 = 17 \cdot 109$	$1859 = 11 \cdot 13^2$	$1883 = 7 \cdot 269$
$1891 = 31 \cdot 61$	$1897 = 7 \cdot 271$	$1903 = 11 \cdot 173$	$1909 = 23 \cdot 83$
$1919 = 19 \cdot 101$	$1921 = 17 \cdot 113$	$1927 = 41 \cdot 47$	$1937 = 13 \cdot 149$
$1939 = 7 \cdot 277$	$1943 = 29 \cdot 67$	$1957 = 19 \cdot 103$	$1961 = 37 \cdot 53$
$1963 = 13 \cdot 151$	$1967 = 7 \cdot 281$	$1969 = 11 \cdot 179$	$1981 = 7 \cdot 283$
$1991 = 11 \cdot 181$	$2009 = 7^2 \cdot 41$	$2021 = 43 \cdot 47$	$2023 = 7 \cdot 17^2$
$2033 = 19 \cdot 107$	$2041 = 13 \cdot 157$	$2047 = 23 \cdot 89$	$2051 = 7 \cdot 293$
$2057 = 11^2 \cdot 17$	$2059 = 29 \cdot 71$	$2071 = 19 \cdot 109$	$2077 = 31 \cdot 67$
$2093 = 7 \cdot 13 \cdot 23$	$2101 = 11 \cdot 191$	$2107 = 7^2 \cdot 43$	$2117 = 29 \cdot 73$
$2119 = 13 \cdot 163$	$2123 = 11 \cdot 193$	$2147 = 19 \cdot 113$	$2149 = 7 \cdot 307$
$2159 = 17 \cdot 127$	$2167 = 11 \cdot 197$	$2171 = 13 \cdot 167$	$2173 = 41 \cdot 53$
$2177 = 7 \cdot 311$	$2183 = 37 \cdot 59$	$2189 = 11 \cdot 199$	$2191 = 7 \cdot 313$
$2197 = 13^3$	$2201 = 31 \cdot 71$	$2209 = 47^2$	$2219 = 7 \cdot 317$
$2227 = 17 \cdot 131$	$2231 = 23 \cdot 97$	$2233 = 7 \cdot 11 \cdot 29$	$2249 = 13 \cdot 173$
$2257 = 37 \cdot 61$	$2261 = 7 \cdot 17 \cdot 19$	$2263 = 31 \cdot 73$	$2279 = 43 \cdot 53$
$2291 = 29 \cdot 79$	$2299 = 11^2 \cdot 19$	$2303 = 7^2 \cdot 47$	$2317 = 7 \cdot 331$

Appendix B

A List of Primes

2	3	5	7	11	13	17	19	23	29
31	37	41	43	47	53	59	61	67	71
73	79	83	89	97	101	103	107	109	113
127	131	137	139	149	151	157	163	167	173
179	181	191	193	197	199	211	223	227	229
233	239	241	251	257	263	269	271	277	281
283	293	307	311	313	317	331	337	347	349
353	359	367	373	379	383	389	397	401	409
419	421	431	433	439	443	449	457	461	463
467	479	487	491	499	503	509	521	523	541
547	557	563	569	571	577	587	593	599	601
607	613	617	619	631	641	643	647	653	659
661	673	677	683	691	701	709	719	727	733
739	743	751	757	761	769	773	787	797	809
811	821	823	827	829	839	853	857	859	863
877	881	883	887	907	911	919	929	937	941
947	953	967	971	977	983	991	997	1009	1013
1019	1021	1031	1033	1039	1049	1051	1061	1063	1069
1087	1091	1093	1097	1103	1109	1117	1123	1129	1151
1153	1163	1171	1181	1187	1193	1201	1213	1217	1223
1229	1231	1237	1249	1259	1277	1279	1283	1289	1291
1297	1301	1303	1307	1319	1321	1327	1361	1367	1373
1381	1399	1409	1423	1427	1429	1433	1439	1447	1451
1453	1459	1471	1481	1483	1487	1489	1493	1499	1511
1523	1531	1543	1549	1553	1559	1567	1571	1579	1583
1597	1601	1607	1609	1613	1619	1621	1627	1637	1657
1663	1667	1669	1693	1697	1699	1709	1721	1723	1733
1741	1747	1753	1759	1777	1783	1787	1789	1801	1811
1823	1831	1847	1861	1867	1871	1873	1877	1879	1889
1901	1907	1913	1931	1933	1949	1951	1973	1979	1987

1993	1997	1999	2003	2011	2017	2027	2029	2039	2053
2063	2069	2081	2083	2087	2089	2099	2111	2113	2129
2131	2137	2141	2143	2153	2161	2179	2203	2207	2213
2221	2237	2239	2243	2251	2267	2269	2273	2281	2287
2293	2297	2309	2311	2333	2339	2341	2347	2351	2357
2371	2377	2381	2383	2389	2393	2399	2411	2417	2423
2437	2441	2447	2459	2467	2473	2477	2503	2521	2531
2539	2543	2549	2551	2557	2579	2591	2593	2609	2617
2621	2633	2647	2657	2659	2663	2671	2677	2683	2687
2689	2693	2699	2707	2711	2713	2719	2729	2731	2741
2749	2753	2767	2777	2789	2791	2797	2801	2803	2819
2833	2837	2843	2851	2857	2861	2879	2887	2897	2903
2909	2917	2927	2939	2953	2957	2963	2969	2971	2999
3001	3011	3019	3023	3037	3041	3049	3061	3067	3079
3083	3089	3109	3119	3121	3137	3163	3167	3169	3181
3187	3191	3203	3209	3217	3221	3229	3251	3253	3257
3259	3271	3299	3301	3307	3313	3319	3323	3329	3331
3343	3347	3359	3361	3371	3373	3389	3391	3407	3413
3433	3449	3457	3461	3463	3467	3469	3491	3499	3511
3517	3527	3529	3533	3539	3541	3547	3557	3559	3571
3581	3583	3593	3607	3613	3617	3623	3631	3637	3643
3659	3671	3673	3677	3691	3697	3701	3709	3719	3727
3733	3739	3761	3767	3769	3779	3793	3797	3803	3821
3823	3833	3847	3851	3853	3863	3877	3881	3889	3907
3911	3917	3919	3923	3929	3931	3943	3947	3967	3989
4001	4003	4007	4013	4019	4021	4027	4049	4051	4057
4073	4079	4091	4093	4099	4111	4127	4129	4133	4139
4153	4157	4159	4177	4201	4211	4217	4219	4229	4231
4241	4243	4253	4259	4261	4271	4273	4283	4289	4297
4327	4337	4339	4349	4357	4363	4373	4391	4397	4409
4421	4423	4441	4447	4451	4457	4463	4481	4483	4493
4507	4513	4517	4519	4523	4547	4549	4561	4567	4583
4591	4597	4603	4621	4637	4639	4643	4649	4651	4657
4663	4673	4679	4691	4703	4721	4723	4729	4733	4751
4759	4783	4787	4789	4793	4799	4801	4813	4817	4831
4861	4871	4877	4889	4903	4909	4919	4931	4933	4937
4943	4951	4957	4967	4969	4973	4987	4993	4999	5003
5009	5011	5021	5023	5039	5051	5059	5077	5081	5087
5099	5101	5107	5113	5119	5147	5153	5167	5171	5179
5189	5197	5209	5227	5231	5233	5237	5261	5273	5279
5281	5297	5303	5309	5323	5333	5347	5351	5381	5387
5393	5399	5407	5413	5417	5419	5431	5437	5441	5443
5449	5471	5477	5479	5483	5501	5503	5507	5519	5521
5527	5531	5557	5563	5569	5573	5581	5591	5623	5639
5641	5647	5651	5653	5657	5659	5669	5683	5689	5693

Index